Seasons of Grace

RELIGION IN AMERICA SERIES

Harry S. Stout
General Editor

A Perfect Babel of Confusion
Dutch Religion and English Culture in the Middle Colonies
RANDALL BALMER

The Presbyterian Controversy
Fundamentalists, Modernists, and Moderates
BRADLEY J. LONGFIELD

Mormons and the Bible
The Place of the Latter-day Saints in American Religion
PHILIP L. BARLOW

Religion and Social Order in Albany, New York 1652–1836
DAVID G. HACKETT

Seasons of Grace
Colonial New England's Revival Tradition in its British Context
MICHAEL J. CRAWFORD

Seasons of Grace

Colonial New England's Revival Tradition in Its British Context

MICHAEL J. CRAWFORD

New York Oxford
OXFORD UNIVERSITY PRESS
1991

Oxford University Press

Oxford New York Toronto
Delhi Bombay Calcutta Madras Karachi
Petaling Jaya Singapore Hong Kong Tokyo
Nairobi Dar es Salaam Cape Town
Melbourne Auckland

and associated companies in
Berlin Ibadan

Published by Oxford University Press, Inc.
200 Madison Avenue, New York, NY 10016

Oxford is a registered trademark of Oxford University Press

Library of Congress Cataloging-in-Publication Data
Crawford, Michael J.
Seasons of grace : colonial New England's revival tradition in its
British context / Michael J. Crawford.
p. cm. — (Religion in America series)
Includes bibliographical references and index.
ISBN 0-19-506393-7
1. Revivals—New England—History—17th century.
2. Revivals—New England—History—18th century.
3. Evangelical Revival—Great Britain.
4. New England—Church history.
5. Great Britain—Church history—17th century.
6. Great Britain—Church history—18th century.
I. Title. II. Series: Religion in America series
(Oxford University Press)
BR520.C72 1991
269'.0974'09032—dc20
90-42077

1 3 5 7 9 8 6 4 2
Printed in the United States of America
on acid-free paper

To Elva

ACKNOWLEDGMENTS

Nearly two decades ago at Washington University in St. Louis, work on my undergraduate honors thesis on the millenarian thought of New Englanders during the Great Awakening made me aware of the close cooperation between revivalists in Great Britain and America during the eighteenth century. I subsequently undertook to investigate the transatlantic connections among the revivalists by attempting, first, to find out what the revivalists understood a revival to be. Out of that inquiry grew my analysis of the evolution of the concept of a revival of religion between the mid-seventeenth and mid-eighteenth centuries. I owe a debt of gratitude to Professor John M. Murrin of Princeton University, whose courses at Washington University first alerted me to the rich diversity of early American history and introduced me to Puritan and evangelical religious thought.

Professor Richard L. Bushman, of Columbia University, directed the dissertation at Boston University upon which much of the present work is based. His steady belief in the value of my findings gave me the confidence to see them through to publication. For that, as well as for lessons in the process of intellectual discourse, wise counsel, and example of boldness and originality of thought, he has my sincere thanks. Professor David D. Hall, of Harvard University Divinity School, suggested many useful lines of inquiry. Professors John D. Walsh, of Jesus College, Oxford University, and W. R. Ward, of the University of Durham, directed me to important studies of eighteenth-century British revivals. Professor Ned Landsman, of the State University of New York at Stony Brook, who read the draft manuscript, brought to my attention recent studies of eighteenth-century Scottish religion and made many useful recommendations. Other intellectual debts are evident in the text and notes.

My colleagues at the Naval Historical Center have more than tolerated a historian of religious thought among them; they have provided a climate conducive to intelligent discussion of a wide variety of historical topics. Dr. William S. Dudley, senior historian at the center, has encouraged my pursuit of interests outside naval history. Charles Brodine, Barbara Dailey, Marycarol Hennessy, and Patrick and Tamara Melia loaned me books to which I otherwise had difficult access. The members of the Maryland Colloquium on Early American History, hosted by the History Department of the University of Maryland at College Park, discussed several of my ideas presented to them in papers in 1983 and 1989. Those were opportunities any scholar would have found constructive, and which as a historian working outside of an academic environment I found particularly valuable.

An Engelbourg Memorial Traveling Fellowship, awarded by the History Department of Boston University, helped fund a research trip to Great Britain in 1977 during which I gathered primary materials for this study. A Michael Kraus Research Grant in History awarded by the American Historical Association in 1987 enabled me to obtain library privileges at Georgetown University.

Librarians and archivists at the following institutions granted me access to manuscript materials: Boston University, Boston, Special Collections; Calvinist Methodist Archives, National Library of Wales, Aberystwyth; Connecticut Historical Society, Hartford; Dr. William's Library, London; Massachusetts Historical Society, Boston; and New College Library of the University of Edinburgh. Librarians at Houghton and Widener Libraries of Harvard University; the Boston Public Library; the Boston Athenaeum; the Congregational Library, Boston; the British Library; the Bodleian Library of Oxford University; the Library of the University of Glasgow; and the Library of Congress were helpful, as well.

Scattered portions of this study were published in a different form in "Origins of the Eighteenth-Century Evangelical Revival: England and New England Compared," *Journal of British Studies* 26 (1987), copyright (1987) by The North American Conference on British Studies, and are published here with the permission of that organization.

Finally, and most important, I want to express my heartfelt gratitude to my wife, Elva, and our young son, Evan, for the time and encouragement I needed to complete this work. Elva has been my constant source of comfort, counsel, and wisdom.

October, 1990 M. J. C.

CONTENTS

Seasons of Grace

INTRODUCTION

Revivals and Revivalism

Open an American newspaper to the religion pages and, as likely as not, you will find an announcement that one Protestant congregation or another is holding a revival. Such a revival usually consists of a series of services for preaching, prayer, and hymn singing intended to reinvigorate the religious commitment of the members of the church and to promote evangelical conversions within the congregation. In its modern use, the term *revival* has two religious meanings, a series of special services whose purpose is to revitalize evangelical piety, and such revitalization itself. Modern churches hold revivals praying that God will favor their meetings with an outpouring of his Holy Spirit. They plan revival services in hopes of a revival of true piety. The concept of a revival as an outpouring of the Holy Spirit on a community emerged fully elaborated from the Great Awakening of the 1740s. Not until the Second Great Awakening at the start of the nineteenth century was the term *revival* used to refer to special services to promote piety. Yet, the techniques of revivalism developed in the earlier period along with the idea of an outpouring of the Spirit.

This monograph traces the evolution of the idea of a revival of religion and the emergence of revivalism in colonial New England. Because thinking about revivals as communal events appears most explicitly in the writings of New England's evangelists, the focus here is on that region. New England's understanding of revivals, however, did not develop in isolation. Her religious culture was a subculture of Great Britain's. Ideas, practices, and institutional forms continuously crossed the Atlantic in both directions, while evangelical leaders maintained a lively transatlantic interchange of letters, publications, and personal visits. Consideration of the British context provides an essential perspective for a study of the evolution of the idea of a revival. Contemporaneous with New England's Great Awakening were revivals in the Middle Colonies, England, Wales, and Scotland. Each of these movements was a regional manifestation of the broader, pietistic Evangelical Revival that reached westward from Germany and forward from the mid-seventeenth century. The near simultaneity of these eighteenth-century movements suggests both a similarity of cause and a mutuality of influence. Viewing New England's Great Awakening in the context of the other contemporaneous revival movements in the British world affords a more accurate understanding of its origins and makes possible a determination of its distinctiveness.

The origins of evangelical revivalism in America lay in the years between the Restoration of the British monarchy in 1660 and the end of the Great Awakening in the 1740s. In that period, Puritanism as a movement came to an end. This is true if we consider it as a political attempt to reform church and state, for the English Act of Toleration in 1689 transformed nonconformists within the Church of England into dissenters from it, while the loss of Massachusetts's charter in 1684 and the acceptance of religious toleration for Protestants by all New England governments effectually put an end to the Puritan experiment in America. It is also true if we consider Puritanism as a particular kind of piety. Perhaps Christianity will always have its Puritans, but as a religious persuasion, the Puritan movement splintered into evangelicalism and rationalism, neither of which was Puritanism. When social, political, and intellectual conditions ceased to support Puritanism as a viable British subculture, evangelical leaders found themselves compelled to discover and devise new ways of sustaining and promoting piety. Over the years, the clergy experimented in search of workable solutions to the problem of piety. Their devices influenced the revivals' form, and, when the revivals appeared in strength, evangelical leaders seized on them as the solution they sought. The course of revivals and ideas about revivals mutually influenced one another, for, while revivals arose out of real psychological needs of the people, evangelical leaders had a major role in shaping them.

Historians of the Great Awakening have closely examined the evolution of ideas about the action of the Holy Spirit on the individual in conversion.[1] The present work examines another, less formal, "theology of the revival," a set of ideas about the action of the Holy Spirit in reference to entire communities. Religious revivals in New England's towns and congregations were not new in the eighteenth century. Periods of intensified displays of piety and concentrations of applications for church membership can be found from the beginning of church settlement, often associated with the early years of a town's founding, the first years of a new pastorate, and periods of high mortality.[2] New in the eighteenth century were the consideration of these communal experiences as discrete entities, their close analysis, and standardized expectations about the patterns they follow. Just as Reformed theologians described a "morphology of conversion" in the individual soul, theorists of the Great Awakening delineated a morphology of revival of religion in the community.[3] Paralleling the conversion narrative, the revival narrative emerged as a religious genre.

The language evangelicals used to talk about revivals and the revival of religion is the focus of most of the chapters of this book. J.G.A. Pocock pursues the history of political thought as the study of the changes in the languages of political discourse current within a given culture. In his view, those languages or "modes of discourse" demarcate the meanings available to political thinkers, and change in political thought occurs as thinkers modify the language or overthrow prevailing modes of discourse and replace them with new paradigms.[4] One could argue that the history of religious thought, too, is the examination of changes in the languages of religious discourse of a given culture, and that the historian of a culture's religious thought seeks the meanings available to its religious thinkers and explanations of the changes in their paradigms. This present study examines how one element in the thought of Anglo-American evangelical Protestants in the period between the Stuart Restoration and the Great Awakening fit into the structure of their thought. It examines how evangelicals in Great Britain and America thought about the revival of religion by studying the language they used to express those thoughts.

The emphasis is placed on language not merely because, as Pocock says, "people think about what they have the means of verbalizing."[5] More important for the study at hand are two observable characteristics of perception: First, people tend to perceive what they expect to perceive, as illustrated by the fact that authors are usually poor proofreaders of their own compositions because they see what they think they have written rather than their slips of the pen or typographical errors; and second, people tend to expect to see that for which they have words—for example, a Floridian may see only snow where an Eskimo, with a greater vocabulary, recognizes a half dozen different frozen aqueous forms, or when an untrained eye may be sure that a bolt of cloth is green, an interior decorator will note that the same bolt is emerald, hunter, kelly, lime, or olive green. By developing a vocabulary to describe revivals, the evangelical clergy made it easier for themselves as well as the laity to recognize revivals and revival-related phenomena.[6]

Language affects perception and, hence, understanding. How one understands what is going on naturally influences how one reacts to or participates in the phenomena perceived. Thus, the language available to describe phenomena may affect the very phenomena. This is what happened in the evolution of the religious revival. What people thought a revival of religion should be affected the form the revivals actually took. Such an interpretation could lead to the dilemma of the chicken and the egg. Edmund S. Morgan confronts a similar dilemma in his study of the morphology of conversion: To what extent did the morphology described by the theologians reflect the actual experience of conversion, and to what extent did people experience conversion according to the morphology because that is how they had been taught to describe conversion?[7] From this dilemma there are two useful, valid, and logically related escapes. One is John Owen King's response to Morgan:

People use language to order experience. "It is not a question of choosing between words and experience, or between a language of conversion and conversion itself. Prescription is not opposed to experience; prescription is language with which to order and craft experience."[8] The other escape is through Pocock's admonition: A dichotomy between thought and reality is false. "The paradigms which order 'reality' are part of the reality they order, . . . language is part of the social structure and not epiphenomenal to it, and . . . we are studying an aspect of reality when we study the ways in which it appeared real to the persons to whom it was more real than to anyone else."[9] It is not to suggest hypocrisy of observer/participants in the revivals to argue that they ordered their understanding of the revivals, and thereby influenced the structure of the revivals, according to patterns they had come to expect, while at the same time the course of revivals modified the observer/participants' understanding of the revivals.

Examining the high culture and the international context of indigenous popular local movements, this study develops two interrelated themes: the evolution of the idea of a revival of religion in Great Britain and British America during the years 1660 to 1750; and the implementation of these ideas in practical revivalism in different ways in Great Britain and New England while the evangelical movements influenced each other. The result intended is a clearer understanding of how revivals and revivalism became integral parts of American culture.

The Constituency of the Evangelical Revival

Just as the union of fuel, oxygen, and a spark produces fire, the conjunction of three major elements explains the coming of the Evangelical Revival to Great Britain and British America in the eighteenth century: constituency, leadership, and a catalyst. These three elements account for the emergence, the form, and the timing of the Revival. Part I of this study examines the evangelical leadership and its development of a revivalist ethos. Part II begins with a look at the principal catalyst that sparked the Great Awakening, the internationally itinerating Methodist preachers. The success of these men depended on the constituency who responded to their leadership. Revivalism developed out of the efforts of pastors to adapt to changing relations with, and to serve the changing needs of, their flocks. To help us understand the leaders, then, we shall first take a look at their constituency.

Dread of death and damnation are important forces for explaining the Revival of the eighteenth century. Revivalists elicited powerful responses to their portraits of hell's torments, awakening the sense of danger that was a first stage in the process of conversion. Since death faces everyone, however, and the dangers of epidemics, warfare, natural disasters, and economic hardships were not peculiar to the early eighteenth century, the heightened sensitivity to the preaching of terror at that time remains unexplained.[10] Feelings of guilt for sin underlay the power of appeals to unexpected death, for it was not just death but punishment thereafter that frightened congregations. The power of Jonathan Edwards's "Sinners in the Hands of an Angry God" derived from the auditors' application to themselves of the appellation "sinners." The sermon worked because its audience identified with the loathsome spider that could drop any moment into the fire. While sin, like death, may be a constant among men, the sense of guilt for sin seems to have an ebb and a flow within a society. The eighteenth-century Evangelical Revival appears to have marked a high tide of consciousness of sinfulness. The consciousness was evinced by specific segments of the populations of Great Britain and America, and it is by identifying those groups most susceptible to revivalism that historians have tried to discover the roots of the anxiety that fueled the revivals.[11]

One characteristic of the constituents of the Revival on which historians have focused is their youthfulness. That the Revival in Great Britain and New England attracted large numbers of young persons supports a theory that problems of resolving sexual tensions fed the anxiety manifested in the revivals.[12] Young persons' distress over their own failure to control sexual drive most likely arose from the loss of proscriptive powers by the churches, neglect of civil authorities to punish transgressions of the moral law, and decline of parental influence[13]; however, the emergence of adolescence as a more difficult stage in life was but one facet of a larger social transformation, and the Evangelical Revival ought to be seen as more than the manifestation of a particular generation's crisis of transition to adulthood.

Analyses of the socioeconomic and geographic origins of early Methodists in England have led several scholars to argue that evangelical revivalism was a part of a psychological adjustment to new social arrangements that accompanied changes in the economy.[14] The great preponderance of early Methodists came from the laboring classes. The movement hardly reached above the middle class and failed to attract the poorest members of society. Occupation was as important as class. Those connected with the landed interests, such as personal and domestic servants, rural laborers, and tenant farmers, were unreceptive to evangelicalism, whereas those connected with commerce and manufacturing, such as hand loom weavers, miners, quarrymen, and fishermen, were disproportionately represented in the movement. Methodists came from the north and west, from pastoral highlands, extraparochial tracts and wastes, and industrial towns and villages.[15] What sense can be made of these characteristics? Not only were these groups inadequately served by the

Church of England, but the redistribution of the population into these areas and occupations multiplied situations in which the traditional authority structure of English society did not operate. Among these groups, the manorial system was weak, the paternalism and social expectations of the common-field agricultural community were absent, and churches were few and parishes extensive. Among them, systems of voluntary church membership operated more freely than in the arable countryside, where the establishment power structure retained its hold. Methodism spread where there was the least resistance from the establishment.[16]

Two scholars, from very different perspectives, argue that Methodism grew among these groups not only because the establishment was unable to deter evangelicalism among them, but also because those groups among whom the dependency system was weakest had the greatest psychological need for evangelical religion. Alan Gilbert contends that evangelical religion spread rapidly among the industrial workers of eighteenth-century England because it rescued them from "anomie and social insecurity" created by separation from the intimate relationships and moral oversight of the common-field village community. The evangelical churches provided clear directives concerning behavior and lifestyle, while they offered "the feeling of belonging to a cohesive social group, of being integrated into a complex network of primary relationships."[17] Bernard Semmel argues that the evangelical Arminianism of Wesleyan Methodism, with its "revolutionary message of liberty and equality—of free will and universal salvation," appealed to masses of men alienated by the new industrial society from their patrimonies and expected vocations and helped them toward autonomy. Just as Calvinism in the sixteenth and seventeenth centuries enabled members of the middle class to become "new men," giving them the internalized values necessary to flourish in commercialized society, so Methodism enabled the working classes to make a similar transformation in the eighteenth century, integrating them into the industrialized state. Methodism "transformed men, summoning them to exert rational control over their own lives, while providing in its system of mutual discipline the psychological security necessary for autonomous conscience and liberal ideals to become internalized."[18]

Historians of Scotland have suggested that there, as in England, the economy produced a social context favorable to revivals. During the first half of the eighteenth century, Scottish commerce, industry, and agriculture underwent important changes. These developments were gradual and uneven, so that by mid-century they had not caused serious widespread social disruption. Agricultural improvements, such as more scientific methods of crop rotation and cultivation and consolidation of farms, began to appear at the start of the century, but "improving landlords" were scattered and the agricultural revolution did not yet transform the countryside. By the eve of the revivals, these long-term economic changes were not prevalent enough to have produced notable apprehension. One result of the changes in the Scottish economy during this period was not an increase of "disinherited" or of "anomie" but rather a slow rise in the standard of living. A short-term disruption was a more immediate source of anxiety. Poor harvests in 1739 and 1740 brought great scarcity and high prices from the autumn of 1739 until the summer of 1741, leading to grain riots and a high death rate, especially among children. The abundant harvest of 1741, bringing down the price of grain, preceded the Scottish revivals.[19] Yet, Callum G. Brown sees the popular religious discontent of the second quarter of the eighteenth century, variously manifested in the evangelical Secession, occasional revivals, and opposition to a patron's choice of minister, as evidence of "unease with change in rural society." As he sees it,

> improvement, whether agricultural or industrial, was giving birth to a commercially-oriented society in which social competitiveness and differentiation was weakening the communal inter-dependence upon which the agrarian parish church had been founded.[20]

In particular, full-time occupation in the clothing trades, such as weaving, spinning, dying, and bleaching, loosed ties with landed society and led these workers to develop a sense of separateness. In the latter half of the eighteenth century, among the rapidly expanding denominations of evangelical dissent, both Methodist and Presbyterian, craftsmen and nonagricultural workers such as fishermen, but especially weavers, would be conspicuously prominent.

In contrast to Great Britain's, New England's social structure lacked any extensive dependency system. Yet students of the Great Awakening have similarly concluded that the Evangelical Revival there, as in Great Britain, was part of a psychological adjustment to social transformations related to economic change. Widened markets and increased opportunity for investment in frontier lands and in waterborne trade, consolidation of the wealth of an elite, and the beginnings of a permanent class of landless agricultural workers and urban wage laborers increasingly meant impersonal economic relationships, greater social distance between people, clashes between individuals in the market place and in the courts, and conflict between factions in politics. New England was becoming a more open society in which the community's hold on the individual was reduced, and in which a broader range of behavior was accepted. In the midst of these changes, New Englanders continued to value harmonious community as the ideal.[21] Richard Bushman and James Henretta contend that converts emerged from the revivals with transformed personalities, as "new men" capable of living in the competitive society of

Yankee America and willing to oppose their own consciences to traditional authority.[22]

Scholars disagree over whether to describe the Evangelical Revival as a movement of the middling sort or of the poor. Gilbert's anomie—disorientation and sense of isolation in the absence of normative standards of conduct and belief—seems too extreme a term to describe the condition of the early adherents of the Revival in England. Analyzing the social origins of early Methodist preachers, Michael Watts asserts that these were people of "modest means and humble circumstances" but not the "disinherited."[23] Anomie seems even less applicable to the condition of the majority of subjects of the Great Awakening in New England. Several studies of the Great Awakening in rural localities have found that its subjects came principally from families of church members. These families tended to be "stable, rooted, and closely identified with the community."[24] James Walsh, looking at rural Woodbury, Connecticut, states it was not the "disinherited" who produced the Great Awakening; in contrast, Gary Nash, looking at Boston, Massachusetts, says the crowds who followed the radical evangelists such as James Davenport and Andrew Croswell "seem to have been primarily composed of these dispossessed," including slaves, servants, and the impoverished.[25] Harry S. Stout and Peter Onuf find that the supporters of the radical revival in New London, Connecticut, consisted not of the "idle poor," not of the "lowest rank," but rather of those "not well entrenched in the religious or civil hierarchies of town life."[26] While the Revival's origins seem to have lain in the middling sort in both Great Britain and New England, it drew a following as well from those who were strangers to economic, social, and political power. The latter were attracted by the movement's critique of those in authority and its condemnation of contemporary morality and materialism. As among Virginia's Baptists later in the century, participation in an evangelical community emboldened those alienated from mainstream culture to condemn prevailing mores.[27]

The converts of the Scottish revivals had a profile similar to those of England and New England, mainly young, predominantly common working people: tenant farmers, small craftsmen, women in service. "The far greater number" of the some seventy awakened parishioners of Golspy, in the north of Scotland, said their pastor,

> are of ages from twenty to fifty years, few of them below twenty, and four only from sixty to seventy. They are of the farmers and tradesmen, or their wives and servants, and but few of their children; and amongst them are seven widows in low circumstances.[28]

The best primary source for characterizing the constituency of the Scottish revivals is a manuscript collection of personal conversion narratives of 106 converts, recorded in their own words by William McCulloch, minister at Cambuslang[29] (see Appendix 1). There appears no reason to believe that this group was not representative of the hundreds, coming from near and far, awakened at Cambuslang in 1742. The sex, age, marital status, and occupation of most of these converts can be derived from their accounts. The sample consists of twice as many women as men (71 and 35, respectively). Most of the Cambuslang converts were single, which was a factor of age, the median of which was 23 years. On average, the men were slightly older than the women (males averaged 26 years, women 24). Four of the narrators were less than 15 years of age, the youngest being 13 years old; only one testator was older than 55, a 65-year-old widow. The majority, 60 percent, were youths between the ages of 16 and 25 years; more than a third (37 percent) of the total were concentrated in the age group 18 to 21 years. Of the 23 men whose occupations are clear, 5 were in apprenticeship or service, mainly as weavers, 13 were craftsmen (4 shoemakers, 4 weavers, a dyer, a gardener, a mason, a tailor, a "tradesman"), 2 were tenants, 1 "old soldier" was a collier, and 1 was a municipal official. It is reasonable to assume that many of the men whose occupations are not obvious from their accounts were craftsmen, tenant farmers, or in service. Few of the women's occupations are clear. Two were seamstresses, several were in service, and a number mention spinning at the wheel. The occupations of either their father or husband, and in one case of both, is known for 42 of the women. Several of their fathers held higher than average social status: a gentleman, two merchants, a small independent landowner ("portioner"), and a schoolmaster. One woman convert was the wife of the bailie of Hamilton; however, the majority of fathers and husbands of the female converts were of ordinary status: tenant farmers (12), and craftsmen and tradesmen: bleacher, carter, cooper, gardener, maltman, peddler ("packman"), sailor, ship carpenter, shoemakers (4), smith, tailor, weavers (4), and "workman." There was a soldier's wife, and the humblest end of the social scale was represented by a collier's daughter and a day laborer's wife. Propertied Scots in general kept themselves clear of what they viewed as the delusion and enthusiasm of the revival; but it was people struggling at the margins of respectability, not the "dispossessed," who embraced it.[30]

Demographic, economic, and geographical descriptors are not sufficient to identify the constituency of the Revival in Great Britain or New England, for there were persons susceptible to the movement living in the same communities and of the same social circumstances with others who were unresponsive to or repulsed by it. Upbringing within a pious family and the influence of pious educational institutions were crucial. Those persons who had been taught the catechism from the cradle, who had been reared in the fear of God, and who had internalized the values of pious parents were readier than others to feel guilt for moral transgressions, to interpret anxieties in a religious context, and to find solace in submission to a loving God. Nearly all of the Cambuslang converts whose experiences McCulloch recorded noted their pious upbringing, which included family and private prayer, Scripture reading, learning the catechism, and regular church attendance. The revivals in the north of Scotland during the 1740s occurred in the shires of Ross and Sutherland, areas of historic Presbyterian strength. The charity schools of the Society for the Propagation of Christian Knowledge and the Welsh-language charity schools founded by Griffith Jones, along with the societies for reformation of manners (i.e., for public morality), and the Anglican religious societies, all active in the first half of the eighteenth century, were important in preparing a British clientele for the Revival. The kind of religious teaching these institutions provided was as important as the fact of religious training alone. By upholding difficult, if not impossible, standards of external behavior and internal holiness as requisite for salvation, the High Church intensified the anxiety from which Methodism's free grace offered escape. Pious men and women who found themselves continually falling short of their goal, despite days of fasting, nights in prayer, and frequent alms giving, discovered relief when they placed their burden at the foot of the cross.[31] In New England, the Revival was most effective among families in which parents inculcated in children the necessity of the conversion experience.[32] Institutions similar to the English societies for the reformation of manners and especially religious societies were in widespread use in New England communities. Some religious societies were organized around neighboring families and others as youth groups. The formation or revitalization of these societies was often the first signal of a religious revival in a New England community. In America, as in England, a background in strict piety seems to have predisposed many toward the Revival.

Philip Greven believes that a particular pattern of child-rearing was crucial and fundamental in the formation of the evangelical temperament in early America. Those reared in the evangelical pattern were much more susceptible to evangelical conversion than those who were not.[33] Greven's analysis has implications for the Methodist Revival in England. He draws much of his evidence for the character of evangelical child-rearing from the case of Susannah Wesley, mother of John and Charles, founders of Methodism. He does so even though his concern is early America and the Wesleys lived in England. And he does so without indicating that the Wesley household was High Church Anglican, a persuasion not usually considered evangelical. The comparison between this High Church household of the Wesleys and evangelical households in America indicates an important common element in the evangelical and High Church temperaments. That common element was the desire for self-annihilation.[34] Works of piety in High Church circles, such as Thomas a Kempis's *Imitation of Christ*, William Law's *Christian Perfection*, and Jeremy Taylor's *Holy Living*, prescribed a rigorous and methodological program of self-abasement, self-denial, and prayer. High Church piety called for nothing less than complete subjugation of one's will to God's in imitation of Jesus Christ's obedience to the Father.[35] If Susannah Wesley's child-rearing practices—the attempt to break the will of the child as early as possible, followed by a continuing expectation of unquestioning obedience, a regulated and regular daily schedule, and denial of worldly pleasures—were typical of High Church households, then a background of High Church family life and education would have provided a psychological basis for conversion similar to that which, according to Greven, was provided by an evangelical upbringing. Conversion, which often came during the transition to adult responsibilities, recapitulated the infantile subjugation of the will to the parents. Indeed, many High Church Anglicans in mid-eighteenth-century England replicated the Wesley brothers' journey from discipleship of William Law's High Church precisianism to membership in the Methodist fold.

The question remains, why were the traditional practices of the existing religious institutions insufficient to serve the psychological needs of so many pious evangelicals and High Church Anglicans so that those persons were susceptible to revivalism? An explanation prevalent at the time of the revivals and appealed to in varying degrees of sophistication in the recent literature is that the vitality had gone out of the piety of existing institutions. Religious observance had become cold and formal. In its very essence, supposedly, Pietism is a reform movement, only existing in opposition to a prevailing religious norm.[36] One important lesson we have learned since Perry Miller, however, is that to speak of a revival of evangelical piety in the eighteenth century is something of a mistake. The term revival implies that a piety that had been vital in the mid-seventeenth century had become dead or dormant by the eighteenth. Recent studies document the continuation of a healthy strain of lay piety into the eighteenth century.[37] Studies of specific localities indicate that the core constituency of the Great Awakening in New England were members of families in which traditional Reformed piety had been kept alive.[38] Marilyn Westerkamp suggests that the same is true for the revivals among Presbyterians in the Middle Colonies and in Scotland.[39] Older studies affirm the strong lay leadership in the evangelical movement in Scotland.[40] Nor were revivals as communal events, themselves, altogether new with the Great Awakening. The Scottish revival tradition stretched back to the 1620s, if not earlier, and New England's congregations had known periods of intense religious concern and church growth since their founding. New to the eighteenth century were the revivals' extent, intensity, and concentration in time, as well as the ways in which revivalists promoted them, and these are what essentially constitute what has come to be known as the Great Awakening. The eighteenth century Evangelical Revival had more to do with adaptations to structural changes in society, in particular, those conducive to religious voluntarism, than with the death and rebirth of evangelical piety. Our puzzle is not so much the rebirth of piety as why piety expressed itself in the form of revivals.

The sense that piety had declined in the churches arose less from changes in those churches than from changes in the circumstances of religious commitment. Voluntarism in religion was of especial importance to the emergence of revivalism as a major religious phenomenon in the eighteenth century. Puritanism had always depended on the voluntary participation of those who cared about piety; however, whereas under the Tudors and Stuarts Puritanism grew in the face of authorities vigorously enforcing conformity to the Book of Common Prayer, in the eighteenth century the Evangelical Revival flourished amid the relative indifference of the state to religious diversity. The threat of civil penalties to those who participated in evangelical activities was minimal. Because, on the one hand, it was easy to avoid performance of any religious duties and, on the other hand, the penalties for becoming involved with any particular Protestant group became less severe, church growth came to depend more and more on voluntary commitment that took place in less emotionally charged circumstances. The contagious enthusiasm of the public revival provided the emotional impetus and the occasion for the ordinary lay person to make that commitment. For the ordinary clergyman, the revival came to be a means by which to stimulate voluntary commitment and, in the absence of other forms of official affirmation, to validate his ministry. On both sides of the Atlantic Ocean, the Revival was a response to the decline of the influence of traditional social authorities and the rise of independent attitudes, voluntarism, and secularization.

Why were many persons in different parts of Britain and America ready to give powerful emotional responses to the evangelical message in the second quarter of the eighteenth century? Scholars have offered similar interpretations for Britain and America by seeking psychological sources of the emotional energy released in the Evangelical Revival. They have isolated similar factors: anxiety produced by fear of death and guilt for sin; demographic and economic changes that increased the number of persons troubled by anxiety; religious upbringing and pious education that reinforced anxiety; rationalism and formalism in the churches that left anxiety unattended to; and threats to traditional piety that made layman and clergyman alike eager for confirmation of that piety's power. These factors are supposed to have created a constituency susceptible to a new preaching style that presented a theology to explain and a mechanism by which to relieve those disturbing feelings and that replaced them with more comforting feelings of forgiveness, belonging, and joy. The Revival provided an opportunity in both Great Britain and America for anxious persons to reaffirm their attachment to the ideal and to rediscover some of the reassurance of the tightly knit, loving, Christian community, despite changed circumstances that proscribed return to earlier conditions. In England, the Methodist Societies served as substitute communities for the uprooted. In Scotland, the revivals affirmed the attachment of the church to evangelical religion in the face of enlightenment latitudinarianism in the General Assembly and schism by evangelical radicals. In New England, towns and villages touched by the Revival sought to recapture the sense of consensus and unity enshrined in their historical identity.

Revivalists themselves were part of the constituency susceptible to the evangelical call. While members of the clergy were subject to the same psychological stresses from changing social and economic relationships as were others, their profession made them acutely sensitive to those changes. As the intellectual and spiritual leaders of their communities, they felt a responsibility to interpret social conditions, to judge the moral climate, to oppose prevailing evils, and to promote religion and virtue. Alarmed by increasing evidence of behavior they saw as unchristian competitiveness, avarice, and irreligion, numbers of these clergymen looked for ways to effect moral reformation and spiritual revival.

There is another reason why these clergymen welcomed a crusade for reform and revival: they sought to uphold the status and prestige of their calling. Secularization and anticlericalism struck at their position as community leaders. By rallying their flocks with the call to religious renewal, some thought they could restore the respect and influence traditionally theirs. In the circumstances of de facto voluntarism in religious and moral affairs, persuasion came to be the only practical means for the clergy of Great Britain and America to combat secularism and to attract followers. From this situation, there derived a role for aggressive evangelism and revivalistic techniques. Similar dynamics were at work on both sides of the Atlantic to call forth a religious leadership eager as well as able to reach the emotional core of masses of men and women.

I

COVENANT AND REVIVAL, 1660–1739

1

The Outpouring of the Holy Spirit 1660–1690

"As particular persons, so Churches and Congregations have their *day of Grace*."

INCREASE MATHER, *Returning unto God the Great Concernment of a Covenant People* (Boston, 1680).

Throughout the seventeenth century, the English used the word *revival* to refer to a renewed interest in branches of literature and the arts.[1] A renewed concern for the things of religion they usually called *reformation*. When they applied the verb *to revive* in a religious context, as suggested by such scriptures as "wilt thou not revive us again that thy people may rejoice in you" (Isa. 57:15), and "revive thy work in the midst of the years" (Hab. 3:2), reformation—of morals, administration of the sacraments, or church government—was the object intended: "If God make this Ministry a *Converting Ministry*," Samuel Torrey of Massachusetts prophesied in 1674, "the Work of Reformation will be again revived."[2] In the first decades of the eighteenth century the phrase *revival of religion*, meaning something distinct from reformation, became increasingly common. For example: "We Pray, we Fast, make Laws, and dispute about Reformation, but yet we are in Affliction, and the Hand of God goes out against us: And so it will, unless there be a reviving of Religion."[3] The growing prevalence of the term revival of religion at the end of the seventeenth century underscores a significant development in British and British/American thought, the evolution of the idea of a revival of religion.

Some Theological Assumptions

The idea of a revival of religion in the later seventeenth century rested on two basic assumptions about the way God dispenses his grace, as well as on biblical precedents and promises concerning the church. An understanding of the development of the idea of a revival begins with these foundations.

The assumption that God offers his grace in an inconstant and inscrutable manner is central to the concept of a revival of religion. In the Roman Catholic tradition, divine grace is continually available to those who seek it through the sacraments, the sacred liturgy, sacramentals, indulgenced prayers, and charitable acts. In the Catholic view, Christ merited an abundant store of grace, to which his saints add, and over which he gave his church power and authority. In this view, divine life enters the soul at baptism and is nourished by the means instituted by Christ and administered by his church. In the Reformed tradition, in contrast, divine life enters the soul with conversion, which baptism may or may not be instrumental in initiating. The chief means of grace are reading of Scripture, hearing the word of God preached, and participation in the Lord's Supper. Grace comes through these means not to all who resort to them, but only to the elect. It comes to those elect only at God's appointed time. Although prayer, Scripture, and the Lord's Supper nourish the divine life of the soul of an elect person after conversion, the elect person is likely to experience the divine favor that is channeled through them at intervals and as special seasons of grace.[4]

The second assumption on which the concept of a revival of religion rests is that God deals with entire communities as discrete moral entities. Puritans conceived that God chastised and encouraged, punished and rewarded, a people according to their moral condition. They called this relationship between God and people the "national covenant." The model of a national covenant was the relationship between Yahweh and his chosen people, the Israelites, recounted in the Old Testament. When the Israelites obeyed Yahweh's commandments, they prospered; when they forgot their mission or worshipped other gods, Yahweh sent prophets to recall them to their duty and disastrous wars and captivities to remind them of their obligations as his people. When the Israelites repented their sins and reformed their lives, Yahweh restored his favor. Puritans extended this kind of relationship to every body-politic and elaborated it to apply to every body-social, from nation down to family. Every social and political entity stood accountable before God. The welfare of each depended on his favor.

The nature of a people's relationship with God differed from that of an individual's, for "good works" had a function in the national covenant they did not have in the process of the salvation of souls. Communities never entered an afterlife wherein they could be rewarded or punished: They had to receive judgment or mercy on earth. In addition, whereas individuals were either saved or damned, communities were composed of saints and sinners together. Hence, a community's relationship with God was not determined so much by the inner graciousness of the people as it was by their external compliance with his law. In Reformed theology, compliance with God's law is not meritorious toward salvation, for a person without divine life in his soul cannot obey out of love of God. His obedience emanates only from self-centered causes. Divine life, which enables one to love God, is a free and unmerited gift from God. A saint obeys out of love because he has been saved, not in order to be saved. The national covenant, however, does not directly concern eternal salvation; rather, it concerns temporal prosperity. Obedience does merit favor, and disobedience, disfavor. Obedience emanates either out of love or out of fear. True love of God operates only among the saved; but fear—of temporal punishment, of social ostracism, of divine wrath—could influence everyone. Therefore, civil authorities could work to retain or regain divine favor for their people by encouraging virtue and punishing vice. Anyone in authority, be he governor, justice of the peace, pastor, or head of family, bore the responsibility of enforcing God's law within his sphere, for the good of nation, town, congregation, and family.

If it were possible for a people to fulfill their part in the national covenant by their own efforts, they would have no need for divine intervention. The Puritans understood that without divine aid no people could, in fact, long sustain external virtue, avoid backsliding into selfish behavior, and prevent the emergence of prevailing sins. They believed that it was God who enabled a people to fulfill their obligations to him. Whenever the Old Testament Israelites returned to Yahweh, it was Yahweh who turned them. God upheld both sides of the covenant. Just as God granted his grace in an inconstant and inscrutable manner to his elect, so did he shed it at intervals on a people in covenant with him to preserve them as his people.

Like the Old Testament Israelites, a Christian people stood in relation to God in two capacities, as nation and as church. Although in Reformed thought members of the church made up only a portion of the citizenship, the nation's fulfillment of the covenant depended on maintenance of the true church in purity. Religious as much as moral declension would lead to withdrawal of divine favor. Without a solid core of practicing saints, the nation would be lost. Thus, the state had a stake in the church's health. Nevertheless, the state, alone, could not preserve the church, for only the Holy Spirit could create the saints who constituted it. Even were the state to turn against it, the church could be the nation's salvation when God fulfilled his promises to revive her through the outpouring of the Holy Spirit.

During periods of religious persecution, Reformed religious leaders comforted and reassured themselves and their people with two scriptural promises. One, Christ's true church would survive, and the powers of evil never prevail against it. Two, Christ's true church would, at last, be victorious on earth; it would enjoy a millennium of peace and prosperity before the end of time. God had promised outpourings of his Holy Spirit to sustain Christ's church and to make it triumphant.

The pouring out of the Spirit was the metaphor most commonly used in the late seventeenth century to describe the action that God took in rescuing a people from its sins. "When God doth restore his people from any deep and general defection, and renew his Covenant with them," said Samuel Torrey, "he doth promise, and actually pour out abundance of converting grace, and so revive and renew the work of Conversion. . . . If ever these Churches be thoroughly recovered, it will be, it must be by such a dispensation of converting grace unto an unconverted generation." There was "no hope no possibility of the Resurrection of Religion otherwise than by such a dispensation of the Spirit and Grace of God, *until* God shall pour out his Spirit from on High."[5] The metaphor was biblical: "Until the spirit be poured upon us from on high, and the wilderness be a fruitful field" (Isa. 32:15); "I will pour my spirit upon thy seed, and my blessing upon thine offspring" (Isa. 44:3); "Neither will I hide my face any more from them: for I have poured out my spirit upon the house of Israel, saith the Lord God" (Ezek. 39:29); "It is time to seek the Lord, till he come and rain righteousness upon you" (Hos. 10:12); "I will pour out my spirit upon all flesh" (Joel 2:28). The Reformed interpreted these passages as promises from God that he would preserve the true church through all trials and declensions by periodic rescues.

The Old Testament relates how God preserved the Israelites as his people despite their repeated backsliding. Through occasional enlarged dispensations of his grace, again and again he brought them back to his pure worship and to obedience to his law. The leading biblical precedent of an outpouring of grace on God's people, however, was found in the New Testament's Acts of the Apostles. The descent of the Holy Spirit on the apostles gathered in the upper room on the day of Pentecost gave them the courage to preach the crucified and risen Christ and the power to convert thousands. The author of Acts believed that the outpouring of the Holy Spirit on that day was a fulfillment of Old Testament promises, in particular Joel 2:28–32:

> This is what is spoken by the prophet Joel: "And in the last days it shall be, God declares, that I will pour out my Spirit upon all flesh, and your sons and yours daughters shall prophesy, and your young men shall see visions, and your old men shall dream dreams; yea, and on my menservants and my maidservants in those days I will pour out my Spirit; and they shall prophesy. And I will show wonders in the heaven above and signs on the earth beneath, blood, and fire, and vapor of smoke; the sun shall be turned into darkness and the moon into blood, before the day of the Lord comes, the great and manifest day. And it shall be that whoever calls on the name of the Lord shall be saved." Acts 2:17–21

The evolution of the idea of an outpouring of the Holy Spirit for the reformation of a people into the modern conception of a religious revival began in the period before 1700, following the Stuart Restoration, when Protestant ministers of every persuasion in Great Britain and British America concurred that religion was dying. Convinced that the nation was on a back-sliding course, the ministers felt called on to become Jeremiahs and preach reformation. If the people did not reform themselves, God would send the nation calamities to call them back to their duty. The last third of the century witnessed a variety of techniques designed by the ministers to promote reformation. Yet, balancing the calls to the people to change their ways could be heard with equal clarity reminders that the power to reform was not theirs, but God's. After 1700 the balance shifted to the latter message. All attempts at reformation had been frustrated, preachers announced, because God's Spirit had not accompanied them. A lasting reformation could come about only as consequence of a revival of religion through the outpouring of God's grace.

Great Britain

The return of the Stuarts to power in 1660 introduced an era of adversity, "the Great Persecution,"[6] for nonconforming Protestants in Great Britain. For encouragement under persecution, Reformed preachers drew on ideas about the national covenant, the promises to the church, and the outpouring of the Holy Spirit. In the process of expounding these ideas, three British preachers, Robert Fleming, John Howe, and John Owen, described both of what would later be called a revival and a great awakening. New Englanders cited their writings occasionally over the ensuing decades, down to the Great Awakening, when attempting to predict and to explain a revival of religion.

After the Restoration, the double-headed axe of episcopacy and patronage cut down the presbyterian structure of the church in Scotland. All clergymen were required to attend their bishops' synods, and every clergyman occupying a living filled since 1649 was required to apply to the lay patron for presentation to the bishop for admission to the benefice. Conscientious Presbyterians found it difficult to submit to these nonpresbyterian authorities; 270 ministers refused and were deprived of their livings and forbidden to reside within 20 miles of their former charges. Lay persons loyal to Presbyterianism suffered, too, from ruinous fines for nonattendance at parish churches and from the violent suppression of conventicles. These troubles lasted until after the Glorious Revolution, when prelacy and patronage were abolished, Presbyterianism established, and the ejected ministers restored to their parishes.[7]

In *Fulfilling of the Scripture*, in 1669, Robert Fleming (1630–1694), one of the ejected Scottish Presbyterian ministers, made use of the millennial promises to encourage Scottish Presbyterians at a period that looked bleak for their church.[8] "It is a *dark* time now with the Church of Christ," with atheism at home, churches abroad decaying, religion everywhere wearing out, while sin increases, Fleming observes. Men professing the doctrine of sanctification ridicule those who seek to put it into practice. The doctrine that Christ's righteousness is imputed to the elect is openly denied. Even ministers oppose piety.[9] No matter how dark the times become, however, the church has cause for hope. The past fulfillment of the Scripture prophecies is reason to trust in the promises yet unaccomplished. The Scriptures promise not only that the church will survive, but that it will triumph. In fact, writes Fleming, things are so bad for the church, it is probably the last assault of Antichrist before his destruction and the last night before the dawn of the millennium.[10] Fleming's version of the millennium is not a literal reign of resurrected martyrs. Rather, it is one in which the church would enjoy calm, outward prosperity, and peace; people and nations would convert; and "pure ordinances, a more universal oneness amongst the worshippers of God, the walk of Christians with a discernable lustre of holiness, will be made to commend the Gospel."[11]

Fleming argues that the occurrence of "more solemn times of the Spirit" is proof of the reality of the work of God on the souls of men, and of his care for the church. "The more eminent extraordinary outpourings of the Spirit witness this truth." When ministers and people at certain times appear more diligent in the performance of their religious duties, more successful in the use of their talents, and more powerful and lively in their religious experiences than at other times, God's promise to pour out his Spirit is being fulfilled.[12]

Fleming offers several examples from Scottish history of remarkable outpourings of the Spirit. First, there was the so-called Stewarton Sickness in the west of Scotland about the year 1625. This religious stirring began in the parish of Stewarton and spread throughout the area.

> For a considerable time, *few Sabbaths* did pass without some evidently *converted*, and some convincing proofs of the *power of God* accompanying his *word*: yea, that many were so *choaked and taken by the heart*, that through TERROUR (the SPIRIT in such a measure convincing them of *sin* in *hearing of the word*[)] they have been made to FALL OVER and thus CARRIED OUT OF THE CHURCH, who after proved most *solid* and *lively* Christians.[13]

Another outpouring of the Spirit occurred at a Communion service held at Shotts, near Glasgow, 20 June 1630. The several decades preceding the Shotts Communion had witnessed the gradual development of the Scottish Communion season, a practice that expanded rapidly in the 1620s. Congregations would hold solemn Communion services at which several ministers from a vicinity preached over a period of from three to five days, and which hundreds of persons from the surrounding parishes attended. Over the course of a summer, when the Communions were usually scheduled, an individual might have the opportunity of attending such occasions with several congregations in a vicinity. Those who gathered at Shotts for such a service in 1630 spent the night following ordinances in prayer for grace. During the sermon the next day, 500 persons were thought to be converted. "It was the sowing of a seed through *Clidesdeal*," writes Fleming, "so as many of most eminent Christians in that country, could date either their *conversion*, or some remarkable *confirmation* in their case, from *that day*."[14] The Stewarton Sickness and the Communion at Shotts became part of the collective memory of pious Scots, who looked back at those occurrences as models of divine intervention in the life of the church.[15]

In England, those who could not in good conscience submit to the authority of bishops or make use of the Book of Common Prayer in public worship suffered persecution for nearly two decades after the Restoration. None who would not receive Communion according to the rites of the Church of England could be a member of a municipal corporation. Heavy penalties were levied for attendance at worship not according to the Book of Common Prayer. Only clergy ordained by a bishop could hold benefices. Nonconforming ministers were forbidden to live or visit within five miles of any incorporated town or any place where they had served as ministers. Without a licence from a bishop, no one could legally teach in a school or a private family. Some 1,700 clergy lost their livings under these laws.

To comfort and sustain the English Nonconformists amid their disappointments and trials in Restoration England, John Howe (1630–1705), like Robert Fleming, invoked the divine promises to preserve and glorify the true church and those who remained faithful to it. Howe, a Presbyterian who had served both Oliver and Richard Cromwell as domestic chaplain at Whitehall, was ejected from his living at Great Torrington in 1662. In 1678 he preached a series of fifteen sermons on Ezekiel 39:29, "Neither will I hide my face any more from them: for I have poured out my Spirit upon the House of Israel, saith the Lord God."[16] The sermons, published under the title *The Prosperous State of the Christian Interest Before the End of Time, By a Plentiful Effusion of the Holy Spirit*, comprise a defense and analysis of two doctrines drawn from Ezekiel's prophecy:

> First, That there is a state of permanent serenity and happiness appointed for the universal church of Christ upon earth.—Secondly, that the immediate original and cause of that felicity and happy state, is a large and general effusion or pouring forth of the Spirit.[17]

The millennial state, according to Howe, is to be brought about by a communication of the Spirit to mankind more rapid and abundant than ever before. The Spirit will work through means: civil magistrates who will be nursing fathers to the church; heads of families who will maintain family worship, instruction, and discipline; and ministers of the Gospel who will deliver awakening sermons and deal effectively with souls. Howe laments that "there is a great retraction of the Spirit of God even from us [ministers]; we know not how to speak living sense unto souls, how to get within you. . . . The methods of alluring and convincing souls, even that some of us have known, are lost from amongst us in a great part." With the promised effusion of the Spirit, however, preachers "shall know how to speak to better purpose, with more compassion and sense, with more seriousness, with more authority and allurement, than we now find we can." The Spirit will revive religion not only through means, but also by immediate operation on souls, bringing about conversions and increasing the holiness of believers, as it had at Pentecost.[18]

In Howe's vision, the millennium is not inaugurated by an apocalyptic cataclysm. Rather, it begins gradually and gains momentum as it spreads across the earth. Howe does not predict an imminent millennium. Instead he comforts his auditors with the hope of some harbingers, urging them "to wait patiently and pray earnestly, that of so great a harvest of spiritual blessings to come upon the world in future time, we may have some first-fruits in the mean time. As it is not unusual, when some very great general shower is ready to fall, some precious scattering drops light here and there as forerunners." In Howe's scheme, localized revivals will precede the general awakening.[19]

Howe asserts that the pouring forth of the Spirit is the only means for bringing about a revival of religion and producing the millennium. Nothing can mend the world except what mends human hearts, and this the Holy Spirit alone can do. Preaching the Gospel, a very exact form of church government, and even miracles will have no good effect unless the Spirit is poured out on men. God dispenses his grace arbitrarily in the times and places he chooses. The duty of the church is to wait, pray, and preach the Gospel.[20]

John Owen (1616–1683) was another English Nonconformist minister who sought to inspire hope through reference to divine promises of the outpouring of the Holy Spirit. Owen, a prominent Independent theologian, held important posts during the interregnum, including chaplain to Oliver Cromwell, preacher to the council of state, and vice-chancellor of Oxford University. With the Restoration, he was ejected from his living at the deanery of Christ Church, Oxford. A few years later he was indicted under the Conventicle Act, but avoided imprisonment.

Though the 1670s and until his death, Owen repeatedly preached jeremiads.[21] He excoriated England for its sins, warned of approaching punishments, and called for repentance and reformation. He delineated the duty of the people to forsake their evil ways, of civil magistrates to discourage vice and to encourage piety, and of ministers of the Gospel to promote vital religion. Yet, again and again he returned to the notion that the vicissitudes of religion are in the hands of Providence. Religion decays when God withdraws and revives when he restores his Spirit.

Having said that the only remedy was an outpouring of the Spirit, Owen continually assured his hearers that there were grounds to hope that a remedy would come. He felt it especially important to offer these assurances to the Protestant Nonconformists during and immediately after the Exclusion Crisis, 1678–1681, when the danger to true religion from the Catholicism of James, Duke of York, seemed great. In April and May of 1680, Owen preached a series of sermons on "The Use of Faith" in times of trial.[22] He reminded his hearers of God's promise the church would be preserved through all adversity. In addition, he said,

> God hath yet *the fulness and residue* of the *Spirit*, and can pour it out when he pleases, to recover us from this woful state and condition, and to renew us to holy obedience unto himself. . . . I believe truly, that when God hath accomplished some ends upon us, and hath stained the glory of all flesh, he will renew the power and glory of religion among us again, even in this nation.[23]

Following the defeat of the Exclusion Bill, Owen published *An Humble Testimony Unto the Goodness and Severity of God in His Dealing with Sinful Churches and Nations* (London, 1681). "The truth is," he wrote, "the land abounds in sin,—God is angry, and risen out of his holy place,—judgment lies at the door." Reformation was needed to avert judgment, but the only cause of reformation would be an outpouring of the Spirit. Could England expect such an outpouring? "At present there seems to be no other hopes of it," Owen answered, "but only because it is a sovereign act of divine grace." Yet, "sovereignty can conquer all obstacles," and this was the way God had healed the church in the past.[24] On a fast day in December, 1681, Owen spoke "Seasonable Words for English Protestants" specifically to offer evidence that God had not forsaken England: "In this woful state there is yet an intimation made of a covenant interest of Judah in God, and that God did yet own them as his in covenant."[25]

Owen spoke of two principal signs of the coming of an outpouring of the Holy Spirit: prayer for the millennial kingdom, and the revival of the power of evangelical preaching. He wrote, "It is our duty to pray for the accomplishment of all the promises and predictions that are on record in the book of God concerning the Kingdom of Christ and his church in this world." "Prayer for the accomplishment of promises hath been the life-breath of the church in all ages."[26] Repentance, reformation, and revival would come to England, he also insists, only if "there be, through the providence of God, provided another manner of administration of the word throughout the nation than at present there is; which is the only means of conviction and conversion unto God." Evangelical preaching is necessary for religion to revive.[27]

Pious British Nonconformists in the era of persecution during the reigns of Charles II and James II found in the vision of a revival of religion hope for better times brought about by the pouring out of the Spirit of God on the British nation. Patient faith, prayer for the church, and revival of the power of evangelical preaching would all be signs of the approach of a time of grace for the nation. Revival of religion would mean not only greater personal holiness and more numerous conversions, but also moral reformation in the nation and the purification of the national church. It would be a day of grace for the people as a whole.

In his book *The American Jeremiad*, Sacvan Bercovitch posits a sharp distinction between the European and the New England jeremiad, or political sermon, of the seventeenth century. The former, he says, simply combined the summoning of the nation to moral reformation to avoid divine chastisements, on the basis of the dictum that national sins deserve national judgments, with a lament at human frailty. The latter, on the other hand, contained in its summons the promise of inevitable reformation through divine intervention because the nation, that is, New England, was God's chosen people. In the rhetoric of the New England Puritans, national calamities became not mere punishments for transgressions, but proofs that the nation was elect. They were awakenings sent by God to recall his people, demonstrations of his care. The examples given from the works of Robert Fleming, John Howe, and John Owen illustrate that the promise of divine favor and the "unshakable optimism" that Bercovitch discovers in the American jeremiad were not the exclusive possession of New England. Many Englishmen and Scots held a belief as firm as any New Englander's that theirs was God's elect nation and would not be abandoned. The conviction that good would inevitably defeat evil, the essential vision of the Book of Revelation, underlay Christianity itself. In combination with the Reformed teaching that God would save whom he called and reform whom he saved, it heartened the pious in England and Scotland as well as in New England.[28]

New England

Like the British Nonconformists, after the Restoration, New England Puritans turned to the promises of the pouring out of the Spirit at times when they were particularly aware of their inability to help themselves.

Two interrelated calls to action dominate the sermons published in New England between the Restoration and the Glorious Revolution. One is the call for reformation in doctrine, worship, discipline, and morals. The other is the call for the revival of religion, the resurgence of vital piety through an increase in conversions and in the sanctification of the saints. The two calls were related, for only when love of God fired the souls of the people with a zeal to implement his will would a thorough and lasting external reformation take root: "The great Apostacy of New England," wrote Increase Mather,

> is in respect of the Spirits of men, and general decay as to the power of godliness. Were a *thorough Reformation* in this matter, we need not fear any evil, but without this, no other *Reformation* will continue the Lord's Presence with us.[29]

Reformation, the preachers taught, was essential if New England were to fulfill its national covenant. According to their mythology, New England had come into existence by virtue of a solemn agreement with God. The people had engaged themselves to fulfill a special mission. Increase Mather explained:

> The Lord intended some great thing, when he planted these Heavens, and laid the foundation of this Earth, and said unto *New England* (as sometimes to *Zion*) *Thou art my People:* And what should that be, if not that so a *Scripture-Pattern of Reformation*, as to *Civil*, but especially in *Ecclesiastical* respects might be here erected.[30]

As long as the nation fulfilled its part of the bargain, as long as magistrates suppressed vice, the people conducted their market transactions justly, the Gospel was preached, and the sacraments kept pure, God would protect the nation and make it prosper. From the beginning of settlement, New England's leaders had warned of the dangers of neglecting her holy mission. By the second generation, the charge of "declension" rang from the pulpits. In 1663, John Higginson found it necessary to remind the people that "New England is originally a plantation of Religion, not a plantation of Trade."[31] The people, alleged the preachers, were forgetting their high purpose, they were becoming worldly, vice increased, and even the public worship was threatened with corruption. Now every drought, shipwreck, epidemic, and Indian raid was judged to be a warning from God that unless the people undertook a general reformation, God would abandon New England, allowing it to sink under the weight of its own sins.

The prevalence of sin, the sins of church members, and especially the dearth of conversions among the young led the preachers of second and third generation New England to believe that religion in New England was dying. Yet, the ministers remained confident that "there is still hope, that it may

Revive and *Live*."[32] Hence, they exhorted the people to "labour to get a new Heart and new Spirit."[33] A spiritual reformation would save not only souls, but also society, for the new heart would provide the repentance required for a serious external reformation.

The religious literature of the entire era was a plea for reformation, spiritual and moral. Titles reflected this central concern: *An Exhortation unto Reformation* (Samuel Torrey, Cambridge, 1674), *The Necessity of Reformation* (the "Reforming Synod," Boston, 1679), *Reformation the Great Duty* (Samuel Willard, 1694). Yet, while the people were being instructed to be up and doing, they were reminded of their impotence to effect reformation. "We must labour in it," they were told, "under a sense of our own insufficiency to Reform";[34] "The way of man is not in himself, so perverted are we by nature, that until a day of power comes down upon us, our hearts will not incline to God";[35] and, "When we have done all that we can, *Religion* must be revived, not without the greatest Miracle of Divine Power, Grace and Mercy by God himself."[36] The people were obliged to obey God's law, in their hearts as well as in their actions, yet the power to obey was God's alone to grant. This paradox lay at the center of Reformed theology: man was fully responsible for his sins, God fully responsible for man's salvation.

When urging reformation, the preachers relied on a particular set of scriptural passages that emphasized the necessity of man's repentance before God would show mercy:

> And therefore will the Lord wait, that he may be gracious unto you. Isa. 30:13.
>
> Return ye backsliding children, and I will heal your backslidings. Jer. 3:11.
>
> Therefore now amend your ways and your doings and obey the voice of the Lord your God, and the Lord will repent him of the evil that he hath pronounced against you. Jer. 26:13.
>
> Repent and turn yourselves from all your transgressions. Ezek. 18:30.
>
> Thus saith the Lord of hosts; Turn ye unto me, saith the Lord of hosts, and I will turn unto you, saith the Lord of hosts. Zech. 1:3.
>
> Return unto me and I will return unto you, saith the Lord of Hosts. Mal. 3:7.
>
> Remember therefore from whence thou art fallen, and repent and do the first works. Rev. 2:5.

The preachers used these passages to emphasize man's initiative in the process of reconciliation with God: The offended God would turn his face and smile on man when man repented and turned toward him. "It is in vain for any sinful persons to expect God turning to them in mercy and favour, who do not turn unto him with true repentance and reformation";[37] "This is the way to recover the presence of God, to return to him, and therefore turn from all sin";[38] "God is waiting to be gracious. . . . Hence, if we Reform, there is hope";[39] "Let us amend, and God will repent."[40]

To make it clear that the initiative in a reconciliation was God's, the ministers cited a contrasting set of scriptures:

> Turn us again O Lord God of Hosts, and cause thy face to shine and we shall be saved. Ps. 80:19.
>
> I have seen his ways, and I will heal him. Isa. 57:18.
>
> Turn us unto thee O Lord and we shall be turned. Lam. 5:21.

In the former passages, man turns to God; in the latter, God turns man. Since reformation depended on the revival of religion, and revival on God's free grace, the primary role of men was to pray:

> When the matter requires more then ordinary help from the Lord himselfe, and yet the Lord seems to withdraw and to be asleep, then the Servants of God must awaken our Saviour, saying, *Lord save us or we Perish*.[41]

God's way was to channel his grace through his chosen means, especially the preaching of the Gospel and distribution of the sacraments.[42] Impotent man could only make use of those means while praying to God to make them effective.

The doctrine of the covenant allowed the preachers to ask the people to do that which they lacked the power to do. God saved those nations that kept the national covenant and those souls that took hold of the covenant of grace. In both cases, God had promised to save those who fulfilled certain tasks, and to give the power to fulfill those tasks to those whom he would save. Nations were required to obey; souls were required to believe. Samuel Willard attempted to make this doctrine clear:

> When God hath a purpose of mercy to a rebellious People, and is resolved to exalt his Grace upon them, and also makes known these purposes in those discoveries which we call absolute Promises, he doth it so as not to cross this Rule of his Covenant with People, and therefore engageth not only to give deliverance from his Judgments; but also to do it in such a way as withall to give them the condition.[43]

A people turned to God only because God turned them. John Davenport justified his exhortations to reformation by referring to the conditions of the covenant:

> It is true, that we cannot turn unto God of our selves, till he turn us unto

> himself, . . . yet, when God calleth the vessels of mercy to turn unto him, and addeth his Promise for their encouragement thereunto, his Spirit in and by the promise, works converting grace in them.[44]

Although salvation, both temporal and eternal, resulted from divine mercy, the people must still act in order to fulfill the conditions of the covenant. Thus, Samuel Torrey could go so far as to assert, "surely it lyes much, yea most, and almost onely in you, to retain the gracious Presence of God in these Churches, by taking hold upon him by his *Covenant*".[45] Man's responsibility to act had to be balanced by the caveat that his actions did not earn God's favor. Reformation was both a result of God's favor and a condition of the covenant.[46]

The interpretive framework for the revivalism of the eighteenth century, the pouring out of the Spirit, found expression in this period in New England, just as in Great Britain. Alongside jeremiads urging the necessity of reformation, preachers urged *The Necessity of the Pouring Out of the Spirit from on High upon a Sinning Apostatizing People*.[47]

The Indian uprising in 1675–1676 known as King Philip's War reconfirmed for many of New England's Puritans their own insufficiency and ultimate dependence on God for survival and prosperity. That uprising touched every New England family. One out every sixteen white New England males eligible for military service died, half of their towns suffered some damage, and twelve were destroyed. In each of the three succeeding years, 1677, 1678, and 1679, a New England minister delivered and published a sermon that focused on the outpouring of the Holy Spirit as the one necessity to save the people from the sins that had called down on them such devastating judgments.

In *Righteousness Rained from Heaven* (Cambridge, 1677), Samuel Hooker delivered "a serious and seasonable discourse exciting all to an ernest enquiry after, and continued waiting for the effusions of the Spirit." He described the outpouring of the Spirit in general terms. By powerfully influencing the means of grace, God would turn the hearts of the people to him and "make us a religious, godly, righteous people."[48]

In his *Pray for the Rising Generation* (Cambridge, 1678), Increase Mather exhorted his hearers to cry to God "that his Spirit may be poured on the now risen (a multitude of sinful men that are risen up in their Fathers stead) and upon the after rising Generation." Mather described "what is implyed in this *Pouring* of the Spirit" as a great amount of grace infused into the individual soul, not as grace granted to a great number of souls. Yet, he went on to describe past events wherein the outpouring of the Spirit had converted multitudes within a short span of time: "In the last age, in the days of our Fathers, in other parts of the world, scarce a Sermon preached but some evidently converted; yea, sometimes hundreds in a Sermon." He cited Robert Fleming's recently published account of the communion service at Shotts, Scotland, in 1630, and his own acquaintance with a day of prayer and fasting at Milford, Massachusetts, which had been followed by the conversions of many of the young.[49]

William Adams asserted *The Necessity of the Pouring Out of the Spirit from on High upon a Sinning Apostatizing People* (Boston, 1679). Adams presented a detailed analysis of a pouring out of the Spirit. "By *Spirit* here which is said to be poured out upon them," he began,

> we may understand . . . the saving gifts and graces of the Spirit, by the donation of which, persons are enlightened, & regenerated, sinners converted, & Saints more & more sanctified. And by *pouring out of the Spirit*, we may understand a plentiful effusion, or giving forth of the gifts and graces of the Spirit to the sound Conversion and thorough Sanctification of a People.[50]

The two immediate effects of an outpouring would be the conversion of the unregenerate and the growth in grace of the saints. General happiness would increase because of the number of people seeking or growing in holiness. "When the Spirit is poured out upon a people," Adams explained, "all, or the generality of them, or at least very many among them will be either enquiring for, or walking in the way to Zion with their faces thitherward."[51]

These preachers used the promises of the outpouring of the Spirit to prevent despair of reformation. "Be not discouraged," urged Samuel Hooker, "to the Lord belongeth righteousness and salvation, although to us confusion of face." New England had experienced

> the weakness and utter insufficiency of all means in themselves considered to keep us with God, or reduce us to him when turned away. But it may be Christ will shortly come down, and then all will be mended.[52]

"We have had experience of the *inefficacy* of means upon us to bring us into order and to a good frame," admonished William Adams. Warnings, punishments, deliverances, and renewal of punishments had not brought New England to reform.

> Truly we are grown so irrational so as to spiritual concernments, that it is not Gods working upon us only in a rational way that will reclaim us, or bring us to his will.[53]

The growing frequency of references to the pouring out of the Spirit in the last decades of the seventeenth century conformed to a general change in pulpit rhetoric. In his 1975 study of *Power and the Pulpit in Puritan New*

England, Emory Elliott documents this shift during the 1680s, as pastors addressed the psychological needs of New England's second and third generations of Puritans. The men of the first generation were independent, self-reliant, and iconoclastic; however, they did not raise their children to be like themselves. Seeking to control the society they established, they instituted a rigid patriarchy. They raised barriers to church membership to discourage newcomers from gaining control, and they held onto the land, keeping their children, even as mature adults, subordinate. They repressed their children, breaking their wills in infancy and burdening them with guilt. The children, holding themselves in low self-esteem, grew up unsure of themselves and hesitant to make the public confession of faith their parents had made a prerequisite for church membership. The jeremiads of the 1660s and 1670s, by condemning the sinfulness and ungratefulness of the "rising generation" and exalting the founding fathers as heroic saints, objectified and helped the second and third generations face their self-doubts and fears. Yet, the ministers who entered the pulpits in the 1680s realized the need of the second and third generations, as they came into maturity and assumed positions of responsibility, for self-confidence. They shifted their emphasis from guilt to assurance. Instead of stressing the sinfulness of men and the judgment of God, they focused on the power of the Father to save sinners and the forgiveness merited by Christ the savior. They encouraged the people in their social roles by emphasizing duties expected of true Christians rather than the repentance required of sinners. They suffused their sermons with a spirit of optimism, reassuring their hearers that Christ's special protection embraced New England because of the special place it had in God's plan.[54]

By the end of the seventeenth century, the belief that God poured out his grace in abundant measure at particular places and times had become an important devise for maintaining hope for a revival of religion. Samuel Torrey defended this belief in his Massachusetts election sermon in 1695:

> At some times, and in some cases God doth in Sovereign Mercy, Save his Churches and People, both from Sin and Judgment, in a more immediate and extraordinary way of working by himself. . . . *There are certain times and extraordinary cases wherein God thus Saves his People by himself* . . . when their Condition . . . is altogether *hopeless* and *helpless*.

God works more immediately when, by his Word, "*he make it mighty and powerful, to all the saving Effects and Ends of it*; especially in the *Conversion & Reformation* of his People." The promises of the pouring out of the Spirit, said Torrey, "hath been and ever will be in successive Accomplishment to the Church; it is that whereby the Church hath lived and been upheld in all Ages."

Torrey remarked on the labors that had been exerted to effect reformation in New England and concluded, "What remains then but that we believe, pray and wait."[55]

In the later seventeenth century, pious men in both Great Britain and America looked for a great spiritual awakening. They believed that on occasion God granted extraordinary dispensations of his grace to communities, ranging from individual congregations to entire nations.[56] They based this belief on both Scripture and experience. One need not, they declared, restrict application of the Scripture promises to the era of the millennium, "but there is many a *like dispensation* of God in accomplishing these promises under the Gospel." After all, Peter had cited Pentecost as a fulfillment of Joel 2:28 (Acts 2:16–18).[57] William Adams quoted Robert Fleming with approval:

> There is no particular Church where the light hath shined but hath had its special times, some solemn day of the pouring out of the Spirit. . . . Whence a great tack of souls to Christ hath followed, besides the reaching of the Conscience, and stirring the affection of many others under a common work of the Spirit.

Adams concluded, "Oh therefore pray that New England may have such [or another such] a solemn day, before her Sun go down."[58] Likewise referring to Fleming, Increase Mather taught that "as particular persons, so Churches and Congregations have their *day of Grace*."[59]

The Anglo-American concept of a revival of religion in the later seventeenth-century was more theological than anthropological. It was less an understanding of how men act than of how God acts. It was theology in its basic sense, the study of God and his relation to the world. The idea of the revival developed out of the paradoxical logic of the national covenant. On the surface of the teaching of the covenant, as expressed in the jeremiad, the relationship between God and communities was dependent on the actions of men. When they obeyed, God blessed them. When they disobeyed, God punished them. But ultimately, God was the sole arbiter. Without his grace, men could not obey. With and through his grace, they obeyed. Since his grace was both arbitrary and irresistible, the spiritual condition of a community revealed not merely the acts of men but also the acts of God. If religion thrived and morality reigned, God was pouring out his grace. If religion languished and immorality flourished, God was withholding his grace.

Between about 1690 and 1720, many of the clergy in the Reformed tradition in Great Britain and America stressed the understanding that men could do nothing to merit divine smiles, that they must seek his undeserved grace. Sacvan Bercovitch, again, remarked on this shift in emphasis in New England:

> "*Unless God will pour out His spirit on us*": the qualification, repeated over and again, directly and indirectly, grows in volume as the jeremiads proceed into the eighteenth century.[60]

The same is true of the sermons of Reformed preachers in Great Britain. Their long-continuing efforts at reformation proving ineffectual, Reformed preachers in Great Britain and New England evaded despair by shifting the burden of change to God. They came to say that the people had been trying to accomplish too much through their own powers and had not relied enough on God. God had allowed reformation to fail so that his people might recognize their absolute dependence on him. Meaningful reformation would not come about until God poured out his grace in abundant measure. This delving into the theological side of the covenant resulted from the clergy's realization of the failure of their efforts at reformation; however, it also followed from the clergy's loss of practical power to effect changes and to stem undesired ones in their communities.

2

Religious Revival as Deus Ex Machina 1690–1720

> Some think the Zeal of Rulers, and Faithfulness of Officers may put a stop to Sin. But how shall these things be come at, if the Spirit of Religion do not revive among us?
>
> SOLOMON STODDARD, "The Benefit of the Gospel," in *The Efficacy of the Fear of Hell, to Restrain Men from Sin* (Boston, 1713).

After the Glorious Revolution, the Reformed clergy in Great Britain and British America had to redefine their relationship with society. During the succeeding generation, supported by the civil authorities with much less vigor than they had experienced before the Stuart Restoration, they anxiously experimented with means and methods to influence the piety and morality of their communities. Disappointment with the effects of those experiments brought some clergymen on both sides of the Atlantic to look to descent of the Holy Spirit as their sole source of any meaningful improvement of religious conditions.

In England, while the worst aspects of persecution were removed under William and Mary, blasted were any hopes of comprehension of the Nonconformists within the established church. Instead, Presbyterians, Independents, Baptists, and Quakers became tolerated outsiders. They were now dissenters outside the establishment, rather than nonconformists to it and stabilized in numbers at a minority of about 6 percent of the population of England and Wales. Unrecoverable were the days of the interregnum when the Reformed hoped to introduce a temporal as well as spiritual reign of saints. Excluded from political power, Dissenters could not directly influence moral legislation or public policy. The attempt by Presbyterian and Independent ministers to increase their influence through a union based on the "Heads of Agreement" of 1691 foundered on the issue of synodical authority. At the beginning of the eighteenth century, when even the clergy of the established church were losing prestige, the Dissenting ministers were well aware that their sole instrument of influence was moral suasion.

Despite the reestablishment of Presbyterianism following the Glorious Revolution, the authority, both practical and moral, of Scotland's Presbyterian clergy was now but a shadow of what it had been under the earlier Presbyterian establishment. The Scottish state no longer reliably supported church-initiated reform. Civil penalties for excommunication were abolished in 1690. After political union with England in 1707, the legislative power was in London, in Anglican hands. The Toleration Act of 1712 legalized episcopal worship in Scotland and forbade magistrates to enforce kirk censures and summonses. That same year, Parliament restored the right of lay patrons to fill pastoral vacancies.

The Glorious Revolution brought about the fall of the Dominion of New England and restored home rule, with significant modifications in Massachusetts, to the New England colonies; however, the restoration of home rule did not mean a return to religious monopoly by Puritan Congregationalists. Toleration of most varieties of Protestantism became the law. By the 1670s the original close cooperation between clergy and magistracy had broken down. After Massachusetts lost its first charter, her governors refused permission to call a synod, which might have repeated the supposed success of the Reforming Synod of 1679. Even the Saybrook Platform in Connecticut failed to increase the influence of the clergy. New England's law courts occupied themselves with property and debt cases and the enforcement of contracts and concerned themselves less with the punishing of moral transgressions. Because of the increasing heterogeneity of the populace, churches became less important than civil law courts as tribunals for resolution of local controversies.[1]

Secularization made a change in clerical strategies necessary. *Secularization* has many meanings, the most basic among which is a lessening of societal expectation that appeals to the supernatural will affect daily existence.[2] Although the disappearance of witchcraft prosecutions, the popularity of the writings of Enlightenment thinkers, and the interest of the educated in scientific experimentation reflected a lessening of expectation of supernatural intervention in daily life, eighteenth-century Britons and Americans were still highly religious. Secularization in two other senses, however, was significant in England and New England: decline in the practical influence of the clergy over public policy and social behavior, and growth of indifference of civil authorities to the promoting of religious uniformity and performance of religious duties. The churches in Great Britain and New England turned to voluntary organizations such as societies for prayer and pious consultation and societies for the reformation of manners, that is, of public morality, in large part because of their loss of prescriptive power.

Great Britain

Secularization diminished the authority of the clergy of the Established Church in England, and it was they who took the lead in organizing voluntary efforts at spiritual and moral reform in the latter part of the seventeenth and beginning of the eighteenth century. As Tina Isaacs explains, not only was the Toleration Act of 1689 interpreted by many as sanctioning voluntary church membership, but also by disallowing presentments in church tribunals for nonattendance, the act resulted in a dramatic fall in all other ecclesiastical indictments. With the rapid decline of canon law as a means of controlling the laity, Anglican clergymen interested in religious and social reform variously sought to revive the church's religious jurisdiction, supported the Societies for Reformation of Manners—which prosecuted immorality in secular courts—and promoted religious education.[3] By the early eighteenth century, according to Alan Gilbert's analysis, changes in attitudes toward religious practice removed many barriers to participation in voluntary religious agencies. At the same time the clergy's own social and political influence were declining, civil sanctions against Dissent, Recusancy, and irreligion diminished. New forms of religious expression spread because of the declining powers of the religious establishment, and ministers resorted to those new forms because of the increased ability of the populace to ignore formal religious observance. The growth of voluntary institutions associated with the Church of England, such as the Religious Societies, the Society for the Propagation of Christian Knowledge, and the Societies for Reformation of Manners, reflected the erosion of the authority of its ministers.[4]

Appalled by the moral decadence that had spread among the highest ranks of society under the example set by the Stuart Court in Restoration England, several pious Anglicans took steps to restore vigor to religious life. In London, in 1678, a group of devout young tradesmen agreed to meet once a week for religious exercises as a means of counteracting the influence of the deist and Socinian clubs. Under the direction of Anthony Horneck, a German educated at Oxford University, they modeled their society on Germany's Collegia Pietatis, the Pietist prayer groups that appeared first at Frankfurt about 1670 under the leadership of Philipp Jakob Spener. Soon the Religious Societies multiplied and spread widely in England, Wales, and Dublin. By 1710 there were forty-two in and about London. They adopted strict regulations. Members were to be Anglican males, of more than sixteen years of age, and confirmed. Each society was to choose an Anglican clergyman to direct them and to assign the practical divinity to be read. With their charitable collections they kept poor children in school, distributed pious books to prisoners, sailors, and soldiers, and helped support foreign missions. In reply to James II's public celebration of the Catholic Mass, the Anglican Religious Societies organized nightly public prayers and special vigils and fasts.[5]

Anglican and Dissenter alike prayed that the overthrow of their Catholic monarch in 1688 would be a Glorious Revolution not only politically, but also morally. With Protestant William and Mary on their thrones, the government could once more promote sober Christian behavior. Responding to these expectations, the new king and queen issued several proclamations ordering the responsible authorities to enforce the laws against public vice. Several Englishmen, disturbed by public immorality, organized to insure compliance with the proclamations. About the year 1690, a group of four or five members of the Church of England formed a society in London to see that the city officials enforced the laws, particularly those against prostitution, drunkenness, and profanation of the Lord's Day. Their program spread. By 1699 eleven, and by 1701 nearly twenty such Societies for the Reformation of Manners, made up of Anglicans and Dissenters, operated in London and its suburbs. They hired informers, initiated prosecutions, printed blank warrants, and distributed pamphlets with abstracts of the laws and penalties concerning public vice. Societies soon appeared in neighboring counties, as well as in Scotland, Ireland, Jamaica, and New England. The Dutch, the Germans, and the Swiss printed translations of accounts of the societies. The royal proclamations and the propaganda of the reformation societies were designed to motivate their readers with the national covenant: national sins deserve national punishment.[6]

After 1710 the Anglican Religious Societies and the nondenominational Societies for the Reformation of Manners lost their prominence. The Religious Societies sank rapidly into obscurity. Some churchmen objected to the societies' invasion of the ministerial office by their members' visiting the sick and reproving and admonishing one another. Critics accused the members of spiritual pride, of erecting a church within a church, and of destroying parochial communion by their special services. There were two chief causes of collapse. Because of their High Church character, they were suspected of Jacobitism and of influencing the charity schools for political purposes. And, despite the vast differences between the two groups, they were confused with the reformation societies. The Societies for Reformation of Manners were never popular, for the bulk of the people found their use of informers obnoxious. Despite some initial success, measured by the number of prosecutions initiated, after 1710 the societies became much less active. Although the reformation societies survived in London until 1738, their main activity seems to have been reduced to the sponsoring of yearly reformation lectures, one for Anglicans and one for Dissenters.[7]

By the 1720s, the reformation societies' Dissenting lecturers were explaining the inefficacy of the movement with the observation that reformation would not be lasting until God poured out his Spirit. "The Blessing of God, and the Influence of the Holy Spirit, are absolutely necessary to the Reformation of the wicked World," said Daniel Neal. Samuel Price told the societies that "the Reformation of our Country will abundantly countervail the greatest Cost. But we must remember, that it will not be effected *by Might, nor by Power*, without *the Spirit of God.*" Neal and Price called for prayer for an outpouring of the Spirit. In 1735 John Guyse made this explanation the very doctrine of his reformation lecture. By his text, Hebrews 9:10, "until the time of reformation," he implied that reformation must await God's appointed time. "The corruptions of mankind will prove too hard for all human laws and restraints," he said, "unless there be a change upon the heart." Moral reform depended not so much on the execution of good laws, as on the motion of God's grace, issued on the land at God's pleasure. Internal reformation, the change of the heart, would be the only source of lasting external improvement.[8]

In Scotland in the early decades of the eighteenth century, the notion that dying religion would revive only at God's pleasure became a recurrent refrain. Robert Wodrow, author of *The History of the Sufferings of the Church of Scotland*, lamented to Cotton Mather in 1722 that "the Gospel has very little fruit, and the ignorant and unholy continue so still, and matters will continue worse and worse, until the Spirit be poured down from on high." John Maclaurin began his ministry in Glasgow in 1723 with a sermon on "The Necessity of Divine Grace to Make the Word Effectual." When in 1730 Ebenezer Erskine, who would shortly lead a secession of radical evangelicals from the Church of Scotland, proclaimed that "we might look for a work of reformation to be revived in the land . . . when the Spirit shall be poured out from on high," he was merely repeating the accepted wisdom.[9]

Convinced that efforts at reformation would be fruitless without the concurrent action of God's Spirit, the British reformers shifted their emphasis from promoting morality to praying that God would revive religion through a pouring out of his grace. The bleaker the prospects for reform, the greater their emphasis on the power of prayer. One such effort occurred during the reign of High Church Queen Anne, when a sense of foreboding and helplessness before imminent disaster pervaded Protestant Dissent in England. The year 1712, in particular, seemed a time "of more than ordinary Distress," when the Protestant interest appeared "to Ly bound upon the Altar." Dissent was under pressure in England, episcopacy winning toleration in Scotland, and Britain making a peace with France that left Huguenots unprotected. A number of London Dissenters issued "A Serious Call from the City to the Country" for an agreement to set aside an extra hour weekly for private prayer that God "would appear for the Deliverance and Enlargement of His Church." Protestants everywhere, they said, seemed on the brink of destruction, and England was inundated by all sorts of wickedness. "We exceedingly need those fresh Effusions of the Spirit of Grace upon us, which nothing but a *Vehement Cry* to the God of all Grace, will be likely to obtain for us," wrote the authors of the call. The union of Christians throughout the British isles in this agreement might save Britain, or, since "*prayer* will do mighty Things," even signal the coming of the millennium.[10]

New England

A change from emphasis on efforts to effect reformation to emphasis on prayer for religious revival can be traced in the careers of four leading New England ministers, Increase and Cotton Mather, Samuel Danforth, and Solomon Stoddard, between the years 1690 and 1720.

During the late seventeenth century, published sermons preached in New England on public occasions, such as election days, public fast days, and days of thanksgiving, often contained a formulaic exhortation in which each segment of society would be reminded in turn of its particular role in effecting reformation. Civil magistrates were urged to enact and enforce wholesome laws for the suppression of vice, the encouragement of virtue, and the exercise of true worship. Citizens must elect zealous rulers, uphold the law, and cooperate in its enforcement. Gospel ministers must be guiding fathers, admonish sinners, encourage the weak, and preach for conversions. Churches must choose orthodox, holy ministers and keep themselves pure. Parents must train children in the ways of holiness, and children and servants must be tractable and obedient.[11] Boston minister Increase Mather (1639–1723), New England's leading Jeremiah in the last third of the seventeenth century, continually promoted the use of these means to reformation. Mather, however, became progressively disillusioned with their effectiveness until, sometime after 1700, he came to base his hope almost exclusively on the promises of a glorious state of the church on earth in the latter age.

During the 1670s, Increase Mather preached up three prerequisites of reformation: 1) civil magistrates zealous for reformation; 2) public renewal of church covenant; and 3) the maintenance of the requirement of evidence of regeneration for full church membership. As early as the 1670s, however, he began to doubt the zeal of magistrates. He looked to the people as the source of reformation. Not only should they elect godly rulers, but the people should reform themselves voluntarily by renewing their church covenants. Renewal of covenant was a means to involve large numbers in reformation without depending on civil authority.[12]

By 1700, Mather gave up any reliance on the state. His confidence in the pure church and in covenant renewal remained strong, for he now had to rely solely on the elect remnant to save New England. He argued that both the requirement of probable evidence of regeneration for full church membership and renewal of covenants by congregational churches were absolutely necessary for a revival of religion.[13]

In the final fifteen years of his life, Mather concluded that even the pure church would not save New England. Instead, he turned more and more to the promises of a glorious state of the church in the latter days. He closely compared world events with scriptural prophecy and gathered reports of the conversion of Jews and of the work of Dutch and German missionaries in the Far East. No longer relying on the pure New England churches to bring about the millennium, he now relied on the millennium to restore the purity of those churches. He read the once pure churches as a "Specimen of what shall more Generally obtain *in the Glorious Times Approaching*."[14] The millennium was imminent, at which time "the Members of the Church shall be *Converted* Persons. . . . And there will be Purity, both as to the Worship and the Worshippers."[15] Mather, disappointed with human efforts, now placed his reliance on prayer. In God's appointed time, which Mather expected shortly, he would bless the preaching of the Gospel with numerous conversions the world over. The duty of the faithful was to see that the Gospel was preached and to pray for conversions and the millennium.

> We should therefore Pray that there may be a plentiful Effusion of the Holy Spirit on the world. Then will Converting work go forward among the Nations, and the Glorious Kingdom of Christ will fill the Earth.[16]

"I don't doubt," he wrote in 1719, "but there are some here living will see the beginning of those glorious days."[17]

Members of the younger generation of ministers at this time, men who had less firsthand knowledge of cooperation with the state in reformation, underwent a similar change of attitude. Several of these men involved themselves in the spiritual and moral reform movements of the day. Disappointment with these voluntary organizations brought them to look to the outpouring of the Holy Spirit to bring about success. Among these men were Increase Mather's famous son, Cotton, and Cotton Mather's friend, Samuel Danforth.

Proclamations, sermons, and religious and reformation societies in New

England evince the influence of the English movement for the reformation of manners. In March 1690 the governor and General Court of Massachusetts ordered the faithful enforcement of the laws against vice, especially those against blasphemy, cursing, profane swearing, lying, gambling, Sabbath breaking, idleness, drunkenness, and sexual immorality. The colony asked the ministers of the Gospel to read the proclamation from their pulpits and to adjoin admonitions against spiritual sins. The colony had the proclamation printed, and Cotton Mather also printed it as an attachment to his sermon on *The Present State of New England* (Boston, 1690). In his election sermon that May, Mather called for the "Reformation of Manners" and encouraged the civil authorities to execute wholesome laws for the suppression of vice. In 1699 the colony issued a proclamation concerning the observance of the Lord's Day, and in 1704 a declaration against profaneness and immoralities. Inspired by a proclamation of Queen Anne, John Norton made his 1708 Massachusetts election sermon *An Essay Tending to Promote Reformation* (Boston, 1708). "Let her Majesties Royal PROCLAMATION for Punishing Vice and Immorality, be attended," he said.

> It might be great Advantage, were Her Majesties Royal Proclamation follow'd in *New England*, as it was in *Old-England*, with REFORMING SOCIETIES. Let the good Laws Enacted be duly Executed.

Grindal Rawson repeated these exhortations in his 1709 election sermon, *The Necessity of a Speedy and Thorough Reformation* (Boston, 1709). "May it, therefore Please your EXCELLENCY!" he addressed the governor,

> to copy after the illustrious Example of Her most Excellent Majesty, whose repeated PROCLAMATION against Vice, & Immoralities, have abundantly convinced the World, that Her Majesty is at an irreconcilable Enmity with every thing that has a tendency to draw down the wrath of an Almighty and Sin hating GOD, on the Kingdoms and Plantations over whom She Sways Her Imperial Scepter.

Rawson urged the Council, the House of Representatives, the clergy, and the people to help enforce the laws against vice and impiety in order to avert God's wrath. In 1716, upon being asked by the colony to read King George's proclamation "for the Encouragement of Piety and Vertue, and for the preventing and punishing of Vice, Prophaness and Immorality," Benjamin Colman delivered *A Sermon for the Reformation of Manners* (Boston, 1716). In the sermon, Colman praised "the late Zeal of the *Justices of the Peace* in this Place to find out and suppress Vice."[18]

Cotton Mather (1663–1728) was not content simply to exhort. In 1702, in imitation of the London Societies for the Reformation of Manners, he founded a society of men to help the authorities in Boston suppress disorders. The following year he published a manual for these societies. In 1704, to aid the enterprise, he published an abstract of Massachusetts's laws against vice, along with a sermon urging the people to reprove error, the rulers to punish disorders, and the citizens to help by informing against wrongdoers. He elaborated on these reforming societies in *Bonifacius* (Boston, 1710), as one of many ways of doing good. In 1705, Boston had two such societies. By 1713 the Boston societies seem to have disappeared, and Mather's attempts to revive them over the next five years failed.[19]

If reformation societies combatted vice, another type of voluntary endeavor could promote virtue. The Puritan founders of New England brought with them from England the practice of meeting privately for religious exercises. Throughout his ministry, Cotton Mather sought to renew the practice. In 1693, he organized a religious society of black servants. He preached often to religious societies of young men. In 1694, he published rules for the government of such societies, suggesting they meet two hours weekly for prayer, repetition of sermons, psalm singing, and discussion of practical Christianity. In 1706, he urged that reviving "the ancient practice of lesser societies, formed among religious people, to promote the great interests of religion" would be an effectual method to advance the power of godliness. He enumerated several sorts of societies: family groups of a dozen men with their wives, meeting once or twice a month at each others' houses; societies of women; and societies of young men. He repeated these suggestions, along with rules for such groups, in *Bonifacius* in 1710, and in 1724 as a proposal "For the REVIVAL of Dying Religion."[20]

Cotton Mather's strenuous attempts to bring about a moral reformation and a revival of piety through voluntary organizations produced a profound disillusionment with human schemes. He ended up depending on an imminent millennium and the direct influences of the Holy Spirit as completely as any of his colleagues.

Robert Middlekauff concludes that sometime after the year 1710 Mather tacitly admitted that the societies had failed to reform New England or to revive her piety. Nevertheless, in Mather's view, they continued to serve an essential function. Efforts to do good served as a basis of unity for true Christians. The union of Christians on basic principles would prepare the way for the Second Coming, which would bring about the ultimate reformation. When the faithful of all the Protestant churches united on the basis of piety, and the elect were converted, the world would be ready for the return of Christ.[21] Like his father, Cotton Mather came to rely on the faithful remnant awaiting the millennium, but who, while waiting, would preach the Gospel, promote piety, and work for conversions.

A chiliasm as fervid as Increase's fired Cotton Mather's zeal. He expected the Second Coming in 1697, and then in 1716, and then from year to year. By the 1720s he believed that all the signs prophesied to precede Christ's return had been given. By the summer of 1724 he had rejected his father's teaching that the conversion of the Jews must come first.[22] He expected the Second Coming, resurrection of the martyrs, and conflagration of the earth to be literal. The oppression of the Protestant churches would end, the impurities in their worship would be cleansed, and the errors of doctrine would be removed; external peace would reign, and holiness would prevail. Mather had always held that the millennium would bring about the desired reformation. The change in his thinking in the second decade of the eighteenth century was that the millennium alone could bring about any significant improvement. Human efforts were in vain. All that was left for Christians to do was to prepare and pray for the pouring out of God's Spirit promised to the Church in the latter days.[23]

The evolution of the thinking of Samuel Danforth (1666–1727), minister at Taunton, Massachusetts, about reformation and revival neatly paralleled that of Cotton Mather. During the second decade of the eighteenth century, he too evinced disillusionment with human efforts to bring about the desired changes, and resignation to awaiting prayerfully on God for his own good time to revive religion.

During the first decade of the century, Danforth had urged two principal means for effecting reformation. His *Piety Encouraged* (Boston, 1705) called on all segments of society to encourage virtue; and his *Duty of Believers* (Boston, 1708) called on all to discountenance vice. Following Cotton Mather's example, he urged organized efforts for these goals: societies for prayer, and regularly scheduled neighborhood family worship meetings. He referred to the example of the Reforming Synod of 1679, suggesting that "ministers should meet at stated times, either in lesser Conventions, or more general Assemblies" to set forth instructions "in order to the reforming of Evils prevailing among us, and for the Reviving of Religion in the Land."[24]

Like Cotton Mather, Danforth organized as well as exhorted. Having read accounts of the English Societies for the Reformation of Manners, he brought together in 1704 several reputable persons of Taunton in monthly meetings to plan methods for suppressing disorders among the youth of the town. Their decision to encourage family worship met a favorable response. Then, in imitation of the young men's religious societies in London, the young men formed themselves into regular meetings for prayer and repetition of sermons, adopting rules of the young men's meeting organized by Cotton Mather in the north of Boston. Danforth reported that "the disorderly concourse of youth was now over." Persons of all ages, but particularly young men and women, visited Danforth concerning their spiritual distress. On 1 March 1705, 300 inhabitants, men and women over sixteen years, renewed the covenant of reformation taken in 1676 during King Philip's War. Many wept for joy during the ceremony. Several "children of the church" (i.e., persons baptized on the strength of a parent's membership in the church) in neighboring towns joined in the ceremony. On succeeding Sabbaths more persons bound themselves to the covenant. On 4 March, fourteen persons "were propounded to the Church, some for full communion, others for Baptism, being adult Persons."[25]

By the 1710s, Danforth changed his emphasis: His principal message came to be that all human efforts toward moral reformation and spiritual renewal were doomed to failure unless God were to bless them. His Massachusetts election sermon in 1714 called on sincere believers to pray for "a gracious visit from heaven . . . as the only and sufficient relief against impending ruin by sins and judgments." "Nothing short of such a divine visitation," he said, "can recover the church from apostacies and calamities." "All attempts for reformation are unsuccessful," he insisted, "when God is departed from Israel. The means of grace are evanid [without effect] when the Spirit of God is withdrawn."[26] The doctrine of a sermon he delivered in 1717 was that

> the Revival and Progress of Religion, is a work accomplished by God himself, in answer to the Prayers of his People, when they are deeply humbled for their Sins, and sensible that they are destitute of all Help for the promoting of the work, except the help of God alone.[27]

In a letter to the ministers of Boston in 1720, he explained that their endeavors met with frequent disappointments so that God could

> maintain in us an entire dependence on our Lord Jesus Christ for success in our essays for the enlargement of the bounds of His vineyard on earth; finding by our own experience that all our projections and essays of accomplishing any thing that is for the glory of God and the good of the souls of men will avail nothing, till the Lord Himself appear in His glory for the building up of Zion, and he be pleased to work with us.[28]

The need of prayer for an outpouring of grace became a central concern in Danforth's sermons. His 1714 Massachusetts election sermon was "An Exhortation to All to use Utmost Endeavours to Obtain a Visit of the God of Hosts, for the Preservation of Religion, and the Church, upon Earth," and the sermon of 1717 quoted earlier was entitled "The Building of Sion carryed on by Praying."

During the first two decades of the eighteenth century, Solomon Stoddard (1643–1729), minister at Northampton, Massachusetts, in the Connecticut

Valley, underwent a similar change in attitude toward methods to effect a reformation of morals and revival of religion. During the latter part of the seventeenth century, Stoddard had promoted his particular program for the reformation of New England as vigorously as had the Mathers theirs. Like them, by the 1710s, he had reevaluated his methods. Stoddard came to look for a shower of grace—but not for the millennium.

The Mathers and Stoddard, addressing the problem of how to bring potential saints into the church, resorted to different versions of "sacramental evangelism."[29] The Mathers responded to the decline in admissions of new church members by emphasizing growth in grace within the nursery of the church. They spoke of baptism, the Word preached, and the church covenant as the instituted means of grace intended for the use of as many as possible. They offered baptism to anyone who had the barest suspicion of a work of grace and urged people to the table of the Lord's Supper by stressing how small an amount of grace was needed to qualify, and by holding forth the spiritual nourishment the sacrament provided. Stoddard, owning a much more strenuous concept of conversion, believing individuals could know its very moment, rejected the Mathers' strategy. He believed it encouraged people to stop short of true conversion. Stoddard resolved the ministers' dilemma by redefining the relationship between conversion and the sacraments. He dropped the requirement of evidence of conversion for church membership, which had been the hallmark of the New England Way, and contended that the sacraments were ordinances instituted as means of conversion.[30]

While Increase Mather was defending the ancient polity of the New England churches, Stoddard was attacking it as a major obstacle to reformation. According to Stoddard, the power of religion had declined in the land not because the churches had been defiled by unregenerate members, but because too many people had been excluded from the means of grace. The sacraments, baptism, and the Lord's Supper, were converting ordinances. Salvation came through the church, and it was important to bring as many people into it as possible. Stoddard thus upheld a national church, rather than congregational polity, in which all but the openly scandalous would be urged to partake of the sacraments. He defined visible sainthood as historical faith and outward conformity, denying the possibility of anyone's identifying the regeneration of another. Stoddard's Presbyterianism, that is, his belief in a national church and in the authority of its ministers and his denial of the validity of individual congregational church covenants, was an adjunct of his commitment to the promoting of conversions.[31] In *An Appeal to the Learned* (Boston, 1709), he sought to show that the duty of the unregenerate to come to the Communion table was "a Truth . . . whereby a door is opened for the revival of Religion."[32]

Stoddard believed, like the Mathers, in the national covenant and in the duty of the civil authorities to enforce the moral law. By the beginning of the eighteenth century, however, he recognized, like the Mathers, the failure of the state to bring about a moral reformation. Coercion had failed. Some sort of compelling, inward persuasion was required. Stoddard found the solution in *The Efficacy of the Fear of Hell, to Restrain Men from Sin* (Boston, 1713). "There have been great endeavours to promote Reformation," he observed. "Laws have been enacted, Sermons have been preached, Covenants have been made, but all endeavours have had a miscarrying Womb; there has not been one Sin generally reformed these twenty Years," because people are not afraid of hell.[33]

The preaching of terror was part of Stoddard's larger plan. He concluded that a lasting moral reformation could be achieved only through a revival of religion, only if God were to shower New England with his grace. "Some think Family Government may put a stop to Sin," he wrote.

> Some think the Zeal of Rulers, and Faithfulness of Officers may put a stop to Sin. But how shall these things be come at, if the Spirit of Religion do not revive among us? . . . We Pray, We Fast, make Laws, and dispute about Reformation, but yet we are in Affliction, and the Hand of God goes out against us: And so it will, unless there be a reviving of Religion.[34]

By the second decade of the eighteenth century, as Paul Lucas has suggested, Stoddard had recognized the failure of his argument for the sacraments as converting ordinances to win support.[35] After 1710, Stoddard began to argue that the Gospel was the sole means of conversion and ceased publicly defending the sacraments as converting ordinances. During the following years, until his death, Stoddard advocated a new program for the promoting of vital religion, a program that centered on the minister as preacher and as pastoral guide to awakened sinners seeking Christ.

For Stoddard, religion would revive when God channeled abundant measures of grace through a conversionist, evangelical, Gospel ministry. But the period of intense religious concern and numerous conversions that would result held no eschatological meaning. In Stoddard's view, the sovereign God was entirely arbitrary with his grace in respect to individuals and to communities. Whenever God revived religion in any place, no one could determine his intentions.[36]

Stoddard was an exception. Most New England ministers shared the Mathers's interest in the millennium and associated local revivals of religion with the millennial prophecies. Increase Mather expected Christianity to spread through the nations of the world by means of the pouring out of the Spirit of God in numerous conversions through the preaching of the Gospel. He did not look for this to begin in any remarkable manner until the appearance of Christ visibly to the Jewish people, and their conversion, but that he expected shortly. After the conversion of the Jews, "converting work will go on all the world over, even among *Indians and Infidels*." "The Kingdoms of the World will *Gradually* become *the Kingdoms of Christ*, until the whole World is become *Christianized*."[37] Samuel Danforth associated Taunton's reformation in 1705 with the millennium. During that religious excitement, he was moved to "think some times that the time of the pouring out of the Spirit upon all Flesh may be at the Door." In 1717 he suggested that whenever there was a revival of religion in any particular place, the prayers of the people should be for an increasingly wide outpouring of grace so that the millennial promises might be fulfilled.[38]

Increase and Cotton Mather, Samuel Danforth, Solomon Stoddard, and others conceived of New England's history in terms much like those of a classical Greek drama in which the players bring themselves to such a pass that only a deus ex machina, a god descending from the heavens, can bring the plot to a resolution. These men prayed that the God of Christians would descend on New England, like a deus ex machina, producing a harvest of converts and a moral reformation, if not the millennium. One of New England's most prominent clergymen, Samuel Willard, (1640–1707), minister at Old South Church, Boston, referred directly to this pagan theatrical device in this very context in 1695. He acknowledged that "when Good men in all Orders find all attempts to recover a Backslidden People awfully to be frustrated, needs must they sink in dispondency, and lose their Faith and Hope," if they did not know that God often saved a people in such a condition so that they might ascribe their salvation entirely to him: "When times are never so black, second causes afford little or no relief . . . yet Caelo restat iter, and there is THEOS APO MECHANES."[39]

Theos apo mechanes is the Greek for *deus ex machina*. *Caelo restat iter* is an allusion to the story of Daedelus and his wings in Ovid's *Art of Love*, Book 2, line 37. Daedelus had but one remaining avenue of escape from the isle of Crete: "a way remains by the heavens." Similarly, taught Willard, for New England, having tried all other means, heaven was the one remaining hope.

In the interval between the Glorious Revolution and the Great Awakening, reform-minded clergy in Great Britain and America undertook a major reevaluation of the best means to reverse the decay of religion and morals. *Reformation of manners* was a phrase repeated endlessly during the 1690s in Great Britain and America. By the 1720s, *revival of religion* had replaced it as the catchword of the reformers. Before about 1690, ministers on both sides of the Atlantic emphasized the need for moral reformation. God had withdrawn his Spirit, they taught, because of the sins of the land. Therefore, the nation must reform. The ministers told the people that if they reformed, God would revive them spiritually, and they promoted organized efforts to encourage virtue and discourage vice. By 1720, reformers were preaching that if God would revive the people, then they would reform, and they promoted concerted prayer for the return of the divine Spirit. Moral reformation, they concluded, would take place only when God chose to restore vital religion. The clerical reformers believed that God would employ themselves as the principal channels through which he would pour his grace.

3

Preaching Christ: 1710–1730

The Spirit of the Lord must be poured out upon the People, else Religion will not revive. But when the Spirit is upon Ministers, it is a very hopeful sign.

SOLOMON STODDARD, "The Benefit of the Gospel," in *The Efficacy of the Fear of Hell, to Restrain Men from Sin* (Boston, 1713).

It is a truism that a new affective homiletic style was a major factor in the coming of the Revival movement to England and America. By no mere coincidence did that preaching style appear on both sides of the Atlantic Ocean by the 1730s. It had evolved under shared influences. Negatively, those influences included the rise into fashion of moralistic, rationalist preaching and the spread of rationalist heterodoxies against which the evangelically inclined orthodox reacted. Positively, they included new psychological theories and German Pietism. The first decades of the eighteenth century found Reformed clergymen throughout the British world self-consciously assessing their place in society. They believed they had failed, and that spiritual decline continued unabated. Some, searching for an explanation, found a solution in orthodox, conversionist evangelism. The 1720s and 1730s witnessed parallel campaigns in Scotland, England, and America for a zealous, diligent, and conversionist ministry. The evangelically oriented clergy pressed their campaign for an evangelical style of preaching and pastoral ministry all the more urgently because of the rival clerical style of Enlightenment rationalism, which they asserted stood in the way of a revival of religion.

By the turn of the eighteenth century, Scottish, English, and American clergymen who were convinced that piety was expiring found themselves nearly powerless to do anything about it. They comforted themselves with the doctrine of human impotence. If, as John Howe complained in 1678, "the methods of alluring and convincing souls, even that some of us have known, are lost from amongst us in a great part," it was because God had retracted his

Spirit. Without God, men could accomplish nothing.[1] The doctrine of human impotence made sense to clergymen who were in actuality losing influence within their respective societies. The concept of a revival of religion offered a mechanism by which they might regain influence. The ministers believed that God dispensed his grace through means, in particular the preaching of the Gospel. If the doctrine of human impotence explained the ministers' pastoral disappointments, the doctrine of means elevated the importance of the ministers, for as preachers they were the instruments through which God would channel his grace. Religion would revive when the ministers regained the power to allure and convince.

England

The spread of nonevangelical preaching and liberal theology in England early in the eighteenth century split the ranks of Dissent and brought evangelicals together into cooperation to promote a revival of religion through passionate conversionist preaching and ministering.

The rationalist, homocentric spirit of the age affected the clergy of the established church first. The pulpit eloquence of Archbishop John Tillotson (1630–1694) became the standard of excellence. Preachers cultivated the plain, clear, distinct expression of the fashionable essayists. A contemporary Anglican wrote, "Studied expressions and close arguings, neat thoughts and fine turns . . . acceptable to fine wits, . . . *this is the way of our times, and he who excels in the most celebrated* [preacher]."[2] Eventually, these literary developments of the late seventeenth and early eighteenth centuries affected the preaching of Dissenters, particularly Presbyterians and Congregationalists.[3] The Dissenters reduced the number of subdivisions in their sermons and developed their thought more simply and directly.[4] In a 1707 treatise originally written in Latin for the information of Continental clergy, an Anglican described these changes. Most Nonconformists, he wrote, had left off their din and ravings, their shouting and physical demonstrations, "extempore effusions" and "canting phrase and expression. . . . Now their discourses are sober and correct; they study and compose them; they have purged out the old, musty, obsolete words; they take care not to be abrupt and incoherent." Their manner of reasoning no longer resembles the "dotages of Baxter," but the clear method of Tillotson.[5]

Content of sermons changed with style. Among Tillotson's favorite texts had been "My yoke is easy" (Matt. 11:29). The standard message of the English sermon of the Age of Reason was that moral behavior is conducive to temporal happiness. Much in the new preaching resulted from the attempt to defend Christianity against the deistic attack by demonstrating the reasonableness of the Christian religion. Thus, sermons became rational discourses. The clergy tended, in the phrase of Horton Davies, to reduce "revelation to reason, faith to philosophy, and Christian ethics to prudential morality."[6] Dissenters as well as Anglicans responded to the call for Christian apologetics.

The rationalist spirit widened the breach in the ranks of Dissent. Those most in accord with the rationalist spirit moved toward Arminian, Arian, and Pelagian positions. They tended to become Presbyterians. Those who rejected the rationalist spirit moved toward a High Calvinism and tended to become Congregationalists.[7] Between the two extremes stood a group who believed Enlightenment thought and heart-religion still reconcilable.[8] These men became the backbone of evangelical Dissent. They disliked cant, enthusiasm, and narrow party spirit. They tended toward a moderate but orthodox Calvinism.[9]

Although the spread of liberal heresies had alarmed the orthodox since the Restoration, the debate among Dissenters in 1719 at Salters' Hall seems to have been a catalyst for concerted action against the growing popularity of Arminian, Arian, and Pelagian doctrines. The debate itself concerned no doctrine, but whether subscription to the Athanasian Creed should be a test of orthodoxy. Many who were sound on the Trinity opposed making a human composition the basis of ministerial communion. Even so, the debate underscored the existence of fundamental doctrinal disagreements. A number of evangelically oriented Dissenters, fearing that the practice of preaching Christian precepts from the grounds of natural reason had been carried too far, that heathen philosophy had forced out of the pulpit everything peculiar to the Gospel, mobilized to restore the motives of faith in Christ and love of God.

In the 1720s and 1730s "preaching Christ" became the rallying cry of the evangelicals, and Dissenters argued in public over the meaning of the phrase. The meat of the evangelical campaign was an attack on "legalistic" and rationalist preaching that ignored grace and the Scriptures in favor of "mere morality" and reason, preaching, they said, that taught philosophy rather than the Gospel.

The archetype of English Dissenting tracts on evangelical preaching during the first half of the eighteenth century was John Jennings's *Two Discourses: The First of Preaching Christ; The Second of Experimental Preaching*, published posthumously in 1723 with a preface by Isaac Watts.[10] Jennings (d. 1723) conducted a Nonconformist academy at Kibworth, where he was minister of the Independent church. Jennings argued that making the person of Christ the focus of each sermon would distinguish one's preaching "from mere Discourses on Natural Religion." He championed conversionist evangelism: The purpose of preaching was the saving of souls. "Let Christ be the

Matter of our Preaching," his mediatorial office, his life, death, and resurrection, the covenant of redemption, and the sufficiency of his righteousness, urged Jennings.

In 1729 John Guyse (1680–1761), minister to the Independent congregation in New Broad Street, London, delivered two sermons on "preaching Christ," in which he repeated the exhortations of Jennings's *Two Discourses* and stated that the greater number of English ministers did not preach Christ. By decrying the absence of the preaching of Christ, Guyse implicitly challenged those Dissenting ministers who did not stress the atonement to prove they were Christian preachers. Samuel Chandler (1693–1766), minister at the Old Jewry, in London, accepted the challenge. He replied to Guyse in a public letter of 94 pages, accusing him of using "preaching Christ" as a Calvinist slogan to keep up animosities and divisions, "making zeal for a party pass as a certain mark of salvation and so recommending dependence and meer faith, to exclude the necessity of virtue and substantial piety." Guyse answered with a 168-page letter, *The Scripture-notion of Preaching Christ*. Chandler issued a *Second Letter*, which Guyse also answered.[11]

Devout laymen supported the movement to reinvigorate evangelical Dissent. They established several series of lectures in London to promote orthodoxy, with some of the city's most distinguished ministers participating.[12] In 1730 a group of Dissenting laymen founded the King's Head Society, with the purpose of promoting vital religion and strict morality. The society established and financed orthodox lectures. Concerned about the spiritual quality of Dissenting ministers, they established an academy to which they admitted and subsidized only students who gave evidence of strict orthodoxy and of personal religious faith.[13]

From 1730 to 1732, the Dissenters carried on a public debate over ministerial styles by means of a discussion of the best way of reinvigorating the Dissenting interest. In 1730 Strickland Gough, a Presbyterian preacher in London without a charge, published an *Enquiry into the Causes of the Decay of the Dissenting Interest*. He blamed the decline in the size of Dissenting congregations and in the number of ministers on lack of refinement in the pulpits and bigotry in the pews. Evangelically oriented ministers responded in force.[14] Keenly aware of their minority status, the Dissenters argued among themselves over whether their numbers were declining, and if so, why, and over the reason so many of their clergy defected to the established church. On one side of the debate stood a group who believed Dissent would regain its strength through a regimen of affective, evangelical preaching.

Many things in Gough's pamphlet irritated the evangelically oriented: his depreciation of doctrine; his rejection of practices and teachings many considered essentials; his denunciation of congregations concerned to have evangelical preaching; his concern for gentility and the tastes of people of fashion—his advocating dancing masters at academies must have seemed ludicrous to many. Gough's omissions, however, irritated most of all. Nowhere did he deprecate decline in conversions and loss of religious warmth.

Philip Doddridge replied to Gough with *Free Thoughts on the Most Probable Means of Reviving the Dissenting Interest* (London, 1730). Doddridge (1702–1751) had studied under John Jennings, and recently had been chosen to conduct a new Dissenting academy and to minister to the Independent congregation at Northampton. In *Free Thoughts*, he ridiculed Gough's call to suit sermons to the tastes of the polite and educated. To do so would be the ruin of the Dissenting interest, he said. To restrict sermons to subjects of natural religion and moral virtue might attract a few of polite tastes, some of whom have sunk as low as deism, but it would do so at a high expense. It would mean discarding the peculiarities of the Gospel scheme, neglecting the souls of the common people, and alienating the majority of Dissenters, who are plain people, lacking a learned education. Most Dissenters relish evangelical discourses that "awaken, revive, and enlarge the soul." They expect to be instructed in spiritual combat and the operations of the Holy Spirit on the soul and resent those who seem to pour contempt on these subjects "as if they were all fancy and enthusiasm." The majority of Dissenters are ordinary people who desire to be spoken to plainly. They attend preaching not to be entertained,

> but that their hearts may be enlarged as in the presence of God, that they may be powerfully affected with those great things of religion which they already know and believe, so that their conduct may be suitably influenced by them; and to this purpose they desire that their ministers may speak as if they were in earnest in a lively and pathetic, as well clear and intelligible manner. . . .
>
> If the Established clergy and the dissenting ministers in general were mutually to exchange their strain of preaching and their manner of living but for one year, it would be the ruin of our cause, even though there should be no alteration in the constitution and discipline of the Church of England.[15]

Its evangelical style, argued Doddridge, was Dissent's most distinguishing feature and the source of its strength.

David Some objected to the nonevangelical character of Gough's pamphlet. In explaining *The Methods and Motives for the Revival of Religion* (London, 1730) at a meeting of the ministers of several congregations of Dissenters for prayer for the revival of religion, he proposed that since it was vital religion that was dying, a vigorous pursuit of methods to promote piety was required. First, preaching: the Law and the Gospel, not sublime speculations and abstruse controversies should be spoken of; Christ in his offices of prophet, priest, and king, and the details and purpose of his life and resurrection should be frequently and largely explained; and the nature and necessity of conversion should be pressed. Second, pastoral care: public catechizing of young persons; pastoral visiting; particular notice of those under religious impressions; seasonable application of admonitions and reproofs. And third, "if we desire the revival of religion, we should be much in prayer for the blessing of GOD upon our endeavours."

Isaac Watts stated clearly and simply the evangelicals' explanation of the Dissenting interest's decay in his *Humble Attempt towards the Revival of Practical Religion among Christians, and particularly the Protestant Dissenters* (London, 1731). Watts (1674–1748), was a highly respected Congregational minister in London, author of textbooks in numerous fields, poet, and hymnist. He asserted that "the great and general reason" for the decay of the Dissenting interest "is the decay of vital religion in the hearts and lives of men, and the little success which the ministrations of the gospel have had of late for the conversion to holiness."[16] Watts's primary remedy was evangelical preaching. He urged preachers, when composing a sermon, to consider usefulness rather that eloquence, and to remember they are preachers of Christ, not of philosophy. The practice of the age to neglect evangelical themes, particularly salvation by faith alone, and to preach virtue without reference to Christ, he viewed as most unfortunate, for "it is the gospel alone that is *the Power of God to salvation*."[17]

The phrase *revival of religion*, used by Doddridge, Some, and Watts, possessed specific content for the evangelical Dissenters. Decay meant loss of emotional fervor and decline in conversions. The methods of reviving religion suggested in the replies to Gough can be divided into two categories. First, methods for promoting conversions: preaching Christ and the atonement, elucidating the conversion process, and careful pastoral guidance of those under religious impressions. And second, prayer for the outpouring of the Holy Spirit, without which no endeavors would bear fruit.

Scotland

Anglicization upset many Presbyterian ministers in Scotland in the early eighteenth century. They viewed union with England as a spiritual disaster: Morals decayed; toleration and lay patronage sapped the authority of the presbyteries; and a series of heresy trials revealed the importation of Arianism, Arminianism, and latitudinarianism.[18] To halt the spreading cancer, evangelically oriented ministers called on their colleagues to examine their pastoral practices and to return to orthodox, evangelical conversionism.[19]

Stewart Mechie proposed a fourfold division of the spectrum of theological thought in Scotland in the early eighteenth century: Scholastic Calvinism, evangelical Calvinism, liberal Calvinism, and rationalism. The various views along the spectrum were not absolutely separate, but shaded into each other. This typology, slightly modified, will be useful in summarizing the theological climate in which Scottish revivalism developed.[20]

The first group, the Scholastic Calvinists, sometimes referred to as High or Hyper Calvinists, were wary of any doctrinal statement that could be interpreted as deviating from the Westminster Confession. Placing the doctrine of predestination at the center of their teachings, they held that Christ died only for the elect and that the call to believe went out only to them. Thus, their preaching focused on the operations of the Holy Spirit on those called by grace.

The evangelical Calvinists, in contrast, were fervent conversionists:

> They were popular preachers, zealous for free grace, and though unconscious of any deviation from the Westminster Confession they made a warm personal appeal to their hearers and tried to do justice to the broad statements of an offer of salvation to all.[21]

The "Marrow" controversy (1718–1722) signaled the existence of a radical wing among the evangelicals. Republication in 1718 of the first part of *The Marrow of Modern Divinity*, first published about 1644, by Edward Fisher, an English Puritan, manifested a reaction against both neonomianism and High Calvinism. *The Marrow of Modern Divinity* and those who in the Scottish debate came to be known as the Marrowmen accentuated the fullness and freeness of grace, the universality of the call to believe, and the suitableness and sufficiency of Christ as savior for all sinners. They held neither particular nor universal redemption, yet they taught that Christ procured and offered grace "not only in general to all but in special to every one." All are legatees of Christ's testament, they said, but only to the elect does the testament become effectual.[22] Neonomianism is the treatment of faith under the covenant of grace as if it were a new law corresponding to the role of obedience under the covenant of works. The Marrowmen's objection to neonomianism focused on the use of the term *condition*. They opposed calling faith the condition of justification, as though the law of faith were a new law of works. Rather, they saw faith as only the instrument that effected salvation.[23] To the Scholastic Calvinists this anti*neo*nomianism sounded dangerously like antinomianism.[24] Prominent among the Marrowmen, Ebenezer and Ralph Erskine emphasized three teachings: 1) The call of the Gospel is to all men, and all who hear are obliged to believe; 2) assurance that the mercy of God in Christ applies to one's own soul is of the essence of faith; and 3) good works are the fruit of belief, but not a condition of salvation.[25] Most evangelically oriented ministers did not agree completely with the Marrowmen's open way for sinners to Christ but broke most clearly with them on the necessity of assurance for salvation.[26]

In response to complaints the Assembly of 1719 sent *The Marrow of Modern Divinity* to the Committee on Purity of Doctrine. The committee reported the next year, concluding the book contained antinomian sentiments. The Assembly accepted the report, prohibited ministers from recommending the book in any way, and required them to warn people not to read it. Objecting that the committee had isolated individual passages rather than considered the intent of the whole, the Marrowmen tried to have the Assembly's act repealed. In 1722 the Assembly modified the act but left the Marrowmen dissatisfied.[27]

The third group, the liberal Calvinists, emphasized the virtues of moderatism. Their aversion to religious persecution and valuing of freedom of inquiry made them reluctant to prosecute men suspected of teaching heresy. Nondogmatic, they disliked rationalist theology for being too speculative and controversial, and evangelicalism for being too emotional and for undermining the obligations of morality. Like their English mentors, they preached morality from natural law. While steering clear of heterodoxy, they also avoided consideration of distinctly Reformed doctrines.[28] After 1752 these attitudes in favor of the moral content of religion and of tolerance of theological and philosophical speculation would be taken up and actively promoted by a body of churchmen that came to be known as the Moderates. The Moderate Party would begin as a movement to counter what it saw as "the disorderly and destructive tendencies latent in radical Presbyterianism . . . [by] building a strong, unified, orderly Scottish Presbyterian church . . . an enlightened, tolerant Scottish kirk serving as a bulwark of virtue and stability."[29]

The fourth theological group, the rationalists, consisted primarily of John Simson and his followers. By the first quarter of the eighteenth century, Enlightenment rationalism had affected some of the Scottish clergy, notably Simson, professor of divinity at the University of Glasgow, and some of his students. In 1717 the Assembly found that Simson, who had been reading John Locke, denied original sin in infants, believed in the ability of those without the Gospel to be saved, and taught that there are means within natural ability of obtaining saving grace that God has promised to bless with success. The Assembly reproved Simson for magnifying natural reason and the power of corrupt nature to the deprecation of revelation and free grace. Although prohibited from venting his unorthodox opinions, Simson retained his teaching post. In 1729 the Assembly found him guilty of teaching Arminianism and suspended him from any further teaching. They did not, however, depose him or deprive him of his income.[30]

The liberal Calvinists' "legal strain" of preaching disturbed the evangelically oriented, and several disciplinary actions in which the Assembly appeared to be excessively harsh against the evangelically oriented and inexcusably lenient toward men who had clearly passed the boundary into Arminianism and Arianism widened the breach in the church. Disposition of the proceedings against Professor Simson in particular aggravated divisions among the clergy. The Assembly's leniency scandalized the Marrowmen, for whom the Simson case became the symbol of the church's departure from doctrinal purity. "A flood of erroneous doctrines," announced Ebenezer Erskine,

> is vented in our church. . . . A flood of Arminian errors runs in our day, exalting the freedom of man's will to the prejudice of the freedom and sovereignty of the grace of God, and exalting reason above revelation. A flood of anti-evangelical errors [is come in]. . . . Some instead of preaching the Gospel, entertain poor people with empty harangues of morality, or something worse, even in the hearing of our national assembly; and yet such things pass without censure.[31]

The Arian heresy, believed the Marrowmen, had entered not only the pulpits but also the church courts.

In the aftermath of the Simson suspension, Ebenezer Erskine issued the radical evangelicals' call to combat, *The Standard of Heaven Lifted up against the Powers of Hell, and Their Auxiliaries*, a series of sermons preached in 1730. *The Standard of Heaven* declared that unbelief, hypocrisy, superstition, and patronage were evidence that the Spirit of the Lord had departed the land. Erskine lamented "O what a flood of cursing and swearing, lying, murdering, cheating, stealing, backbiting, malice, envy, covetousness, Sabbath-breaking, doth prevail among us!" Atheism, blasphemy, and books ridiculing miracles abound, he said. Even heresy is left unpunished by the church. The enemy comes in like a flood because the Lord is departed. The Lord has departed because we have grieved the Spirit by disregarding Christ and the Gospel. Magistrates, ministers, and people have disregarded the motions of the Spirit on their hearts.

One thing alone, said Erskine, can make the Gospel successful among a people and destroy Satan's kingdom:

> It is the Spirit of the Lord, accompanying the preaching of the word, and displaying the glory of Christ therein. . . . It is not the flourish of words, it is not the force of human rhetoric, or flaunting harangues of morality . . . no, it is a faithful display of the glory of Christ, a simple proposal of the gospel, an opening of the mysteries of the kingdom.[32]

Only evangelical preaching, accompanied by God's grace, will be effective.

In 1733 the issues of preaching style, doctrine, and patronage united to produce a major secession from the church. The action of the Assembly that brought about the final disillusionment of Ebenezer Erskine and his followers with the Church of Scotland concerned the practice of patronage. In 1712 Parliament restored to lay patrons the right, abolished in 1690, of nominating ministers to vacant posts. Evangelicals opposed patronage as destructive of presbyterian polity, as an infringement of popular rights, and as a tool for imposing nonevangelical ministers on unwilling congregations. The evangelicals, who held that it was the right of the people, represented by all male heads of households, to choose their minister, were not the only ones to oppose patronage. Others, believing patronage to be a part of an attempt of the landed to humble the clergy, advocated that presbytery and kirk session select ministers, with the congregation having only a qualified veto. A third group, which included Francis Hutcheson, professor of moral philosophy at Glasgow University, applying Revolution Whig theories, feared that patronage could operate to make the clergy of Scotland subservient tools of the crown and great lords. They sought to support civic virtue by placing the power of selection into the hands not of the populace at large but of the local landholding gentry. Patronage had the support not only of the patrons, including the crown, but also, by the 1720s, of many of the liberal Calvinists, who found in patronage a means of settling ministers of their own persuasion. Some supported patronage because it was the law of the land and defiance encouraged disorder: In the late 1720s serious disputes over the induction of ministers appointed by patrons but unwanted by the congregations because of the nonevangelical sentiments of the presentees became common, and armed soldiers sometimes had to effect the unpopular inductions. In the second half of the century, convinced that if the Scottish church was to have any real influence it must work within the established power structure and not against it, the Moderate Party would erect the law-and-order point of view into a principle.[33]

The act of Parliament of 1712 that restored to patrons the right to fill parish vacancies provided that if the patrons did not exercise their right within six months of a vacancy, the local presbytery had the right to install a new minister. In practice, presbyteries often submitted the choice to the people. In 1732, in an action that supported the Whiggish objectives of men like Hutcheson, the General Assembly removed the choice from the people by restricting it to elders and Protestant heritors in cases in which patrons failed to act. In October, in a sermon before the Synod of Perth and Sterling, Ebenezer Erskine denounced this act of the Assembly. Previously, patronage had been a system imposed by act of Parliament. Now, he protested, the church not only formally legitimized this infringement on presbyterian polity, but further infringed popular rights as well. The synod censured Erskine for his invidious reflections on the Assembly. In 1733 the Assembly upheld the censure. The issue had now become the right of ministers to speak their consciences against decisions of the Assembly. Erskine and three others protested the censure. The Assembly demanded an apology for the protest. Erskine and his three supporters refused. The Assembly's commission suspended them. The four ministers declared the suspension null and void and continued their ministry. The commission then loosened the four from their charges and declared them no longer ministers of the church. The four announced they would continue as ministers and declared a secession, not from the Church, they said, but from its prevailing party. They would continue in communion with all who would continue in communion with them.[34]

On 15 December 1733, the Seceders formed the Associate Presbytery. They drew up "A Testimony to the doctrine, government, and discipline, of the Church of Scotland," in which they explained the reasons for the Secession. First, the Assembly had subverted the presbyterian polity. Second, the prevailing party was corrupting the doctrine contained in the Confession of Faith. The prevailing party practiced a new and fashionable mode of preaching that consisted of a "sapless and lifeless discanting upon moral virtues," and in which the peculiar doctrines of the Gospel were seldom mentioned. The Church of Scotland was in danger of becoming orthodox only in the sense the Church of England was: subscribing to the Calvinist articles while the clergy taught Arminian doctrines. Scotland was "in the utmost danger of losing the gospel in its power and purity through the prevalancy of a corrupt and unsound ministry." The third reason for the Secession was that the Assembly imposed unwarrantable terms of ministerial communion by restraining ministers from protesting defections from pure doctrine and discipline.[35]

The Assembly sought reconciliation. In 1734 it repealed the offending act concerning patronage, appointed a commission to petition king and parliament for repeal of the patronage act, and empowered the Synod of Perth and Sterling to act for reconciliation. The synod removed the sentence of deposition. The Seceders, however, refused to return, because the corrupt party still controlled the Assembly. Before they would return, the Seceders demanded that the Assembly declare patronage contrary to church principles, require presbyteries to inquire into the power of godliness and work of the Spirit on the souls of candidates for the ministry, and issue a warning against infidelity and gross errors.[36]

Scots soon realized that the Secession was to be permanent. In 1736 the Seceders began actively proselytizing. Other ministers, including Ralph Erskine, joined the Secession, increasing the number of Secession churches from four to fifteen by 1737. People unhappy with the nonevangelical ministers forced on them formed associations for prayer and requested the Associate Presbytery to license preachers for them. By 1740 the Secession increased to thirty-six, and by 1746 forty-five, congregations.[37]

The majority of evangelical ministers remained within the church. While they shared the Seceders' opposition to patronage and to nonevangelical preaching, they rejected Secession as destructive of Christian unity. The views of John Willison, John Maclaurin, and Robert Wodrow reflect those of the evangelical clergy who stayed in the church but continued to criticize it.

John Willison (1680–1750), minister of Dundee, was considered the leader of the popular party from the middle 1720s, because of his opposition to patronage. He was appointed to the commission sent by the General Assembly in 1735 to petition Parliament for repeal of patronage. Throughout the 1730s he led the movement to conciliate the Seceders and effect a reunion. He was a gifted evangelical preacher and diligent catechizer of the young. His treatises on devotional and practical religion were long among the favorites of Scottish religious literature.

Willison's sermon *The Church's Danger, and the Minister's Duty*, delivered before his synod in 1733 at the height of the Secession crisis, illustrates the evangelicals' thinking about the role of evangelical preaching in any revival of religion. In the first half of the sermon, Willison analyzes the evidence that the church was in a backsliding condition: schisms and divisions, doctrinal errors, barrenness of conversions, the Spirit withdrawn from ministers, and the persistence of patronage. In the second half, he states the actions to be taken to remedy the problem: Prayer for an outpouring of the Spirit on the ordinances and assemblies of the land, so that they may be effectual for conversion, and on students for the ministry that they may be acquainted with regeneration and diligent in serving God; and prayer for an outpouring of a spirit of moral reformation and of zeal for piety so that vice and sin might be suppressed. Only an outpouring of the Spirit would restore purity of doctrine and peace and unity to the church. Since a zealous, able ministry would be the means God would use to revive conversion, Willison urged his fellow ministers to pray for

> a time when by the Effusion of the Spirit *a great* and effectual Door shall be opened to us, . . . a Door of *Utterance* opened in Ministers Mouths, . . . a Time when Ministers Minds shall be enlightened, their Hearts warmed, Memories strengthened, and Tongues loosed, so that they shall have a great Facility in uttering their Thoughts, . . . when the Arrows of the Word shall be directed by a powerful Hand to pierce the Consciences of Men.[38]

Effective preaching is evangelical:

> Let all of us study in a special Manner to be found among those who deserve the

> Character of able Ministers of the New Testament, . . . one who is well-skilled in the Mystery of Christ, and him crucified; that knows how to open up the Covenant of Grace, and the Method of our Justification by Jesus Christ and his Righteousness only.[39]

And effective preaching is affective: Let us

> have our Sermons to the People *flowing from the Heart*. . . . That which comes from the Heart, is most likely to reach the Heart. The best Way to bring our People to believe and be affected with our Doctrine, is, to let them see that we believe it, and are affected with it ourselves.[40]

Willison, like other Scottish evangelicals, echoed the English Dissenting evangelicals' contemporaneous campaign for "preaching Christ."

John Maclaurin (1693–1754), one of the ministers of Glasgow, and brother of the noted mathematician Colin Maclaurin, was a celebrated preacher and theologian. He had special charge of the Highlanders in Glasgow, preaching in Gaelic for those who did not understand English well enough to attend regular worship services. He supported the Societies for Reformation of Manners, worked for poor law reform, and promoted the Glasgow Town Hospital for the poor and insane, established in 1733. He was a leader of the opposition to the violent introduction of presentees to unwilling congregations and a vocal critic of doctrinal deviations at the universities. He began his ministry in Glasgow in 1723 with a sermon on "The Necessity of Divine Grace to Make the Word Effectual," and during the rest of the decade repeatedly asserted that "the preaching of Christ, and no other sort of knowledge or wisdom is blessed as the means of turning us to God," and that even the best of evangelical preaching would remain ineffective without the concurrent operation of the Spirit.[41]

Robert Wodrow (1694–1734), minister at Eastwood, took the middle ground between the popular party and their opponents. In the words of his son, Robert, Jr., "He never inclined to Join any Party . . . but . . . Indeavoured to Strike a midle Safe way without Running to any Extremes, By yielding a little to Both Partys."[42] At the accession of George I, he took an active part in the attempt to abolish patronage but opposed every attempt to avoid compliance while it remained law. He defended orthodoxy from antinomianism and Arminianism alike. On the one hand, he sat on the Committee on Purity of Doctrine that condemned *The Marrow of Modern Divinity*. On the other hand, he lamented that Scottish ministers were

> falling in with the fashionable English way of preaching in harangues without heads [*i.e.*, the dividing a sermon into doctrine, reasons, and uses]; and love to call grace virtue, and other ways of speaking which differ much from our good old way in this church.[43]

He agreed with his fellow evangelically oriented ministers that the revival of religion awaited God's time.

Working to reconcile the Seceders, the evangelicals pushed through the General Assembly of 1736 an official call for evangelical preaching. The act instructed ministers to warn their hearers against current doctrinal errors and to insist frequently on the necessity of revelation. Ministers were to preach the depravity of human nature, man's impotence to do good without grace, justification and salvation through union with Christ with the Holy Spirit's working faith in the soul. Ministers were to insist not only on the necessity of faith for salvation, but also on the necessity of repentance for sin and reformation from it. Moral actions were to be taught as acceptable to God only as the result of faith, and as worthless in point of justification, which depends solely on Christ's righteousness. "Conformity to the moral law" is to be urged "not from principles of reason only, but also, and more especially of revelation." Preachers should "rightly divide the word of truth," speaking distinctly to the various cases of the converted and the unconverted. Sermons should be suited to the capacity of hearers, and delivered with "zealous freedom and plainess." Sermons should edify rather than amuse.[44]

On the eve of the Great Awakening, Scottish evangelical ministers were, owing to the political forum provided by the General Assembly, especially self-conscious of their identity as evangelicals. They shared with Anglo-American evangelicals attitudes toward the revival of religion. Revival was needed because vice demanded reformation of morals, errors demanded return to pure doctrine, and conversions were at a standstill. And revival was possible, all things being possible with God. They agreed that revival would come about by the outpouring of the Holy Spirit, and that the means would be a zealous ministry preaching the Gospel of justification by faith alone and the New Birth.

New England

In the first quarter of the eighteenth century, two lines of development were underway in New England that, when they converged, would produce successful evangelical revivalism. One group of preachers, centered in Boston and headed by Cotton Mather, gradually moved away from traditional Puritan emphasis on formal reason and logic and urged the kind of vital piety described in the writings of German Pietist August Hermann Francke.[45] In preaching, Mather's marked Christocentrism in the 1720s exactly paralleled the campaign for "preaching Christ" undertaken by orthodox Dissenters in England. In pastoral work, Mather and his disciples pursued practices, such as catechistic work and regular visitation, advocated by Richard Baxter and German Pietists, that involved them directly in the lives of their parishioners and enabled them to promote personal piety among them.[46] Another group of preachers, centered in the Connecticut Valley and under the mentorship of Solomon Stoddard, was perfecting a homiletic style similar to that advocated by Francke and Jennings and reaping evangelical fruits thereby.

The contemporary debates in English Dissent over "preaching Christ" evidently affected New Englanders. The English evangelicals sent to America the published sermons from the several series of orthodox lectures in London. On their part, the American evangelicals bemoaned the failure to preach Christ and the replacement of faith by "heathen philosophy." In the preface to Thomas Prince's sermon *God Brings to the Desired Haven* (Boston, 1717), for instance, Increase Mather wrote,

> *I cannot but rejoice to see that the Author* Preacheth CHRIST, *ascribing to Him the glory of His Works of Providence as well as Redemption. Many late Preachers have little or nothing* of Christ *in any of their* Sermons *(shall I call them) or* Harangues.[47]

It is more likely Mather was rejoicing that young Prince, who had spent the years 1709 to 1717 in England, had not been contaminated by the rationalist spirit there, than that he was referring to the spread of rationalism among New England ministers.[48] Thomas Foxcroft chose for his ordination sermon in 1718 the doctrine that "Christ is the Grand Subject which the Ministers of the Gospel should in their Preaching mainly insist upon." "Whatever Subject they are upon," he said, "it must be some how *pointed* to Christ."[49] He denounced those who scoffed at the preaching of the mystery of Christ as enthusiasm or cant. That in these remarks the American ministers had an eye on developments in England is evident from comments made by Cotton Mather to the convention of Massachusetts ministers in 1722. Early in the year Mather had heard that "Arminian books are cried up in Yale College for eloquence and learning, and Calvinists despised for the contrary."[50] "Truly," he told the convention, "the fearful *Decay of Christianity* in the World, is very much owing to the inexcusable *Impiety* of overlooking a Glorious CHRIST, so much in the *empty Harangues* which now often pass for *Sermons*." Evangelical truths—human impotence to do good, the atonement, salvation by faith alone, the nature and necessity of the new birth, and so on—were threatened with banishment from some churches. The source of this development, thought Mather, was "such *Books* as have been very much in Vogue among us; *Books*, whereof it may be complained, *Nomen CHRISTI non est ibi*, and the *Religion of a Regenerate Mind* is not there to be met withal." He quoted a recent author's statement that "*the Presbyterian Divines*," of England, "*have been observed of late . . . to Preach after the manner of the Church-of*-England *men*."[51] In September, the defection of Timothy Cutler and several others at Yale to the Church of England proved to Mather that "preaching Christ" was endangered in America by the decay of evangelical truth in England.

New England Congregational and Presbyterian ministers considered themselves fellow United Brethren with the Dissenters and the Scottish Presbyterians. In 1715 a convention of Congregational ministers of Massachusetts and Connecticut met at Boston to draft an address to the new monarch, George I. They also informed the Nonconformists in England of their "ambition to be acknowledged as your *United Brethren*." They sought "such a good correspondence . . . between us, and our *United Brethren* in the Church of Scotland, and the dissenters in England, that they may look on what is done unto us as done unto themselves." They presented the loyal address to the crown through the Body of the Protestant Dissenting Ministers of the Three Denominations in and about the City of London, which had been founded in 1702 by the Baptist, Congregational, and Presbyterian clergy for the purpose of approaching the crown as one deputation.[52]

The Scots and English Dissenters reciprocated this feeling of unity between themselves and the New Englanders. In 1712 London Dissenters distributed their "Serious Call from the City to the Country" for spending a common hour weekly in prayer for the church not only in Great Britain but also in America, "because *they* are all bound up with us in the same *Bundle of Life*, and their Fate is likely to be involved in ours." Cotton Mather welcomed the proposal and sought to bring his religious societies into the agreement.[53] In 1726 the Scot Robert Wodrow proposed to Boston's Benjamin Colman a similar project for transatlantic united prayer.[54]

Before the Great Awakening, New England ministers engaged in a lively dialogue with their British colleagues concerning the prospects for a revival of religion. In the first quarter of the eighteenth century, Cotton Mather was the major conduit of this dialogue. His voluminous correspondence extended beyond Great Britain to the Pietists in Germany, and even to the Dutch missionaries in the East Indies.[55] Through the letters and publications he sent abroad, Mather hoped to promote evangelical pietism as a means to the Christian unity that he believed would be the principal mechanism for introducing the millennium. He viewed the increase of evangelicalism and missionary enterprise as signs of the approaching kingdom of Christ.

During the 1720s, the Benjamin Colman–Isaac Watts axis became a major channel of communication between the evangelically oriented clergy in New England and the evangelical Dissenting ministers of England. Colman (1673–1747), who had spent the late 1690s in England, including two years preaching in Bath, was the original pastor of Boston's Brattle Street Church, whose liberal membership practices distressed the Mathers. Watts served as Colman's literary agent in London and Colman performed a similar service for Watts in Boston. Watts's numerous correspondents in New England included Eliphalet Adams, Zabdiel Boylston, Mather Byles, Thomas Foxcroft, John Greenwood, Samuel Mather, Thomas Prince, Elisha Williams, and Solomon Williams. He was also a personal friend of Massachusetts Governor Jonathan Belcher. Watts presented most of his own works and many works of others to Harvard and Yale Colleges, and his works were greatly admired in New England. While at Yale, Jonathan Edwards studied his sermons, texts, and poems.[56]

In 1728, Watts saw through the London press a series of discourses preached by Colman in Boston called *Some Glories of our Lord and Saviour Jesus Christ*. Focusing on Jesus, the discourses were intended as helps toward a more edifying reception of the Lord's Supper. Colman dedicated the work to the English lay Dissenter Thomas Hollis, in gratitude for his gifts to Harvard College. In the dedication, Colman associated the discourses with English Dissenter William Harris's discourses on the Old Testament types of the Messiah and with Watts's sermon on the propitiation of Christ,[57] as "offerings *to the honor of the blessed Redeemer*." The preface was written by William Harris and signed by the English Dissenters Edmund Calamy, Jeremiah Hunt, Isaac Watts, Daniel Neal, and John Evans. They welcomed Colman's sermons as an American contribution to their campaign for preaching Christ.

The Calvinist clergy in New England held the Church of Scotland in high esteem, viewing it as a bulwark of orthodoxy.[58] New England's most prominent clergy had a direct association with Scotland, for from its universities they received honorary Doctorates of Divinity. In the early eighteenth century, Scots and New Englanders frequently exchanged thoughts on the state of religion. Cotton Mather and Robert Wodrow, for instance, maintained a steady correspondence on the subject.[59]

Through the services of Mather, Wodrow began a correspondence with Benjamin Colman in 1717 that lasted until Wodrow's death in 1734.[60] Early in their relationship the two men agreed that religion was in a state of decay. In a letter of 23 January 1719 Colman answered Wodrow's inquiry "after the state of Vital Religion among us." "To be sure," Colman wrote, it "decreases with our Growth, and as the world grows upon us." Although pride and vanity "eat out the heart of serious Godliness, . . . we have a great Number of serious Godly praying people, both older and younger, here and thro' most parts of the land." The outward profession of religion continues, the sabbath is kept, and family worship is maintained. "But certainly the power of Godliness is much on the Decay."[61] Wodrow's letter to Colman written the same week, 29 January 1719, carries similar sentiments. "Real religion is under a sensible decay, and our sun is a winter sun. Profaneness and sin are dreadfully abounding, and the love of many waxeth cold."[62] Wodrow communicated news of religion he received from Colman and his other correspondents in New England, who included Edward Wigglesworth, Hollis Professor at Harvard College, to a meeting of eight or nine ministers in his neighborhood.[63]

Colman gave Wodrow fairly full explanations of a few ecclesiastical controversies in New England, such as the opposition to the settlement of Peter Thacher at New North, Boston, in 1720, and the formation of a new church by the discontented party. Wodrow's accounts of the ecclesiastical controversies in Scotland were sketchy. His account of the condemnation of *The Marrow of Modern Divinity* in 1721 was simply that several plainly antinomian points in the work had been condemned, and that about a dozen ministers had presented a protest, "some expressions whereof seem to tend towards Antinomianism."[64] In 1726, all Wodrow told Colman about the Simson controversy was that the professor was being investigated concerning his views of the Trinity.

In the spring of 1734, Wodrow's son Robert wrote Colman of his father's death. Young Wodrow, twenty-three years of age and not yet ordained, undertook to continue the correspondence with Colman. He made his descriptions of the Scottish controversies somewhat fuller than his father's. Following in the elder Wodrow's footsteps, he rejected the Secession but identified with the Seceders' evangelical zeal and opposition to patronage. He informed Colman that the four ministers who had protested the Assembly's act of 1732 concerning patronage had been declared no longer ministers and had been loosed from their pastoral relation to their parishes. He offered to send a recently printed narrative of the affair and expressed hope that the next Assembly would rectify many grievances and amend the act of 1732.[65] In 1735, Colman must have written Wodrow about the controversy arising from Robert Breck's heterodoxy, for in August Wodrow wrote Colman,

> I heartily lament over your Dark uncomfortable and Erroneous situation. If the foundations of Religion be saped [*i.e.*, sapped, "undermined"] with Error if your Clergy and Leaders mislead your people and if [from] the paths of truth unto those horrid Errors you mention then Dismal must be the Consequences.

Wodrow then identified the Arminian problems in New England with those in Scotland: "I am grieved you partake of our Errors and Shisms [*sic*]." He quoted the Scripture that had become the evangelical's rallying cry: "But I would fain hope that the Spirit of the Lord will lift up a Standart when the Enemy comes in as a flood." And he continued with an exhortation to pray for a revival of religion:

> Let us Double our ernest Suit to Heaven for the Reparation of our Breeches the Removal of our Errors Shisms Divisions and factions the Redress of our griev-

> ances the Revival of Piety and goodness the spreading of useful Knowledge The Purity of worship and Doctrine.

Colman had also informed Wodrow of revivals in the Connecticut Valley, for Wodrow remarked that "the account of the saving change wrought on the people of Hampshire [County, Massachusetts]" was cause for joy.[66]

Colman also corresponded with John Willison. In 1730 Colman was appointed correspondent of the Society in Scotland for Propagating Christian Knowledge, which supported missions to American Indians. About the year 1730 his associate, William Cooper, commenced a correspondence with John Maclaurin.

The New England and the Scottish ministers must have had only very general ideas of the religious developments and controversies in each other's country. Still, the evangelicals were in contact, they discussed their common belief that religion was in a state of decay, and they sympathized with each other's struggle against the danger posed by antievangelicals. They shared a belief in the need of a revival of religion and an understanding of what such a revival would consist.

The Connecticut Valley

During the first third of the eighteenth century, sets of New England clergymen undertook a campaign like those among evangelical Dissenters in England and evangelical Presbyterians in Scotland for orthodox, affective, conversionist preaching. In extolling the preaching of the Gospel as the principal divinely appointed means of grace, they sought to use their monopoly as preachers of the Gospel as a vehicle for restoring their influence in society. If evangelical preachers were essential to a revival of religion, as they implied, then New England needed faithful ministers now more than ever.

Declining prestige began to disturb New England ministers well before the end of the seventeenth century. Unlike their fathers, second generation clergymen shopped around before accepting a call to settle. They sought communities that would assure them of the obedience and salary commensurate to their station.[67] By the eighteenth century, increasing resentment because of contentions over authority and salary decreased the ability of pastors to lead their flocks.[68] The turning of the sons of the socially prominent to other professions measures the decline of the status of the clerical profession.[69] The ministers responded by objectifying the office of the ministry. They stressed the sacerdotal nature of the office, modified the link between the ministry and the gathered church, replaced lay with clerical ordination, formed ministerial associations, and emphasized their preaching function as the necessary means of grace. In their associations, the ministers asserted their unity as members of their profession and sought to increase their authority. A spate of printed works by ministers about the ministry revealed their consciousness of status. Between 1709 and 1740, New Englanders published eighty ordination sermons, all dealing in some way with the nature and duties of the Gospel ministry.[70] In these works, the ministers insisted that their calling was one of honor and dignity, and deserving of just recompense. They reminded their readers that the ministers' calling demanded faithfulness, diligence, study, and hard work. At the same time that they were insisting on their right to deference, the ministers sought to regain the goodwill of the people by forsaking prophetic zeal for a stance of benignity and meekness.[71]

Solomon Stoddard chose a different method to restore the influence of the clergy. While joining his colleagues in insisting on respect for the ministers, he never forsook prophetic zeal. After 1710 he began to stress the charismatic authority of the minister along with the objective.[72] Stoddard promoted a charismatic evangelism: converted preachers, full of the Holy Spirit, appealing to anxious men and women to flee to Christ for safety, would revive religion while restoring the influence and status of the clergy. By charismatic ministry (*charism*: "gift of the Holy Spirit") here is meant a leadership that relied for its authority and influence on the evidence the minister gave that he was filled with the Holy Spirit and acting on the Spirit's promptings. In 1708 Stoddard had argued "the efficacy of the Word of God doth not wholly depend upon the Piety of him that dispenses it. . . . Men may Preach to others, yet be cast-aways themselves.[73] While "they that are Converted themselves are most likely to be instruments of the conversion of Sinners, Yet it is lawful for men in a natural condition to Preach the Word."[74] After this, however, he began increasingly to emphasize the "need of *experimental* knowledge in a Minister." "It is a great calamity to wounded Consciences," he wrote in 1714, "to be under the direction of an unexperienced Minister."[75] Only someone who had experienced preparation and conversion himself could understand the condition of another engaged in the same process. In 1724 he explained that a good education was not sufficient to make an able minister:

> Every Learned & Moral man is not a Sincere Convert, & so not able to speak exactly and experimentally to such things as Souls want to be instructed in.[76]

If preaching the Gospel and skillful direction of awakened sinners were the means to a revival, then the failure of revival could be ascribed to the ministers. In *The Defects of Preachers Reproved* (New London, 1724), Stoddard attributed the decline of religion to ministers who pampered sinners for fear of offending them. "For hence it is," he wrote,

> that there is so little Conversion. . . . The want of dealing plainly with men is the reason, why there is seldom a noise among the *Dry Bones*. In some Towns there is no such thing to be Observed for Twenty Years together.[77]

The people did not abandon their sins because ministers were afraid to offend their congregations by condemning their immoral liberties. Sinners were not awakened because ministers did not preach about the danger of damnation. "If Sinners don't hear often of Judgment and Damnation, few will be Converted."[78] Few were converted because ministers did not properly understand the nature of conversion. They suggested false signs of grace. They did not teach the necessity of humiliation before faith, and they taught that men were frequently ignorant of the time of their conversion. And finally, ministers were frequently ineffective preachers because they read their sermons.

> The Reading of Sermons is a dull way of Preaching. . . . It is far more Profitable to Preach in the Demonstration of the Spirit, than with the enticing Words of mans wisdom.[79]

Stoddard outlined his program to promote revival by means of conversionist evangelism in a series of seven sermons on "The Benefit of the Gospel to those that are wounded in Spirit," which he preached during a revival at Northampton and published with his essay on *The Efficacy of the Fear of Hell* (Boston, 1713). The doctrine of the first sermon of the series underscored the importance of a charismatic ministry: "Ministers had need have the Spirit of the Lord upon them, in order to the reviving of Religion among his People." He wrote,

> Sometimes Religion is in a withering Condition, but there are means that are serviceable for the reviving of it: And this is one special means, when the Ministers have the Spirit of the Lord upon them. The Spirit of the Lord must be poured out upon the People, else Religion will not revive. But when the Spirit is upon Ministers, it is a very hopeful sign.[80]

The presence of the Spirit with the ministers is conducive to revival in several ways, Stoddard observed. The Spirit gives them a zeal for God's glory and for the salvation of souls, so that they preach on awakening themes, and so that they declare the Word powerfully.

> When they have an holy Zeal, that makes them to be *Boanerges, Sons of Thunder*, they will be earnest and fervent. . . . Zeal will inflame the Heart, and make Men declare the Word of God, so as to awaken others and not lull them to sleep.[81]

Men are so hard-hearted, blind, and self-loving, their consciences can be awakened only by powerful preaching. The Spirit makes ministers able guides to Christ. It gives them the understanding and skill to make men realize their sinfulness, to speak terror to their consciences, to represent "in an affecting Manner" the misery of hell, and to reveal to men the deceits and false hopes of their hearts.[82]

The second through the sixth sermons of "The Benefit of the Gospel" share one message: There is power in the Gospel to save souls. The terror of the divine law must be preached to sleepy sinners, "that they may be sensible of the Terribleness of Damnation. . . . That they may be sensible of the great danger of Damnation. . . . That they may be sensible of the justice of their Damnation."[83] Once sinners are thus awakened, the promise of grace can be offered them, "because such Men are *prepared* to receive the Gospel."[84] Those who have the benefit of the Gospel have the opportunity for salvation. In a *Treatise Concerning Conversion* (Boston, 1719), Stoddard would insist that the Gospel is *the* means of conversion.[85] In the seventh sermon of the series, Stoddard asserts that there are special seasons of revival and urges prayer that God would grant such a season to the whole land.

Over the eleven years following publication of "The Benefit of the Gospel" and *The Efficacy of the Fear of Hell*, Stoddard published three important works designed to aid a revival of religion through conversionist evangelism. *A Guide to Christ* (Boston, 1714) is a practical manual to assist ministers in directing awakened sinners. It closely analyzes the morphology of conversion, the steps of preparation, the varieties of spiritual experiences, the multitude of false hopes, the clever deceits of the heart, and the only sure sign of grace, grace itself. *A Treatise Concerning Conversion* is less practical in intent. It avoids any discussion of false signs of grace in order to focus on what conversion is and "the Way wherein it is wrought." *The Defects of Preachers Reproved* is an exhortation to ministers to adopt his program.

By 1700, Stoddard, a committed moral reformer, concluded that ministers, weakly backed by the state and disempowered through congregational polity that gave the laity control of the churches, had but one useful tool remaining, moral suasion.[86] Stoddard urged ministers to move their auditors by portraying in vivid imagery the horrors of damnation and the ever present danger of dying unconverted. He taught that ministers ought to instruct people in the examination of their hearts for signs of grace and that ministers could not reach men's hearts unless the ministers themselves were converted.

In the Connecticut Valley during the 1710s and 1720s a number of ministers echoed Stoddard's new program for reviving religion. Included in this group were William Williams of Hatfield, Eliphalet Adams of New London, Benjamin Lord of Norwich, Isaac Chauncy of Hadley, and Jonathan Marsh of Windsor. These men agreed that moral reformation could only be accomplished through a religious awakening led by an effective ministry. In 1721, for instance, Jonathan Marsh told the Connecticut General Court that civil rulers could not be blamed for the failure of reformation, since "the Thorough REFORMATION of a *Sinning People* is not to be Expected; however Pious RULERS may be Spirited for the Work, Except the Heart of the PEOPLE be Prepared for it."[87] The responsibility for preparing men's hearts was the ministers'. The ministers must awaken the consciences of the people. The spread of conviction of sin and of conversion to Christ would "pave the way for *a General Reformation among us*."[88] Philip Gura argues that William Williams, probably unintentionally, "placed more emphasis on the minister and his evangelical role than on the state of the souls he was bringing to Christ."[89] These Connecticut Valley clergymen complained that ministers no longer preached conversion, preaching morality instead, so that people did not understand the nature of conversion. They themselves were careful to preach on the nature of saving faith and closing with Christ. They denounced sermons designed to please the ear and urged awakening sermons that condemned sin and called for reformation. The preaching of the Gospel, they emphasized, was the means of conversion. A conversionist appeal to the emotions of the people, they believed, was the most effective means to revive religion.

These ministers also believed that a conversionist appeal was the most effective way to win back the lost prestige of the ministers. As part of his program to promote revival, Stoddard sought to increase the formal authority of the clergy. In his *Examination of the Power of the Fraternity* of 1718,[90] he censured the laity who misused their power through the congregational polity to prevent the implementation of reforms. Because of the obstacles to clerical authority presented by the congregational structure of the New England churches, Stoddard concluded that the chief instrument left to the ministers was their authority as preachers of the Gospel.[91] In 1721, Jonathan Marsh spoke of the struggle for power between the clergy and laity. "Tis easy to observe a Spirit of Jealousy prevailing in the People of this Land against the Ministry of it," he said.

> Hence every Difference, though of a lesser nature from the Primitive Constitution of these Churches, is Suspected by some Religious and Holy men and cryed out against, as Innovation and a Degree of Declension and Apostasy, as the way to invest the Ministers with the Sole power of Church Government, and to cut off the Privilege of the Brotherhood.

The way "to take off the Prejudice of the People against [the clergy]" was through the preaching of conversion.

> Now what is to be done for a Remedy in the case? Lets Labour to come up to and if possible to go beyond those gone before us in a Spirit of Zeal for Practical & Experimental Piety.[92]

A general adoption of their program by the ministers of New England would have many happy effects, explained the Connecticut Valley reformers. Not the least in importance would be an elevation of the status of the ministers. Respect would be based more on the charismatic authority of the preachers than on the authority of the clerical office itself.[93]

Marsh implied that if all the ministers followed the Connecticut Valley formula, revival and reformation would become general. Benjamin Lord asked "How rarely are any converted among us? If there be some remarkable Harvests in some places, yet in the general, may not the Church complain that her Converts are as the small *gleanings of the Vintage*."[94] The remarkable spiritual harvests to which Lord referred followed, he said, from revivalist preaching:

> Sometimes when Faithful Ministers have wisely Timed and Proportioned the Preaching of the Law with the Gospel, They have seen the Travail of their Souls in a Plentiful Harvest of Converts. The Preaching of the Word, has been, and still is the Great means of bringing home Sinners to Jesus Christ.[95]

In the tradition of Stoddard's *Defects of Preachers Reproved*, Lord lectured ministers on their responsibility to preach for conversions:

> Learn Hence, How great the Work of Ministers is, & how solemn their Charge to Adapt their Ministrations to the Necessity of Sinners, and the Important design of their Conversion.[96]

The Connecticut Valley revivalists actively campaigned in favor of their program. Four of their number were responsible for nine of the thirty-six sermons on the ministry published in New England between 1721 and 1730.[97] With the other New England preachers, they stressed the special character and the difficulty of the ministry, the minister's need to study long hours, and to be diligent and zealous, and experienced in guiding sinners through the labyrinths of their own hearts. Their sermons stand out because of the thoroughness and explicitness of their directions for evangelical conversionism. Slighting edification of saints in favor of conversion of sinners, they examine in detail how to preach the law to the secure and how to preach the Gospel to the humbled. They stress the importance of teaching people how to distinguish the true sign of grace from the false signs.

What most clearly sets apart the sermons of the Connecticut Valley reformers as a distinct group is their emphasis on the doctrine of preparation: Ministers should preach terror before they preach peace; they should be sons of thunder as well as sons of consolation. These preachers rejected Cotton Mather's notion that "the desires of grace are grace."[98] "A Man may desire to be justified by Christ's Righteousness," warned William Williams, "and yet not close with him upon the terms of the Gospel."[99] Instead, they followed

Stoddard's view of preparation. Conversion occurred much later in the morphology than Mather taught. Isaac Chauncy urged his reader "that he would embrace the Doctrine of the Preparatory Work that goes before Conversion. . . . Though this is not a fundamental Article, yet the denial of it may lay a foundation for your eternal Ruine."[100] With Stoddard, they believed that, while a thorough external reformation was a prerequisite, "reformation is preparatory to Conversion, but is not it self Conversion."[101] They taught that men must first be encouraged to attempt moral reformation by the preaching of the terrors of the law. Such preaching led persons to see their own impotence to do good, and the need for reliance on Christ's merits alone. Only after humiliation should awakened sinners be made the offers of the Gospel.

Following Stoddard's lead, by the 1720s a group of preachers in the Connecticut Valley had developed a successful revivalist system of preaching.[102] Their style was an affective presentation of the terrors of damnation and of the glorious offer of free grace, spoken in an apparently extemporaneous manner and in plain language understood by the common people. In content, the sermons were directed almost exclusively to the unconverted. They were organized on the basis of the doctrine of preparation.[103] The sinner needed actively to seek regeneration: "Work out your own salvation in fear and trembling" (Phil. 2:12). As Benjamin Lord wrote,

> There is in Natural Men a Power with common Assistance to do much towards their Conversion. . . . We be not capable of Converting our selves, yet we are capable of Endeavours after Conversion.[104]

Since the preparationists' morphology began with awakening, the sinner's realization of his danger of going directly to perdition, the preachers emphasized the continually imminent danger of death: "God threatens to destroy them Suddenly."[105] The revivalists sought to frighten the unregenerate with the terrors of hell. Eliphalet Adams informed ministers that

> we must endeavour to convince [the unconverted] of the Great Danger of continuing in such a condition, we must shew them the Destruction and Misery that is before them, except they repent, and put them upon Flying immediately from the wrath that is to come.[106]

After awakening, came humiliation, the realization that one was without any merit and fully worthy of immediate and eternal punishment, entirely incapable of saving himself. Now, when the sinners are debased, on the verge of despair, the offer of grace through the atonement of Christ is made, as Eliphalet Adams explained:

> We must strike while the Iron is hot, when we have Awakened and Convinced People and *almost perswaded* them *to be Christians*, we must now bestir our-

> selves and follow the matter home till it Issue in a *Sound* and *Saving Conversion*.[107]

Benjamin Lord's treatise *The Faithful and Approved Minister* (New London, 1727) captures the essentials of conversionist preaching as practiced in the Connecticut Valley. Lord believed that to preach for conversions meant to preach first the threats of the law and then the offers of the Gospel. He gives careful instructions on the matters to be handled in sermons. Ministers should see to it that their people learn the essentials of the faith and the moral law, and that they develop the skill to examine their spiritual state once their consciences are awakened in preparation for conversion. Sermons should be directed principally toward two types of persons, unawakened sinners and humbled sinners. To the first, the law must be applied, so their consciences might be touched, so they might see their sins and their danger of damnation. They must be shown the "*Cursed Nature & Damning Fruits*" of sin, and the certainty of their being hurled into the "fiery Gulph," unless they repent. Finally, they must be shown their "utter Impotency to help *themselves*." Once the sinner has been humbled by the law, "when the awakened Soul is Trembling with a sight of his wretched and helpless Condition, acknowledgeth the Justice of GOD in the Ruin of Sinners, and being reconciled to GOD as a Soveraign . . . and lyes at his Mercy," then the calls of the Gospel should be offered. The promise of free grace, the greatness of God's mercy, the sufficiency of Christ's merits to save a sinner "without a spark of goodness to recommend him," and the safety of venturing his soul with Christ should be explained. Lord has only a word about the pastoral care of the soul once converted. "After Conversion . . . there will be need to offer something Consolatory and Confirming," along with instructions about growing in grace.[108]

The Great Salvation Revealed and Offered in the Gospel Explained (Boston, 1717), William Williams's tract on conversion, also illustrates the revivalist teaching of these ministers. Williams focuses first on the importance of regeneration. He explains that men are impotent to save themselves and certain of damnation without Christ. Thus, the tract begins with the first stage in the preparationists' morphology of conversion, awakening. As means to awaken sinners, Williams refers to the horrors of hell: "Can you bear the weight of an Omnipotent One? The anguish of the never dying Worm? The burning heat of those Flames that never shall be extinguished?"

The next stage of the morphology is humiliation, so Williams next reminds the reader that to be saved he must recognize his absolute dependence on Christ. Then Williams teaches the reader how to examine his experiences in order to judge his spiritual state. He rejects as false signs of godliness not only external evidences such as strict conformation to the moral code, but also good affections, alluding to and rejecting Cotton Mather's teaching that the desire for grace is evidence of grace. The Gospel, says Williams, is not only a message of the good news of salvation, but is the instrument, especially when preached, through which salvation comes. One is saved when he accepts the offer of grace revealed in the Word. After having explained the way one is to seek salvation, attending the preaching of the Word, Williams returns to and concludes with an awakening message: "The miseries of Hell are inconceivable! . . . The Neglect of this Great Salvation will expose Men to Great and Unavoidable Misery."[109]

The Great Care & Concern of Men under Gospel-Light (New London, 1721) is a good example of the revivalist sermons of the Connecticut Valley. Jonathan Marsh delivered it on 14 December 1720. The 1721 revival at Windsor proved the effectiveness of Marsh's preaching. The sermon teaches the doctrine of preparation: Every person is under the obligation to seek regeneration. No man can find salvation by himself, yet "in ordinary God gives his Grace to men in the way of painfully Seeking of it."[110] Since it is the Gospel that is "the great Medium God uses for the bringing about of this Work,"[111] the way to seek salvation is to attend Gospel preaching. The greater part of the sermon is devoted to the application, which is divided into three uses, of awakening, of examination, and of caution. "Let this Doctrine be Improved to Awaken secure Sinners," Marsh urges.[112] Here he follows Stoddard's advice to tell the people about damnation:

> *If they Die in their present condition, they will wish they had never been Born.* . . . For their Torment 'twill be intolerable. Tis better not to be at all, than to live under a total Privation of that Good which man is made capable of Enjoying, and without which he cannot be happy. The duration of Being proves but a Torment to a man in such a case. Besides this Privative part of their Misery there is the Positive part of it, which is suffering the Extremity of Wrath for ever, which makes it to be so much the more shocking & amazing to the Soul. . . . The State of Gospel Sinners in another World, is worse than Death or Annihilation.[113]

Under the heading "of Examination," Marsh teaches the hopeful convert how to determine whether his conversion is genuine. Again, he follows Stoddard. First, he rejects the false signs of grace: a high profession of piety, great efforts in religious and moral duties, feelings of comfort and encouragement in one's attempts to lead a righteous life.[114] Then he lists the necessary prerequisites, whose absence would indicate no conversion had occurred: One must be convinced of his own corruption and of his need for grace, and one must be aware of his own absolute unworthiness of grace.[115] And finally, Marsh offers the true sign of grace: "*Try it by your being brought to Believe in*

Christ. . . . Enquire then whether you have accepted of Christ upon Gospel Encouragement."[116] Reflecting Stoddard's teaching that one gracious act is sufficient evidence of a valid conversion, Marsh urges his hearers to test their conversions "*by your Performing of acts of Grace consequent upon Believing*. . . . Now if you can find a holy Life or gracious Actings, these are Evidences of Sincerity; The *Actings of Grace* conclude the *Principle of Grace*."[117] Stoddard taught that since only a converted person can perform a gracious act, if a person knows he has performed one gracious act, a particular act of faith or love, he can be sure of his gracious state.[118] Marsh concludes with his third use, "of CAUTION," in which he encourages his hearers, exhorting them not to give up hoping for and seeking regeneration.

The division between the evangelically and nonevangelically oriented was much less open among New England's Reformed ministers than in Great Britain. New England's Reformed ministers were highly reluctant to admit to serious divisions among themselves, and few of them would have considered themselves nonevangelical.[119] Yet, by the 1720s some non-Anglican New England ministers were clearly moving into the liberal camp.

Take Samuel Whitman, of Framingham, Connecticut, for example. His 1714 Connecticut election sermon fits the conversionist pattern of the Connecticut Valley revivalists.[120] By the 1720s, however, he had moved away from conversionist concerns. No revivalist could fail to use the occasion of a funeral to warn the unregenerate of the danger of dying in that unhappy condition. Yet, in a funeral sermon in 1727, Whitman assumed all his hearers saved.[121] In *A Discourse of God's Omniscience* (New London, 1733), his emphasis is on good works: "If we walk as before God in this world, it will shortly be our Privilege and Happiness, to be and live with Him in Heaven for ever."[122] He was not considered a warm-hearted preacher and seems not to have favored revivals.[123]

One Connecticut minister publicly challenged the campaign of Stoddard's disciples for revivalist preaching. John Bulkley, minister at Colchester, whom some suspected of an inclination toward the Church of England, preached an ordination sermon in 1729 that not only is contrary to all that the Connecticut Valley revivalists were urging, but also differs from any other New England ordination sermon published in the period. Bulkley urges *The Usefulness of Reveal'd Religion* to preserve natural religion, that is, morality. William Williams had argued that

> there is indeed a necessity that Moral Duties should be press'd, as well as Doctrines of Grace laid open: But this is the main thing, that Men be rightly informed and established in the Understanding and Belief of the Great Truths of the Gospel, the Purposes of Gods Grace concerning the Way and Means of Reconciling and Saving lost Sinners.[124]

Bulkley's priorities are the reverse:

> We must Preach the *unsearchable riches of CHRIST*; set before sinners those Treasures of Wisdom and Grace that are laid up in him. We must preach also Duties of Religion *purely positive*: But yet at the same time, we must not so insist on these things, as to neglect the other (I mean the Doctrines and Duties of Natural Religion) and not allow them their due time and place in our Ministry. *Moses* must not be divided from *CHRIST*, the *Law* from the *Gospel*: *Paul* as Evangelical a Preacher as he was, yet was far from this.

Bulkley argues that discourses on morality have their due time and place and are an essential part of efforts to convert sinners. Whereas evangelicals believed morality and natural religion were being preached to the neglect of grace and Christ, Bulkley suggests that the reverse is the case.[125]

The "Arminian threat" that concerned New Englanders in the 1720s and 1730s referred not only to those who taught openly unorthodox doctrine, such as Robert Breck, but also to preachers like Whitman and Bulkley who slighted conversionism. While Breck, Whitman, and Bulkley were exceptions in New England, in the larger context of the British community, the fears of orthodox New Englanders are understandable.

During the first decades of the eighteenth century, non-Anglican British evangelical Protestants identified themselves as successors to the Reformers. They agreed that vital religion had declined since the first half of the seventeenth century. This community of ministers in Scotland, England, and America was aware of a common evangelical orientation, and of a shared emphasis on affective, conversionist, orthodox ministry as the one means, when accompanied by the operations of the Holy Spirit, to effect a revival of religion.

4

Heart Religion

A word coming from the heart will sooner reach the heart.

ISAAC WATTS, *An Humble Attempt towards the Revival of Practical Religion* (London, 1731).

[Let us] have our Sermons to the People *flowing from the Heart*. . . . That which comes from the Heart, is most likely to reach the Heart.

JOHN WILLISON, *The Church's Danger and the Minister's Duty* (Edinburgh, 1733).

You must Preach with Zeal and Affection, that you may command Attention by your Fervency, and Affect the Hearts of your Hearers.

ISAAC CHAUNCY, *The Faithful Evangelist* (Boston, 1725).

The role of emotional appeal in eighteenth-century evangelical preaching was new. There had always been a strong pietistic strain in Puritanism that emphasized the emotions, in particular existential experience of God's love.[1] Puritanism's mainstream, however, had subordinated the emotions to the intellect. Preachers appealed to the affections principally as a means of arousing the hearer to implement that which he had been convinced was the good and the true. As the understanding of psychology changed in the eighteenth century, rhetorical strategies changed as well. Some emphasized man's rationality and denigrated his affections. This trend is evident in latitudinarianism and in the movement toward unitarianism. Others, influenced by John Locke's sensational psychology, adopted a unified conception of the psyche in which the affections, the understanding, and the will could not be separated from each other. These preachers came to believe that one becomes persuaded to embrace the good and the true by a perception of its beauty. Considering the sense of beauty to be a function of the affections, they appealed directly to the "heart" through a use of lively imagery. Here the Enlightenment belief that knowledge comes from experience supported Pietism's belief that knowledge of God comes through personal experience of his grace.[2]

Continental pietistic ideas and practices helped to inspire and legitimize the appeal to the heart. German Pietist influence extended broadly through Great Britain and British America by the 1720s. During the first two decades of the eighteenth century, Anton Wilhelm Boehm (1673–1722), the chaplain of Queen Anne's consort, Lutheran Prince George of Denmark, facilitated British and American knowledge of German Pietism. Boehm vigorously promoted contacts between British Protestants of all denominations and German Pietists. He translated August Hermann Francke's letters and reports into English and selected German Pietists' writings for English consumption. Francke (1663–1727), professor of theology at the University of Halle, was the recognized leader of the Pietist movement in Germany. He was especially interested in the education of the young. In particular, his *Pietas Hallensis*, an account of the charitable orphanage, hospital, and school at Halle, Saxony, first published in an English translation in 1705, interested Britons and Americans. Francke advocated fervent evangelical conversionism and the use of vivid imagery in preaching. His writings influenced several early evangelical leaders in England, including Griffith Jones, founder of the Welsh catechistic schools; Sir John Phillips, Jones's sponsor and philanthropic backer of the Oxford Methodists; Howell Harris, a founder of Welsh Calvinist Methodism; and Dissenters Isaac Watts and Philip Doddridge. New England's Cotton Mather corresponded with Francke over a period of many years about a scheme to use a basic, orthodox piety to unite the faithful remnant of true believers in preparation for Christ's Second Coming. Jonathan Edwards was well aware of Francke's work.[3]

Pietism's direct appeal to the heart attracted ministers frustrated by the inability to halt secularization of their societies, to combat anticlericalism successfully, or to effect any lasting improvement in prevailing moral standards. Its simple faith in Christ's person and its unquestioning commitment to puritanical moral standards proved comforting in the face of the theological and moral skepticism of the age.

A Transatlantic Evangelical Consensus on Affectionate Preaching

By the eve of the great revivals of the eighteenth century, sets of ministers had emerged in Scotland, England, and America who exalted the role of the ministers in any revival of religion. Since the preaching of the Gospel was the principal means God used to convert sinners, good preachers would be needed to bring about a revival of religion. These sets of ministers shared common ideas of what constituted good preaching. They promoted similar versions of the proper and most effective methods of fulfilling the preaching and pastoral functions. The Connecticut Valley group stood out in their emphasis on the doctrine of preparation. Little in these tracts on evangelical preaching cannot be found in seventeenth-century Reformed writings. Indeed, one reason there is such close agreement among these British and American ministers is that all were calling for a return to seventeenth-century Reformed standards.[4]

All agreed with Stoddard that, whenever God is about to revive religion, he increases the zeal and ability of ministers.[5] Zealous and able preachers are central to a revival of religion because the Gospel, especially the Gospel when preached, is the principal means God uses to effect conversion. Benjamin Lord, in a treatise on conversion, wrote that

> other means may be very Serviceable and are of use to promote Conversions: as Afflictions, Mercies, etc., But the Word seems to be the great instituted means of this. . . . Especially the preaching of the word. . . . Faith is wro't by hearing the Word Preached.[6]

William Williams explained that the Gospel not only reveals the way to salvation, but that it is also accompanied with God's presence "so far (as to some at least) that it becomes effectual to convey the privileges purchased by Christ."[7] John Maclaurin, of Glasgow, explained, similarly, that "faith comes by the word of God" because "the doctrine of God's grace is the means of turning our souls to God, and of cleansing us from our filthiness and our idols."[8]

Since the chief purpose of preaching is the saving of souls, according to these men, ministers should preach principally on evangelical subjects, particularly the nature and necessity of regeneration. John Willison, of Dundee, urged the Presbyterian preachers in Scotland to make it "the Scope and End of all their Sermons . . . to lead Sinners from Sin and Self, to *precious* Christ and free Grace displayed in him."[9] A preparationist, Jonathan Marsh argued that the people needed to be taught, and ministers had yet to learn, more about "the Nature of true Conversion in particular as to the Manner how the Soul is prepared for Christ, how tis bro't to Believe & what are the fruits & effects of Faith, *by which it may be known when tis wrought*."[10] In teaching the people, preachers must be careful to distinguish between the case of the converted and that of the unconverted. Isaac Watts complained that

> the general way of preaching to all persons in one view, and under one character, as though all your hearers were certainly true christians, and converted already, and wanted only a little farther reformation of heart and life, is too common in the world; but I think it is a dangerous way of preaching.[11]

Willison called on ministers to "preach to our Hearers differencingly, to distinguish betwixt the Precious and the Vile."[12]

To be effective, the reformers urged, preachers must develop the ability to reach persons in various spiritual conditions. British and American evangelicals commonly interpreted 2 Timothy 2:15, "Study to show thyself approved unto God, a workman that needeth not to be ashamed, rightly dividing the word of truth," in this sense. Watts suggested that the dearth of conversions had resulted from the failure to divide "the Word aright to Saints and Sinners, to the stupid and profane, to awaken'd and convinc'd."[13] John Jennings told ministers to study the variety of the hearts of men and how to reach them. They should apply the Word to the several cases, tempers, and experiences of their hearers.[14] Benjamin Lord exhorted, "Especially be wise to win Souls . . . knowing when to speak Comfort and when to speak Terror, and to whom, . . . being skillful in address to particular Persons under various Circumstances."[15]

The British and American evangelicals rejected rational discourses delivered in an elevated, eloquent style. Too many, complained Jennings, speak with words of "man's wisdom . . . in the Style of the Heathen Sophists." The masters of reason, he said, put the bulk of the audience to sleep, whereas it is the bulk of the audience at whom the preacher should aim. Therefore, preachers should speak plainly and suit their addresses to the capacities of their hearers. Sermons, according to Isaac Chauncy, should be carefully composed in order to be plain, intelligible, orthodox, and convincing.

> The Preacher must not use Enameld Phrases, Flourishes of Rhetoric or oyl'd Expressions, when he would Alarm the Consciences of Sinners.[16]

Since it was the consciences that the preachers sought to reach, their style must be affective, the evangelicals insisted. Jennings called for a "commanding address to the Passions," and "the strongest Fire of the Orator."[17] Isaac Watts told preachers to "endeavour to get your heart into a temper of divine love, zealous for the laws of God, affected with the grace of Christ, and compassionate for the souls of men." His exhortation deserves quotation at length:

> Contrive all lively, forcible, and penetrating forms of speech, to make your words powerful and impressive on the hearts of your hearers, when light is first let into the mind. Practice all the awful and solemn ways of address to the conscience, all the soft and tender influences on the heart. Try all methods to rouse and awaken the cold, the stupid, the sleepy race of sinners; learn all the language of holy jealousy and terror, to affright the presumptuous; all the compassionate and encouraging manners of speaking, to comfort, encourage, and direct the awakened, the penitent, the willing, and the humble; all the winning and engaging modes of discourse and expostulation, to constrain the hearers of every character to attend.[18]

According to Isaac Chauncy, "when Ministers are about to convince and awaken Sinners, their Words must be Daggers and drawn Swords." Sermons should be delivered so as "to move the Affections of men."

> You must Preach with Zeal and Affection, that you may command Attention by your Fervency, and Affect the Hearts of your Hearers. Let your Auditors be awed with your Flaming Zeal, as if they heard a Voice from the Burning Mountain, or the Dark cloud bursting with Flashes of Fire. Such warmth is required that you may pluck Sinners as Firebrands out of the Burning.[19]

Like Stoddard, Watts believed that the preacher should not read his sermon, but deliver it from memory, and enliven it with spontaneous interjections. By so doing, he could animate his voice and countenance "and let the people see and feel, as well as hear."[20] While all did not demand memorized sermons—Jonathan Edwards used notes—many thought that the less reliance on notes the better.[21]

The evangelical reformers concurred that a man who was to lead others to Christ had better have traversed the way first himself. Ministers, said William Williams, have great need of "*EXPERIMENTAL Piety*. A spiritual understanding of the glory and truth of the Gospel, accompanied with an experience of the power of it, begetting Grace in the Soul."[22] A converted man was much more likely to be an effective minister than was an unconverted man. Benjamin Lord agreed:

> They who are to be Instruments of others Conversion, had need be Converted themselves and experimentally well versed in the mysteries of it: or else they are not like to be very usefull to others. . . . Let [the minister] appear as a man well acquainted with the Mysteries of Regeneration. . . . Fervently concerned for Souls, Flaming in his Affections for CHRIST, & zeal for GOD's Honour.[23]

A minister, wrote Watts, should be sure of his own election, and experienced in self-examination. This will give him greater skill in directing others in their spiritual exercises. Experimental acquaintance with the things of religion will enable the minister to "preach more powerfully . . . and talk more feelingly on every sacred subject. . . . A word coming from the heart will sooner reach the heart."[24]

Paul was the evangelicals' model preacher. In 1726, for instance, James Murray (1702–1758), assistant minister at Swallow Field Presbyterian Church, London, published anonymously *The Example of St. Paul*. In the recommendation, Isaac Watts and John Evans call Paul, next to Christ, the brightest and most amiable example for both clergy and laity. "How can Ministers more naturally learn to preach and to behave themselves in the Church of God," they wrote, "than by observing the Subjects on which this excellent Messenger of Christ principally insisted, the Manner and the Spirit with which he spoke and wrote, and his admirable Conversation among all to whom he ministered."[25] Paul was the evangelicals' champion of the doctrine of salvation by faith alone.

No matter how skillful, zealous, diligent, and like Paul a preacher might be, the evangelicals agreed that he would not be successful in saving souls without the concurrent operation of the grace of God. "Let us pray for God's *pouring out his Spirit from on high* upon the Ordinances and assemblies of this Land," cried John Willison, "for our Affairs will never take a Turn to the better, *until the Spirit be poured out from on high*."[26] In his *Humble Attempt towards the Revival of Practical Religion*, Watts expressed the sentiment that the revival of religion would result only from the grace of God delivered through the preaching of the Gospel, and granted in answer to prayer:

> O let us stir up our hearts, and all that is within us, and strive mightily in prayer and preaching to revive the work of God, and beg earnestly that God, by a fresh and abundant effusion of his own Spirit would *revive his own work among us!*[27]

Preaching and the Passions

A fundamental agreement distinguished the evangelical persuasion: Religion is a matter of the heart. Unless the passions are moved, all the convincing arguments to the understanding are useless. The saving operations of the Spirit are an affective, not an intellectual experience.

The evangelical preachers of the 1720s and 1730s justified their attempts to engage the affections of their auditors on two grounds. First, they believed that no degree of logical demonstration of the benefits of virtue and the destructiveness of vice could permanently overcome men's inclination toward sin. Men must be frightened out of hell with the law and lured into heaven with the Gospel. Second, they believed that the conversion of a man from a sinner into a saint involved a transformation of the affections, from love of sin to love of God.

In Solomon Stoddard's theology, conversion involved two separate kinds of operations by the Spirit. First, there is preparatory work, in which the Spirit first frightens the individual into attempting moral reformation by making him sensible of his danger of damnation and then brings the individual to humiliation by showing him he has no power in himself truly to reform. Up to this point, no saving grace has been involved; the soul has done nothing good, but has actually been passive. When the soul has been thus prepared, it is ready for conversion work. By illuminating the understanding with a vision of the glory and excellence of God, the Spirit produces love of God in the heart. The sou is now capable of faith, the accepting of the offer of salvation in the Gospel, the first gracious act of the soul. Since this first closing with Christ is "the greatest change that men undergo in this world," men usually know the time of their conversion. Thus, from first awakening to final assurance, affections—fear, humiliation, love, and joy—are at the center of the conversion process. It is the preacher's job to engage those emotions.[28]

When rationalists rejected emotional religion as "enthusiasm," a mistaken claim of divine inspiration, evangelicals protested that the entire doctrine of the Spirit's operation on the soul was being rejected. The evangelicals did not deny the danger of enthusiasm but insisted that there was a proper and an improper use of the emotions in religion. Two evangelical theorists on the use of the emotions were Isaac Watts and John Maclaurin.

In 1730 Archibald Campbell, divinity professor at Saint Andrew's University, published a *Discourse Proving that the Apostles Were No Enthusiasts*, in which he said that all inner experiences that persons interpret as being supernaturally communicated to their minds "may possible have come about in a natural course and series of things, without any interposing of the Divinity," and that for every person who has had God's manifestation in an extraordinary manner "that has affected [him] with very warm and sensible emotions . . . there are thousands who have had it only in mere pretence, conceit, and delusion."[29] Campbell's thesis, observed Robert Wodrow in a letter to Benjamin Colman, tended to "destroy our doctrine as to the Spirit's work on the souls of his people."[30] Failure of the General Assembly to discipline Campbell was one of the grievances of the Seceders in 1733. In an anonymous publication, a few days before the opening of the General Assembly in May 1734, John Maclaurin complained about the failure of the Church to censure Campbell for his *Discourse* on enthusiasm.[31]

In or about the year 1732, Maclaurin wrote an essay "On the Scripture Doctrine of Divine Grace," evidently as an answer to Campbell and his ilk, but it was published only posthumously in 1755.[32] In this essay, Maclaurin first defends the doctrine of grace, "the doctrine concerning Divine operations restoring the Divine image in the hearts of sinners, and carrying it on gradually to perfection,"[33] and then asserts the necessity of a rational use of the affections in religion.

Maclaurin held that the Holy Spirit is the author of holiness not merely because he is "the author of all the outward instructions and providences that are the means of it," but primarily because he operates inwardly, "immediately upon the heart."[34] Means, such as the Scriptures, preaching, and various dispensations of providence are necessary but not sufficient. "Human corruption, and inefficacy of natural causes to subdue it, has made such interposition [by the Spirit] necessary."[35] In reply to Campbell, he asserted that

> to argue that there are no real operations of the Holy Ghost on the hearts of sinners, because many people deceive themselves in pretending to such things, is as unreasonable as to affirm that there is no true and sincere holiness in the world, because there so many hypocrites.[36]

Maclaurin insisted that devout affections are a necessary part of holiness. He defended "the lively vigorous exercise" of the grace of divine love.[37] One cannot, he asserted, distinguish between the attachment of the will to God and the affections that are included in sincere love to him.

> It is true, indeed that all affectionate devotion is not wise and rational: but it is no less true, that all wise and rational devotion must be affectionate. All suitable divine worship must include the exercise of divine love.[38]

One should use his reason to curb those passions that are hurtful or liable to excess, cautioned Maclaurin. But love to God should have no limits. "As the affections included in divine love are founded on the most reasonable grounds and motives, they are incapable of excess." Nor could one hate sin too much.[39]

Maclaurin championed the deliberate use of means to excite devout affections. That the various means that are useful in exciting divine love are also the means that excite other affections does not prove that divine love is owing solely to the natural means, he argues. There are evil affections, common (not gracious) good affections, and sincere holy affections. Divine grace is needed to make the means effectual for the last of these. Nor is the existence of fraud and of abuse of power to excite the emotions an argument against the validity of good affections or the use of means to excite them.[40] One of the most useful means of exciting devout affections is to give attentive consideration to those affections and to the reasons one should exercise them. Several means are subordinate to this one. First, there is "pathetic or affectionate style." "The turn of thought and style, which is the natural effect of strong affections in one person, is a natural means of exciting the like affections in others."[41] If the affections are good, then the use of pathetic discourse to communicate them is proper. Another means of communicating and exciting devout affections is example. Love in society can be a means of strengthening devout affections, just as devout affections strengthen the bonds of society. Foreshadowing Jonathan Edwards's arguments in *The Distinguishing Marks of a Work of the Spirit of God*, Maclaurin argued that the visible effects of lively affections, such as shedding tears of joy or of sorrow, are proofs neither of sincerity nor of delusion or fraud. The body and the soul are united and mutually influence one another.

> This sympathy between soul and body [is] a general common property of all the affections of human nature [and not] a distinguishing character of devout affections of self-deceivers or imposters.[42]

Powerful divine operations in the soul will have a natural tendency to affect the body; however, since there is great diversity from individual to individual in the external signs of inward love of God and hatred of sin, one cannot judge the validity or the degree of those affections from the signs.

Through his long publishing career, Isaac Watts continually sought to promote the rational use of the emotions in religious experience. In "The Inward Witness of Christianity," a sermon published in 1721, he wrote that "every true christian has a most rational and incontestable evidence of the truth of religion, drawn from the change that is hereby made in his heart."[43] Religion, he believed, is a matter of inward experience.

> It is a witness that dwells more in the heart than in the head. It is a testimony known by being felt and practised, and not by mere reasoning, the greatest reasoners may miss of it, for it is a testimony written in the heart.[44]

Watts wrote volumes attempting to reconcile reason and revelation, but he continually returned to the insufficiency of human reason.[45]

Watts's clearest exposition of the place of the emotions is his *Discourses of the Love of God, And Its Influence on All the Passions: With a Discovery of the Right Use and Abuse of Them in Matters of Religion* (London, 1729).[46] Watts seeks to map a straight path between excessive reliance on reason and blind faith in devout raptures. He concludes "Where the religious use of the passions is renounced and abandoned, we do not find this cold and dry reasoning sufficient to raise virtue and piety to any great and honorable degree, even in their men of sense, without the assistance of pious affections." And where religion consists of a working of the passions without a due exercise of reason, fancy, enthusiasm, and bigotry usually follow.[47]

Like Maclaurin, Watts believed that "Divine Love is the Commanding Passion."[48] And like Stoddard, Watts considered the emotions to be at the center of the conversion process.

> "Religion begins in fear, it is carried on by love, and it ends in joy." Erroneous and unhappy is that philosophy that would banish these affections from human nature, which have so powerful an influence on the religious life, and assist our preparation for death and heaven.[49]

Reason alone is not enough to animate one to flee the wrath of hell, nor to do his duty. Fear is necessary for the former and love for the latter.[50] Just as it is the preacher who "not only instructs well, but powerfully moves the affections with sacred oratory" who will best hold the attention of his hearers, so it is the auditor who is not only convinced in his understanding by the argument, but is also touched in the heart with concern about his sins and fear of divine anger and is filled with zeal and holy purposes who will remember the discourse longest and put it into practice.[51] Hence, it is the duty of the preacher to engage the affections: "When the understanding is enlightened, the passions must also be addressed, for God has wrought these powers into human nature, that they might be the vital and vigorous springs of actions and duties."[52]

Hymnody and Affectionate Religion

Today Isaac Watts is remembered chiefly for his contribution to the development of the English hymn. In his own day, his hymns were loved. Through his hymn writing, and by his encouraging the use of psalms and hymns in worship, Watts sought to encourage the proper use of religious affections.

Clergymen understood the power of psalms and hymns as instruments for expressing pious emotions and made use of them to help create a sense of unity at religious gatherings. Intending to enhance that power, in England and America they led a revolution in the performance, language, and content of church music. The reform of church music began in England late in the seventeenth century within the established church; by the first decade of the eighteenth century, Dissenters, led by Watts, had joined the movement; and in the 1710s it reached New England. By 1730 the revolution there was far advanced. Congregations had set aside a chaotic, rote singing of a small number of tunes in favor of a larger number of tunes, sung by note and in parts. Polemicists in favor of "regular singing," supported the cause with the argument that the more skillful and beautiful the singing the more likely it would serve its primary purpose, eliciting holy affections. The clergy promoted regular singing as a means of restoring religion to its earlier prominence in the lives of their congregations: As Laura Becker concludes, they viewed "religious practice. . . as a means to religious fervor." Clerical interest in retranslating the psalms and composing new hymns accompanied the enlargement of congregations' musical skills. In Scotland, however, the revolution in church music did not begin until the second half of the eighteenth century and was long resisted by many as a challenge to orthodoxy.[53]

High Church Anglicans viewed the singing of psalms as an aid to piety. The Anglican Religious Societies promoted improved performance of psalm singing. The first members of Samuel Wesley's religious society at Epworth were members of his parish choir.[54] Much of the book that would initiate musical reform in New England's churches, *An Introduction to the Singing of Psalm-Tunes*, by John Tufts (1715), a Harvard graduate from Medford, Massachusetts, was lifted from an English psalter by English psalm reformer John Playford.

In 1708, following publication of Watts's *Hymns and Spiritual Songs*, several Independent and Presbyterian congregations in London undertook a cooperative effort to improve the performance of psalm singing. Each of six prominent Nonconformist ministers preached a Friday lecture on the duty of singing God's praises with skill. The six sermons were published together as *Practical Discourses of Singing in the Worship of God* (London, 1708).[55] Among the practices these ministers sought to reform was "lining out." The custom of having the deacon read each line or two so that the people might know what to sing interfered with understanding the sense of the whole. The ministers thought that if the people did not understand what they were singing, the psalms could not effectively help them raise their affections toward God. Realizing that tradition and the expense of supplying everyone with a hymnal would hinder the speedy reform of lining out, Watts tried to accommodate his hymns to the practice by making each line carry a complete unit of meaning.

Reform of the language of the metrical psalm versions meant both modernization and the use of words and phrases meaningful to ordinary people. At the end of the seventeenth century, under the influence of the Royal Society, some reform of the language of the metrical psalter had been accomplished. In 1694 some of the archaisms of the "Old Version" of Sternhold and Hopkins (1562) were replaced. In their "New Version" of 1696, Tate and Brady attempted a more thorough updating.[56]

Some reformers likewise intended to restore meaning to the use of psalms in public worship by changing their content. Just as the affections would not be engaged if the people sang apathetically or if they did not understand what they sang, so their affections would not be engaged if what they sang had little relation to their own situation and experience. Watts used this argument to defend the use of spiritual songs of human composition, that is, nonscriptural, in divine worship, and his revolutionary paraphrase of the psalms, suiting them to the state of English Christians:

> When we are just entering into an evangelical frame, by some of the glories of the gospel presented in the brightest figures of judaism, yet the very next line perhaps which the clerk parcels out unto us, hath something in it so extremely jewish and cloudy, that it darkens our sight of God the Saviour. Thus by keeping too close to David in the house of God, the veil of Moses is thrown over our hearts.[57]

Watts's reform of English hymnody cannot be disassociated from his commitment to the use of the emotions in religious experience. In the preface to his *Hymns and Spiritual Songs*, Watts argued that singing in the worship service should be the action that evokes "the most delightful and divine sensations." Instead, he observed, singing as now managed "doth not only flatten out our devotion, but too often awakes our regret, and touches all the springs of uneasiness within us." The cause of this uneasiness, he believed, was that the words of the psalms often did not correspond to the situation and experience of modern Christians. Thus, he concluded, it is lawful and beneficial to sing new songs, fitted to the circumstances of modern worshippers. All of Watts's hymn and psalm reforms had one objective, to promote the motion of the heart toward God. He pursued this end by replacing everything that he believed served as a hindrance with things more conducive to devotion. Watts developed his argument more fully in his "Short Essay Toward the Improvement of Psalmody," attached to the 1707 edition of *Hymns and Spiritual Songs*. One of the chief ends of singing, he wrote, "is to vent the inward devotion of our spirits in words of melody, to speak our own experience of divine things, especially our religious joy." Thus, like prayer, spiritual songs need not be restricted to inspired forms, fitted to the experience of a people under the Old Testament dispensation. "The improvement of our meditations, and the kindling divine affections within ourselves," Watts listed as another purpose of sung worship. Is not the contemplation of the glories of Jesus Christ, he asked, more likely than the "smoke and sacrifice, bullocks and goats, and the fat of lambs" of the psalms to evoke in us divine affections?[58]

Endowed with a power to move the feelings of great numbers of his contemporaries, Watts's hymns and psalms quickly became favorites among evangelicals on both sides of the Atlantic.[59] In 1731 Philip Doddridge described to Watts the powerful effect the singing of one of Watts's hymns had on a group of "plain country people" in England. "I was preaching in a barn to a pretty large assembly," Doddridge wrote.

> After a sermon . . . we sung one of your hymns; . . . and in that part of the worship I had the satisfaction to observe tears in the eyes of several of the auditory, and after the service was over some of them told me that they were not able to sing, so deeply were their minds affected with it. . . . These were most of them poor people who work for their living. On the mention of your name, I found they had read several of your books with great delight, and that your Hymns and Psalms were almost their daily entertainment. And when one of the company said, "What if Dr. Watts should come down to Northampton?" another replied with a remarkable warmth, "the very sight of him would be like an ordinance to me."[60]

Before the Great Awakening, Watts's hymns were not used in New England meetinghouses, where only scriptural texts were sung. The Reformed, however, did not object to the use of hymns of human composition in private devotion. Cotton Mather sang psalms, hymns, and his own extemporaneous verses in his devotion.[61] Jonathan Edwards during one period had been wont to sing forth "with a low voice my contemplations of the Creator and Redeemer." While he was engaged in contemplating God, "it always seemed natural to me to sing, or chant for my meditations."[62] According to Benjamin Colman, Solomon Stoddard, "when *tir'd* with severe Studies, upon reading one or two of [Watts's] rapturous *Hymns*, he had *returned* fresh to his Work."[63] New Englanders readily adopted Watts's works for private devotion. Cotton Mather recorded in his diary for 2 December 1711 that Watts had sent him a new edition of his hymns, which he resolved to use in the private worship of his family, to suggest to the booksellers that they stock, and to persuade his pious neighbors to purchase.[64] Judge Samuel Sewall was another admirer of Watts's devotional verses.[65] In 1738 Benjamin Colman proposed to his congregation that they make a new collection of psalms and other biblical hymns more suited to the style and diction of the age for use in the worship service. He suggested that the best collection could easily be made from Watts's *Psalms* and from his *Hymns*. "His Poetry is grave and solemn," Colman wrote, "full of Light and Heat, and the *Evangelical* Turn he gives in many Places is wonderfully adapted, in my Opinion, for the Service of Souls."[66] Colman's effort was unsuccessful.

The movement to reform church music and the resulting "singing controversy" in New England followed by a few years the movement in England. By 1700, because of the lack of printed music and trained musicians, psalm singing by New England congregations had become a chaotic singing by rote. The number of melodies in use had dwindled to a handful. Each member of the congregation tended to sing his own version of the tune set by the deacon, at his own tempo, resulting in what Increase Mather called an "Odd Noise." A movement to restore singing by note, promoted by itinerant singing masters and supported by most of the ministers, began about 1715. By 1730 there were many New England congregations musically skilled enough to sing in several parts.[67]

The evidence suggests that the new way of singing contributed to the emotionalism of eighteenth-century revivals. On the whole, the young people and the ministers favored the new way against the predilections of the older members of the congregations. It was the generation that grew up during the singing reform that experienced the Great Awakening. In a letter describing the revival of 1734–1735 at Northampton, Massachusetts, Jonathan Edwards remarked that "no part of the public worship has commonly [had] such an effect on them [the people of Northampton] as singing God's praises."[68] In the *Faithful Narrative* of the 1734–1735 revival at Northampton, Edwards offered a description of the function of singing in engaging the religious emotions:

> Our public praises were then greatly enlivened; God was then served in our psalmody, in some measure, in the beauty of holiness. It has been observable that there has been scarce any part of divine worship, wherein good men

amongst us have had grace so drawn forth and their hearts so lifted up in the ways of God, as in singing his praises.

Edwards went on to extol the technical singing ability of his congregation, associating the increased beauty of the sound with raised religious feelings.

Our congregation excelled all that ever I knew in the external part of the duty before, generally carrying regularly and well three parts of the music, and the women a part by themselves. But now they were evidently wont to sing with unusuall elevation of heart and voice, which made the duty pleasant indeed.[69]

Edwards's observations confirmed the theoretical position of his predecessor in the Northampton pulpit. Solomon Stoddard, promoting the new way, advanced strong claims for the power of singing in public worship. In 1723 Peter Thacher and John and Samuel Danforth preached and published an essay on "Cases of Conscience Concerning the Singing of Psalms," most likely prepared by Stoddard. The piece states explicitly that "singing of Psalms is a Converting Ordinance." The ministers hold that "the sweet and Harmonious Modulation of many Voices of God's Holy Worshippers together, in singing of God's Praises, doth admirably help to excite and raise the affections in the holy Worship of God." Furthermore, "God has vouchsafed to His Saints, much sweet communion with Himself in their Public as well as Private Singing of His Praises." These ministers urge the people to rely on the assistance of grace to transform their singing into genuine praise, "trusting in God the Holy Ghost, for His Influences to Irradiate, Elevate, Invigorate, and Fix our Hearts." The Holy Spirit, working through the psalms, can evoke genuine religious affections.[70]

Cotton Mather shared Stoddard's enthusiasm for the new way of singing and considered the renewed interest in psalmody a sign of the approaching millennium. "It is remarkable," he wrote in 1721,

that when the kingdom of God has been making any new appearance, a mighty zeal for the singing of psalms has attended and assisted it. And may we see our people grow more zealous of this *good work*: what a hopeful sign of the times would be seen in it "that *the time of singing has come, and voice of the Turtle is heard in our land*."[71]

The use of melody to express and evoke religious emotions had, of course, always been part of the Christian tradition. It had an important place in the Nonconformist, evangelical tradition. Richard Baxter (1615–1691), for instance, was, like Watts, in the words of Escott, "a logician, who yet finds room for religious passion."[72] Like Watts after him, Baxter wrote religious verses, hymns, and psalm paraphrases. He defends the use of sung worship to awaken holy passions, from the argument of experience:

> Those that deny the lawful use of singing . . . do disclose their unheavenly unexperienced hearts. . . . Had they felt the heavenly delights that many of their Brethren in such duties have felt, I think they would have been of another mind.[73]

High Church leaders also encouraged the use of singing to express religious feelings. Nevertheless, a decade before the great revivals several individuals felt called on to defend and promote the use of spiritual songs to excite religious passions. In 1728–1729 Nonconformist Isaac Watts, New England Presbyterian Benjamin Colman, and High Church divine William Law each devoted a portion of a major work to this task.

William Law (1686–1761), nonjuror and mystic, alloted Chapter 15 of his *Serious Call to a Devout and Holy Life* (1729), a text that influenced a large number of early Methodists, to the consideration of singing of psalms in private devotions, where he gave it a place of prominence. He laid it down as a common rule for everyone to begin all prayers with a psalm. The psalms should be sung, he insisted, and not merely read over, for the goal is to experience the psalm, so that it might stir up religious feelings. Reading a psalm without singing is like looking at food without eating. Psalms

> create a sense and delight in God, they awaken holy desires, they teach you how to ask, and they prevail with God to give. They kindle an *holy* flame, they turn your heart into an altar, your prayers into *incense*, and carry them as a sweet-smelling savour to the throne of Grace.[74]

Law did not emphasize skill in performance of private devotion; rather, he held that anyone whose heart rejoices in God is capable of expressing that joy through some sort of chanting. Just as religious joy is naturally expressed in song, Law said, so, the act of psalm singing has the natural power to excite in us feelings of delight in God. Singing is a proper means to support devotion. Law explained that psalm singing is among the highest forms of devotion.[75]

In his *Discourses of the Love of God*, Isaac Watts argued that the various parts of public worship were "suited to work upon our senses and thereby to awaken pious passions within us." Especially useful among these devises is singing.

> How happily suited is this ordinance to give a loose to the devout soul in its pious and cheerful affections? What a variety of sanctified desires, and hopes and joys, may exert themselves in this religious practice, may kindle the souls of christians into holy fervour, may raise them near to the gates of heaven, and the harmony of the blessed inhabitants there? Nor are pious sorrows utterly excluded from this ordinance: There are tunes and songs of mournful melody to solace the humble penitent, and to give a sweetness to his tears.[76]

Benjamin Colman entitled the final sermon in his series *Some Glories of Our Lord and Saviour Jesus Christ* (London, 1728), "The New Song," taking as his text Revelations 5:9, "and they sang a new song." In this sermon, Colman defended the use of singing in divine worship, in part, from its usefulness in engaging the emotions. "The tongue of man was made for *worship* and for *tunes*," he argued.

> And the *soul* of man is pleased, enlarged, dilated and profited by singing. Man has an *ear* as well as voice, and every power and pleasure of the mind is sacred to God, and should be improved to his praise.[77]

From similar grounds, he also promoted the improvement of skill in performance:

> Sing better than you have ever done, and it shall be a new song unto the Lord, and to your selves also. Sing now with new light, new affections, new hearts, new life and spirit; with new pleasure and more exalted devotion, so the song shall never grow old, but be always new.[78]

Events of the Great Awakening and Methodist revival bore witness to the importance of the singing revolution. Group singing of hymns in New England's streets became prevalent enough during the Great Awakening to produce public controversy.[79] In 1742 Jonathan Edwards had to compromise with his congregation to prevent them from replacing psalms in divine service entirely with Watts's hymns.[80] The introduction of hymns to divine service constituted a major liturgical innovation in New England Congregational and English Dissenting churches in the mid-eighteenth century, and for a long time thereafter many churches would continue to prohibit nonscriptural hymns during public services. Hymn singing was an important activity among the early English Methodists, who, uniting the English and German hymn traditions, produced some of the most enduring of the English hymns. Singing not only provided an outlet for the expression of pious emotions, but also encapsulated the communal experience of the revivals.

The performance of church music underwent the same pattern of change in Scotland in the late seventeenth century as in New England. By the end of the century the number of tunes commonly in use had been reduced to twelve and many congregations employed only a few of these. Lining out was universal, and precentors were unskilled. The notion that tunes had been written in a specific key was lost and it often happened that the precentor would have to start a psalm over because he had pitched it too high or too low. Congregations sang in the nasal, reedy voice of Scottish folk singing, without regular pulse or rhythm, and embroidered the tunes with improvised melodic decorations. Different parts of the congregation would sing at varying tempos, while some individuals chose a tune different from the one set. Despite the anarchic cacophony, many pious Scots found this form of religious worship spiritually edifying, and some may have continued it in their family devotions after it had disappeared from the churches. In contrast to New England, the movement to reform church music did not produce any effect in Scotland until the 1750s with the rise of the choir movement; and until the middle of the nineteenth century, Scottish Presbyterian churches used only psalms and a few other metrical Scriptural passages and resisted innovations like Isaac Watts's lyrics and Methodist hymns. Nevertheless, the singing of psalms in prayer meetings served the same function in the Scottish revivals as the skilled singing of Watts's hymns did in New England's Great Awakening, the expression of shared feelings and experiences.[81]

In both preaching and singing, the evangelical reformers of the early eighteenth century deliberately chose methods to arouse the passions of their congregations. They focused their reforms on these major parts of the worship service because moral suasion was their principal source of influence, and presiding over public worship was their prerogative and principal function. In so doing, they developed methods that would prove successful during the Evangelical Revival, and prepared a constituency ready to embrace passionate preaching and eager to express their feelings in skillful singing.

5

Revivals: Experience and Experiments 1660–1735

> Entertained thots about projecting some scheme for the reforming the young people of my parish and endeavouring to bring them to a more serious concern about religion.
>
> Diary of Daniel Wadsworth, 5 May 1737.

By the 1730s, evangelically oriented clergymen in Great Britain and America shared a similar theological and pastoral outlook on the revival of religion. This outlook was grounded in the national covenant and had evolved with the changing political and social circumstances. The pastors' experience with actual local revivals also shaped their understanding and expectations about the revival of religion. This experience varied considerably according to the political and social contexts of different parts of Great Britain and America.

In September 1735 Eliphalet Adams, of New London, Connecticut, commented on the reaction of New Englanders to news of numerous revivals in the Connecticut Valley that year. "Some people at a Distance," he observed,

> hearing of this Concern & stirring are quite at a Loss how to account for it, They wonder what should be the occasion, and Nothing less than some prophet or angel sent from above, or some Expectation of the world's sudden Coming to an End, seems to them sufficient to make it Either so great or so Extensive.

Adams answered with a rhetorical question:

> But what is there after all so very unaccountable in this matter? Or what is there so very peculiar in their circumstances who are now, thro' the mercy of God so very much Concerned?[1]

Should not all sinners be so concerned, he asked. The Connecticut Valley awakening seemed extraordinary to many, yet its supporters argued that it was not so very strange, it was unusual only in its magnitude.

Nonconformist ministers Isaac Watts and John Guyse published Jonathan Edwards's account of the Connecticut Valley revivals of 1735 in London in 1737. They entitled it *A Faithful Narrative of the Surprizing Work of God in the Conversion of Many Hundred Souls*. Earlier, having read a shorter version of the account,[2] Watts and Guyse had found the revivals a "strange and surprizing work of God," the like of which they had never heard "since the days of the apostles."[3] In the preface to *A Faithful Narrative* they professed that "never did we hear or read, since the first ages of Christianity, any event of this kind so surprising as the present narrative hath set before us."[4] Even though they found the Connecticut Valley awakening so extraordinary that they could compare it only to Pentecost, Watts and Guyse also wrote of it as if it were a phenomenon not so rare: The year 1735 in the Connecticut Valley was one of those times when

> our ascended Saviour now and then takes a special occasion to manifest the divinity of this Gospel by a plentiful effusion of his Spirit where it is preached: then sinners are turned into saints in numbers, and there is a new face of things spread over a town or a country.[5]

Watts and Guyse professed surprise at the news of the awakening and yet had a ready explanation for it. They found the phenomenon strange and yet recognized it.

Word of the Connecticut Valley awakening produced the same reaction among New Englanders and English Dissenters: both surprise and recognition. Americans and Englishmen were surprised that the occurrence was "either so great or so Extensive." Both Americans and Englishmen associated the awakening with the establishment of Christ's millennial kingdom.[6] Yet, Americans and Englishmen also argued that the awakening was not unique, but that it conformed to the historical pattern of God's dealings with human communities. Many congregations in seventeenth-century England, Scotland, and America had experienced periods of intense concern for salvation among numerous individuals, and more than ordinary numbers of conversions. In several instances, these "revivals" had occurred in the same period among several communities over a wide area. Contemporaries interpreted the Connecticut Valley awakening in this historical context.

England

When news of the Connecticut Valley awakening reached England, the most recent experience with a general awakening among the Dissenters had been in the 1690s. For Dissent as a whole, the episode had been distasteful. The awakening, led by Richard Davis, had contributed to the dissolution of the union of Congregationalists and Presbyterians of 1691 and had reinforced an association of revivalism with enthusiasm and antinomianism. Nowhere in the contemporary literature on the Great Awakening is there reference to the Davis affair, despite the similarities of his techniques and those of the radical revivalists of the later period.

Richard Davis (1658–1714), Welsh born and London educated,[7] upon his ordination as minister of the Independent congregation at Rothwell, Northamptonshire, in 1689, commenced a ministry of energetic evangelism. He preached plainly, directly, and with "thunder." He itinerated widely and sent out laymen—journeymen, shoemakers, carpenters, tailors, and dyers—as evangelists. By 1692 he and his preachers were holding meetings within a radius of eighty miles of Rothwell. Reportedly, by 1696 they had gathered thirteen new churches comprising from 2,000 to 3,000 members.[8]

Davis's practices alarmed many of the United Brethren, particularly the Presbyterians. One Presbyterian considered the lay preachers sent out by the Independents as "but the Devil's Design first to *Debase* the Ministry, and then to *overthrow* it."[9] Davis ran afoul of the United Brethren also on account of his teachings. Davis denied any need of seeking repentance. The elect had been justified from the moment of Christ's death. Thus, Christians need do nothing to be saved. They should not seek pardon, but only the manifestation of pardon.[10] The Presbyterians in the United Brethren, most of whom were moderate Calvinists, construed Davis's High Calvinism as destructive of moral endeavor. Giles Firmin associated Davis's teachings with those of Anne Hutchinson, whose proceedings in New England he had witnessed nearly sixty years earlier.[11] In 1690 Samuel Crisp published the works of his father, Tobias Crisp (1600–1643), some of which had been questioned by the Westminster Assembly for antinomian tendencies. Davis's and Crisp's doctrines corresponded closely enough to cause many of the United Brethren to fear a "plague" of antinomianism.

After several inquiries into Davis's faith and practice, the United Brethren published their findings against him in 1693. They accused him of antinomianism and enthusiasm. They found him guilty of rebaptizing persons baptized in infancy by ministers of the Church of England on the grounds that Anglican priests were not ministers of Christ. They also condemned

> his sending forth Preachers unfit for the Ministry, and unapproved by the Neighboring Ministers; His unchurching such Churches as agree not with his Exorbitant Methods, and Licentious Principles; His wickedly railing against most of the Orthodox, laborious Ministers, endeavouring to the utmost to prejudice the People against their Persons and Labours, as Idolatrous, Legal, and Antichristian; Yea, affirming, That all the Churches are gone a whoring from Christ,

> and that happy is he who is an Instrument in breaking all the Churches, wherein he hath made too great a progress.[12]

Davis reportedly taught that God has ordained all that comes to pass, even sin, and urged his followers to trade only among themselves as far as possible, and never with a conformist. The covenant he had his followers enter reflected an apocalyptic fervor. They considered their opponents to have perverted the Gospel and to be fighting the "last War with the Lamb, in these last days." They, themselves, stood "humbly waiting for [Christ's] Coming in his strength." And they expected a further revelation of the truth.[13] One of Davis's objections to the formation of the United Brethren was that the ministers had acted before God had made known his will. "The Spirit was not poured down from on High," he protested.[14]

The controversy over Davis broke up the United Brethren and revealed a rift in Dissent: The Presbyterians were moving toward Arminianism and eventually unitarianism, while the Congregationalists persisted as Calvinists. Congregationalists objected to the condemnation of Davis, accusing the United Brethren of assuming synodical authority. Some denounced the neonomianism of those who issued a blanket denunciation of the writings of Crisp. Congregationalists began withdrawing from the Union in 1693, and for six years a pamphlet war over the antinomian question raged.[15]

The Davis affair inhibited revivalistic activity for years. It caused revivalism to be associated firmly with antinomianism and enthusiasm at a time when Dissenters were becoming highly sensitive to those charges. Morality was the burden of the era's preaching, and the heresy hunters were swift to attack anything that smelled of antinomianism. Anglicans made use of Davis's behavior and doctrines to condemn Dissent in general and Independents in particular. One Anglican, for instance, used the example of Davis to link the Independents with the Quakers.[16]

In the first half of the eighteenth century, rejecting Davis's revivalist strategy, Dissent turned within itself. It failed to reach out to bring in new members. Rather, it became "tribal," ministering to those families already within the fold. While Calvinist Dissent remained evangelical in doctrine, it ceased to be evangelistic in practice.[17]

The Davis affair made Dissenters wary of revivalist activities but did not cause them to repudiate the concept of a revival. In his *Guide to Prayer* in 1715, Isaac Watts argued that there are occasions when the Holy Spirit makes his presence known powerfully to a congregation. In such cases, he wrote, divine influence has attended the words of the minister, "sinners have been converted in numbers, and saints have been made triumphant in grace." Times of the extraordinary presence of the Spirit "are rare instances," cautioned

Watts, "and bestowed by the Spirit of God in so sovereign and arbitrary a manner, according to the secret counsels of his own wisdom, that no particular christian hath any sure grounds to expect them."[18]

Watts did not associate the Connecticut Valley awakening with the Davis affair. Rather, he seized on the awakening as vindication of evangelical religion against the claims of the deists. The awakening was evidence that God was active among men. In *A Faithful Narrative*, by carefully isolating and rejecting as an aberration anything smacking of enthusiasm or antinomianism, Jonathan Edwards made the Connecticut Valley awakening acceptable to the English evangelicals. Watts and Guyse approved of the teachings of the Connecticut Valley revivalists, such as William Williams, whose *Duty and Interest of a People* they had read. "And if our readers had opportunity (as we have had," wrote Watts and Guyse,

> to peruse several of the sermons which were preached during this glorious season, we [*sic*] should find that it is the common plain Protestant doctrine of the Reformation, without stretching towards the Antinomians on the one side, or the Arminians on the other, that the Spirit of God has been pleased to honor with such illustrious success.[19]

Only a few years earlier, Watts had written his *Humble Attempt towards the Revival of Religion*, and Guyse had recently delivered his lecture *Reformation upon the Gospel Scheme*. Both men had argued that religion would revive when God chose to pour out his Spirit in abundant measure to make effective a course of evangelical preaching. The information they received from Edwards seemed to confirm that God had done just that in the Connecticut Valley.

Scotland

Between the Restoration and the Great Awakening ministers in Great Britain and British America made careful and minute observations of religious conditions in the land in general and in their own cures in particular. They closely analyzed the factors accelerating or impeding decline, and they exchanged these observations with one another in letters. Jonathan Edwards's *A Faithful Narrative* originated as one such letter, addressed to Benjamin Colman. Another example is a letter of 14 September 1709, from Robert Wodrow to John Gib, minister at Cleish, Scotland. The letter is devoted strictly to the conditions of and prospects for religion. This long missive (fifteen pages in his published correspondence)[20] begins with general remarks on religious conditions in Scotland. At the time of the Glorious Revolution, Wodrow states, the Gospel had some success, but since then there has been a growing deadness, facilitated by toleration, the union with England, and controversy over the oath of abjuration. The bulk of the letter describes religious conditions in his own parish. His people are generally obedient to the moral law. On his first arrival at Eastwood, six years earlier, some previously unawakened persons had begun to show concern for their souls, but that concern is now wearing off. Conversions have practically ceased. At Communion services, the people are less diligent and fervent in hearing sermons, there are fewer communicants, and there is less private and public prayer than in former times. Wodrow has some hope for those who were young at the time of the Glorious Revolution and have "had some little convictions and awakenings some years since." But he thinks "the great part of the usefulness of the Gospel at this day is to such as are already converted." When Wodrow seeks a contrast to the present period of religious decline, he turns to the Communion of Shotts of 1630. "When I reflect upon the Communion of Shotts," he writes, "I would fain be more affected than I am with the sad difference between our times and those before us, yea under the darkest times of Episcopacy, as the communion of the Shotts you know was." In the lore of the Scots, Shotts had become the symbol of the good-old-days of the Gospel's power.

Sacramental occasions among the more radical Scottish Presbyterians during the rule of Cromwell and among Presbyterian conventicles during the Restoration period manifested power and popularity on the model of the Communion at Shotts. Approving commentators wrote of several of these events as incidents of special outpourings of divine grace: "Never a greater outletting of God's presence in comunions; tuo congregations befor dead, falling in great love of the ordinances;" "The Lord bowed his Heavens and came down, and displayed his saving Power on that Occasion most comfortably and signally"; "This Ordinance . . . was so signally Countenanced, backed with power, and refreshing Influences from heaven, that It might be said Thou O God didst send out a plentifull rain whereby thou confirmed thine Inheritance when it was weary."[21] During the two decades following the Glorious Revolution and the reestablishment of Presbyterianism, the sacramental season became a regular feature of Scottish religious culture.

Eighteenth-century Communion seasons varied in the depth of piety they exhibited. Ministers lamented when spiritual vitality lagged and rejoiced when religious feeling became enlivened. Not many sacramental occasions were noted as extraordinary times of the presence of God and of the effusive dispensation of his grace. Very few matched the standard set at Shotts. While every Communion might produce some spiritual quickening, only a few qualified as religious revivals.[22]

In the early eighteenth century, some Scottish ministers, using an evangelical preaching style and making use of societies for religious fellowship, had some success in stimulating more than ordinary religious concern. The Erskine brothers are good examples. Ebenezer Erskine (1680–1754) and his brother Ralph (1685–1752) began pursuing an evangelical style of ministry early in the eighteenth century. Ebenezer, settled at Portmoak, in Kinross-shire, in 1703, dated his conversion from 1708. Thereafter, he became a popular preacher. By 1714 crowds overflowed the parish church on Sundays and filled the church even for the Thursday evening lectures. In good weather, therefore, Erskine preached in an adjacent field. He established and drew up rules for praying societies in his parish. Two thousand persons usually attended his annual communion services. Ralph Erskine's settlement in 1711 at Dunfermline, where the ministry had been vacant for five years, sparked a remarkable revival of religious activity. Ralph instituted fellowship meetings and catechism classes and made a regular round of pastoral visits. His Communion services drew crowds of 4,000 to 5,000.[23]

Prayer societies, a practice dating from the seventeenth century, dotted the Scottish countryside. They were composed of devout persons who retained the morality and doctrines of the seventeenth century. These societies helped maintain the piety characteristic of the revival of Shotts into the eighteenth century.[24] Mild awakenings appeared occasionally among Scottish parishes during the 1720s and 1730s. These small revivals are typified by the one at Nigg, East Ross, near Cromarty, in 1739. Here, only a few were under deep spiritual concern at any one time, although many experienced conversion over the year. There were no unusual physical manifestations. The parish society for prayer and conference increased so much in membership that it had to be divided into two.[25]

When the London edition of Edwards's *A Faithful Narrative* was reprinted in Edinburgh in 1737 and 1738, the Scottish evangelicals easily accepted it as an account of God's work.[26] The narrative would significantly influence the interpretation the Scots would give their own revivals in 1742.

New England

From their early days, New England congregations had known seasons in which more than unusual numbers experienced conversions and applied for church membership. Heightened concern for the matters of salvation usually accompanied the founding of a new church. Thereafter, periods in which new admissions to the church were concentrated tended to follow generational cycles. As the children of the founders of the town and church reached maturity in their late teens and early twenties, a portion of them would seek to share in the fellowship of the church. Many of these young adults had been born shortly after the town founding, they matured together, and applied for membership in the same period. Their children, in turn, would attain adulthood and apply for membership within the bounds of a brief span of years. After several generations, the life cycles of town residents would spread out more or less evenly across the years. In the meantime, the wave of each newly maturing generation would bring with it a wave of new converts. Solomon Stoddard felt that something was seriously amiss if a congregation had gone twenty years without "a noise among the dry bones."[27] The phenomena associated with what would come to be recognized as revivals, then, were not unknown to the seventeenth century, and yet, the eighteenth century revivals produced surprise.

On 5 May 1737, Daniel Wadsworth, pastor at Hartford First Church, noted in his diary that he had "entertained thots about projecting some scheme for the reforming the young people of my parish and endeavouring to bring them to a more serious concern about religion."[28] Between the Restoration and the Great Awakening, New England's ministers experimented with various methods to promote "ingatherings" or "harvests" of souls. One aspect all of these methods shared was an emphasis on young persons, those from their late teens into their late twenties.

Robert Pope argues that the halfway covenant, introduced as a response to the mid-seventeenth-century crisis of membership in New England's churches, by the end of the century had become an evangelical tool for bringing in the unchurched. By retaining within the bounds of church discipline the baptized who failed to experience conversion in adulthood, the halfway covenant, it was hoped, would insure that the churches would continue to have significant influence in the community. Originally, a major part of the clergy pressed for this innovation against the reluctance of the laity. During the crisis of King Philip's War, many churches dropped their opposition to the halfway covenant. The jeremiads of the war years accused the churches of neglecting the spiritual needs of the young. Adoption of the halfway covenant was seen as a way of repenting of and reforming this neglect. It was revocation of the Bay Colony charter in 1684, however, that imparted to the halfway covenant its greatest consequence. Since the civil magistrate could no longer be counted on to uphold the national covenant, the need of the churches to spread their influence throughout the community seemed to increase. As Pope explained, "after revocation of the charter, the churches had to reach out into the community for new members and change their attitudes toward the unchurched. The old 'tribalism' no longer sufficed; the churches became evangelical." Well before the end of the century several churches were extending the privileges of halfway membership to adults who had not been baptized as infants.[29]

As Pope has shown, closely related to the halfway covenant was the mass covenant renewal, another innovation used to stir up religious interest. The practice of holding a special observance in which the members of a particular church would renew their covenant to walk together as Christians in the sight of God was introduced during King Philip's War. These covenants of reformation, which had begun as public rituals to appease an offended God, had, by the end of the century, become "a routine part of New England Congregationalism." The covenant of reformation not only involved those already within the church covenant, but often was used as an occasion for encouraging the children of the church to own the covenant. The introduction of the halfway covenant in conjunction with the covenant of reformation was intended to bring the unconverted to a concern about their spiritual state. After 1690, many ministers began inviting the unchurched to join with the church members in covenant renewal, and, hence, to accept halfway membership. In this way, covenant renewal became an evangelical device.[30]

Cotton Mather considered the revival at Taunton, Massachusetts, under Samuel Danforth in 1704–1705 as the successful result of experimental methods that he, himself, advocated. Mather pointed to Danforth's "astonishing harvest" as evidence of the usefulness of "certain particular Methods for Suppressing of Disorder, and for promoting Piety."[31] These methods included a local movement for reformation of manners and young men's religious societies. In a letter, possibly addressed to Mather, Danforth's own description of the Taunton affair sounds identical to the later narratives of the revivals of the Great Awakening:

> *My time is spent* in *daily Discourse* with the *young People*, visiting me with their *Doubts, Fears* and *Agonies*. RELIGION *flourishes to Amazement and Admiration*; that so, we should be at once touched with Soul-Affliction, and this in all Corners of the Place; and that *our late Conversions* should be attended with *more than usual Degree of Horror*; and *Satan* permitted to wrestle with them by *extraordinary Temptations*, and Assaults and Hours of Darkness. But I hope the *deeper the Wound the more sound may be the Cure*: and I have little Time to think of worldly Matters; scarce Time to study Sermons, as I used to do.[32]

The event at Taunton was more a mass covenant renewal than a mass conversion. Only 14 were added to the church, while 300 renewed the covenant. The Taunton episode marked a stage of transition from the moral reformism of the seventeenth century to the revivalism of the eighteenth.

Timothy Woodbridge, minister at First Church, Hartford, from 1683 to 1732, seems to have experimented to produce spiritual harvests. At Hartford First Church between 1686 and 1695 in any one year no more than 23 persons came into full communion, and no more than 19 owned the covenant. The yearly average number of new communicants was 6, and of new covenanters, 8. Then, in the year 1696, a time of "Indian trouble," 203 persons owned the covenant and 45 entered into full communion. Woodbridge gathered another harvest in 1712, when 46 joined in full communion and 5 owned the covenant. Thereafter, however, until his death in 1732, in no year did Woodbridge admit more than 8 persons to full communion. The yearly average number of admissions between 1713 and 1731 was 4; however, it appears that in his later years Woodbridge sought to repeat the experience of 1696. Between 1713 and 1722 no more than 2 persons in his congregation owned the covenant in any one year, yet: on 3 February 1723, 19 men owned the covenant; on 25 March 1725, 11 men owned the covenant; on 7 December 1729, 24 men owned the covenant; on 4 January 1730, 7 men owned the covenant; on 2 August 1730, 23 women owned the covenant; on 27 September 1730, 11 women owned the covenant; and on 11 April, 1731, 15 women owned the covenant. This pattern did not continue under Woodbridge's successor. What produced this pattern? Was Woodbridge holding religious classes, divided by sex, in order to stir up concern for salvation by encouraging persons to declare a commitment to the church covenant? Woodbridge favored and Hartford practiced open Communion until his death. Yet, despite Woodbridge's opposition, the church maintained a distinction between halfway and full communion. Perhaps, if he could not bring many into full membership, the pastor thought he could at least bring the people within the bounds of the church.[33]

The Connecticut Valley

Of all the experimental methods of stimulating interest in matters of salvation, the revivalistic techniques of Solomon Stoddard and his colleagues in the Connecticut Valley had the most success. It was in the Connecticut Valley where the great majority of identifiable revivals between 1710 and 1733 as well as the great awakenings of 1734–1735 took place.

The development of revivalistic techniques in the Connecticut Valley grew out of disappointment with other techniques. The halfway covenant had brought increasing numbers into the church, but had not increased the frequency of conversions. Covenant renewal and other attempts to implement "reformation of manners" had produced only temporary external improvement. The results of the extension of the Lord's Supper under either the Mathers' formula or Stoddard's proved unsatisfactory. The emphasis on early piety and the use of religious meetings for young people brought no dramatic change. The Connecticut Valley revivalists did not reject all of these techniques, but they combined them with a preaching and pastoral emphasis on a recognizable emotional conversion experience. These ministers believed the techniques had failed because they had not been accompanied by the pouring out of the Holy Spirit. They believed that the Spirit was more likely to accompany and make effective evangelical preaching.

Aside from the responses to the earthquake of 1727, the spiritual harvests in New England between 1710 and 1733 that contemporaries took note of are as follows:[34]

1712 Hartford, Conn., First Church, Timothy Woodbridge, pastor;
Northampton, Mass., Solomon Stoddard, pastor;
1715–16 East Windsor, Conn., Timothy Edwards, pastor;
1716 Preston, Conn., Salmon Treat, pastor;
1718 Northampton, Mass., Stoddard, pastor;
Norwich, Conn., First Church, Benjamin Lord, pastor;
1721 Franklin, Conn., Henry Willis, pastor;
Norwich, Conn., Lord, pastor;
Preston, Conn., Treat, pastor;
Windham, Conn., Samuel Whiting, pastor;
Windsor, Conn., First Church, Jonathan Marsh, pastor;
1726–27 Woodbury, Conn., Anthony Stoddard, pastor;
1727 New Milford, Conn., Daniel Boardman, pastor;
1729 Windham, Conn., Whiting, pastor;
1731–32 Hadley, Conn., Isaac Chauncy, pastor;
Lyme, Conn., Jonathan Parsons, pastor;
Windham, Conn., Whiting, pastor.

With the exception of New Milford, all of these towns lay in the Connecticut River Valley.

New England revivalists became aware of the increasing frequency of revivals. In 1714 Samuel Danforth referred to sporadic local revivals of religion as encouragements to pray for a general awakening through the land:

> We have reason to bless God who gives some revivals of his work, sometimes in one church and sometimes in another; and from thence should be encouraged to plead with God for a more general effusion of his spirit upon all his churches, that all parts of his vineyard may flourish.[35]

In 1730 Eliphalet Adams announced:

> Sometimes the Servants of CHRIST are blessed with Success in our Times and that pretty remarkably, The Ignorant are Instructed and made to understand Doctrine, The Wandering are Reclaimed, Sinners are Converted in great num-

> bers, and being Converted they thrive in Grace and Holiness and are built up in their most holy Faith unto Eternal Life, This is the work of God and must be acknowledged to his Praise.[36]

The ministers took particular notice of the confluence of revivals in 1721. We might speak of a "Connecticut Thames River Valley Awakening" of 1720–1722. Within a six-month period in 1721, Samuel Whiting at Windham admitted eighty persons to full communion. Between 1720 and 1722, Benjamin Lord admitted forty-two members at Norwich, just thirteen miles from Windham. In 1721, nearby Preston and Franklin enjoyed revivals, as did Windsor, thirty miles from Windham, on the Connecticut River.[37] In reference to the Windham revival of that year, Adams asserted that the Old Testament had foretold, and the experience of the Apostles had confirmed, that the Gospel would have times of remarkable success. "Three Thousand were Converted by one Sermon, Nations were born at once and Kingdoms in a day." Since the days of the early Christians, religion has much decayed; yet it has revived:

> at sundry times and diverse Places, when the *Spirit* hath been *poured out* more Plentifully *from on high*, then many have been wro't upon by the *Blessing of God* upon the *Means of Grace* and bro't over to the *Acknowledgement of the Truth*.[38]

Jonathan Marsh used the local revivals of 1721 as contrasts to the dull state of religion in general. He warned that:

> tho' God at times is pouring out his Spirit for the carrying on Conversion Work, in a Remarkable manner, in particular places, yet according to the tenour of the *Public Covenant*, as we are a Body Collective by it till we come to Reform and Amend in the General, we ly open to Public Judgments.[39]

The ministers were not satisfied with local revivals. They looked, still, for a great awakening of all New England. Marsh took the opportunity of his Connecticut election sermon in 1721 to publicize to the entire colony the methods he had found successful. His message was that reformation rested primarily in the hands of the preachers, for it was they who must prepare the people by bringing them to a conviction of their sin. To do their part, the preachers must preach awakening sermons:

> Do what in you lies in your ministry to reach the Consciences of men. . . . In a corrupt time there is a great scarcity of Godly men among a People, many are secure and fast asleep in Sin, they stand in great need of Terrour, & are like to perish and go down to Hell without it.[40]

Marsh told the ministers that to "Labour to open the nature of Conversion and to urge the Necessity of it upon your Hearers [is] the way to have the ministry crowned with success."[41]

In the Connecticut Valley, where the doctrine of preparation was stressed and, therefore, the emotional preaching of the terror of hell especially practiced, a growing frequency of sporadic local revivals seemed to confirm the efficacy of the renewed emphasis on evangelical preaching. By application of preparationist theology to their sermons, several clergymen in the valley enjoyed at least one period of religious revival in their congregations before the Great Awakening. Now the Connecticut Valley revivalists began to look for a general religious awakening. They believed that such a general revival of religion would make possible a general reformation of sins, fulfill the national covenant, save New England from divine judgments, redeem lost souls, restore the prestige of the ministry, and, perhaps, inaugurate the millennium.

We know little about the revivals before 1735 because the ministers, wanting to avoid the appearance of pride, were reluctant to describe the successful periods of their ministries in any more than general terms. When Cotton Mather reported in a few sentences Samuel Danforth's revival at Taunton in 1705, he explained that "the Minister and his Happy Assistants would be displeased at my Trespass upon their Modesty, if I should Entertain the World abroad with a full Relation of [the operations of God there]."[42] In 1721 Eliphalet Adams mused concerning the revival in Samuel Whiting's congregation at Windham, Connecticut, "COULD their Reverend Pastor have been prevailed upon so far to have Gratified the Publick, we might have been more curiously Entertained with the Knowledge of many Particulars, which ought not to be forgotten."[43] The reticence of the ministers limits an understanding of the early revivals. To see these seasons through the eyes of the ministers, we can turn to the small number of available contemporary accounts.

Stoddard's revival of 1712 was but one in a series of "harvests" over which he presided (1679, 1683, 1696, 1712, 1718). It was during the 1712 revival, however, that he preached on the phenomenon of the revival. The seventh and last sermon in "The Benefit of the Gospel" teaches that "there are some special seasons wherein God doth in a remarkable manner revive Religion among his People." The body of the sermon consists of a series of general observations on such seasons. Saints are enlivened, sinners are converted, many not converted become more religious, and a great moral reformation occurs. God is arbitrary in choosing where, over how large an area, when, and for how long he will pour forth his Spirit. Stoddard describes revivals in general terms, providing few hints that would enable us to put together the actual experiences. Jonathan Edwards reports that Stoddard said that "in each of them, the bigger part of the young people in the town seemed to be mainly concerned for their eternal salvation."[44] No evidence suggests that anything similar to the crying out or fainting of the 1734–1735 revival occurred, although Edwards does say that in the latter revival persons had "no otherwise been subject to impressions on their imaginations than formerly."[45] All we can say is that Stoddard's harvests consisted of a heightened interest in salvation, particularly among the young, numerous apparent conversions, and a temporary moral improvement.

About as much can be said concerning Timothy Edwards's revivals at East Windsor. Jonathan Edwards reported that during one, in his childhood, he and some classmates built a booth in a swamp where they would pray together.[46] He describes the 1716 revival to his sister Mary in a letter dated 10 May of that year:

> Through the wonderful goodness and mercy of God, there has been in this place a very remarkable outpouring of the Spirit of God. It still continues, but I think I have reason to think is in some measure diminished; yet I hope not much. Three have joined the church since you last heard, five now stand propounded for admission; and I think above thirty persons come commonly a Mondays to converse with father about the condition of their souls.[47]

As a young teenager, Jonathan Edwards was already analyzing the progress of revivals he witnessed. Against such experiences, he would be able to judge the Connecticut Valley awakening.

Writing in 1744, Jonathan Parsons left a concise account of the revival over which he presided at Lyme, Connecticut, in 1731:

> The *Summer following* my Ordination there was a great *Effusion* of the HOLY SPIRIT upon the People. There appear'd to be an uncommon Attention to the Preaching of the Word, and a disposition to hearken to Advice; and a remarkable Concern about Salvation. 'Twas a *general Inquiry* among the Middle aged and Youth, *What must I do to be saved?* Great Numbers came to my Study, some almost every Day for several Months together, under manifest Concern about their Souls. . . . I urg'd them very much to *Works*, and gave it as my Opinion (perhaps too hastily) that such awakened Souls ought to attend upon the *Lord's Supper*: and in less than *ten* Months *fifty-two* Persons were added to the Church. There were *several whole Families* baptiz'd. Many of the *young* People were greatly reformed: they turned their Meetings for vain Mirth into Meetings for Prayer, Conference and reading Books of Piety. There was a Number of them kept a religious Society about *two Years*.[48]

Some historians have used Parson's account as evidence that the New England revivals that preceded Edwards's championing of undiluted Calvinism in the 1730s tended to be tinctured by a reliance on human ability.[49] Parsons, himself, admitted that at that time he had been "greatly in Love with Arminian

Principles." However, Parsons was reflecting on his pastorate in the light of the revivals of the 1740s. His mentioning that he had urged the awakened "very much to *Works*" may be evidence of his rejection of preparationism subsequent to the Great Awakening; his belief that he had probably been too lax in his standards for admitting persons to the Lord's Supper certainly reflects the strenuous standards adopted by New Lights in the 1740s. In any case, the preparationist preaching of other revivalists in the Connecticut Valley was as much within the bounds of Reformed orthodoxy as was Stoddard's. Stoddard's preparationist morphology of conversion gave men no power to save themselves, and Stoddard's God was as sovereign as anyone's.[50]

In the first decades of the eighteenth century a group of Connecticut Valley ministers were developing a viable revivalist tradition, which included a theology to explain revivals and a system of homiletics to promote them. When others outside that tradition, however, also experienced revivals, they sometimes proved unprepared to lead and direct the religious energy into orthodox channels. The revival at New Milford in 1727, for instance, went astray into the paths of enthusiasm.

Daniel Boardman (1687–1744; Yale, 1709) served the congregation at New Milford, Connecticut, from 1712 to 1744. Information on the 1727 revival comes principally from a letter written by Boardman in 1742 and from memoirs of David Ferris.[51] Even though Boardman criticized and Ferris approved the revival, their accounts substantially agree on what happened. Some young companions of Ferris, then twenty years old, who had been accustomed to spending much of their time in "vanity and merriment," suddenly, on the "dying Counsel of a loose young Man," greatly reformed, being brought into serious concern for their souls. They set up private meetings for prayer, reading of books of piety, and singing psalms. The awakened ranged from five years of age to thirty, though most were under twenty-two. Pleased by these developments, Boardman attended many of their frequent meetings. Within a year and a half, more than fifty members were added to the church. About a year after the beginning of the revival, the new converts and Boardman had a falling out. The subjects of the revival asserted that "they had been awakened by the immediate operation of the Holy Spirit on their own minds." According to Ferris, however, Boardman and many of the congregation, asserting that revelation had ceased, denied "the apostolic doctrine of 'Christ within,' and of being 'led by the Spirit of God.'" According to Boardman, the new converts claimed "*a Spirit of discerning and judging the State of others*, so that there were scarce any that escap'd their Censure. . . . Upon this they began to *purge their Meetings*, (to use their Language) and disallow the *unconverted* (as they termed them) to meet with them." Accused of heresy, the young people were brought before the church. The church concluded that their "sentiments were not so heterodox as to prevent communion" with them. Nevertheless, the church discountenanced the separate meetings, which continued. Many considered the young converts Quakers, even though the converts disclaimed knowledge of that people. Heterodoxy held its ground in New Milford. Baptists from Rhode Island wrote the meetings and Rogerenes paid them a visit.[52] Between 1731 and 1734, 19 of Boardman's congregation, most of whom were in full communion, seceded to the Quakers, a serious defection for a church of less than 100 communicants. Nine of the seceders had joined the church during the 1727 revival.[53]

The Connecticut Valley revivalists were aware of the potential dangers in the revivals, but seem to have developed some skill in containing them. Eliphalet Adams, a zealous evangelist, sounded warnings against enthusiastic excesses and Arminian heresies. He emphasized that

> *it will cost a man some pains & Study to be any thing Eminent in the Ministerial Profession.* Ministers in our Times be not Immediately inspired, Qualified & sent forth into the Vineyard, And those few Crazed and soft headed Persons that pretend to be so, do commonly make such wretched Work of it, that we may know from thence their pretence to be Vain; Our Accomplishments must be gained by much Study, Diligence and Prayer to God.[54]

In 1730 he warned that during revivals "we must Carefully avoid both Extremes, On the one hand, we must not run into Enthusiasm and impute every Whim and Notion and odd Impression to the Spirit of GOD, as some are so weak as to do, Nor yet on the other must we deny the Necessity of its Assistance or be backward to acknowledge its Actual Concurrence."[55] When the Quakers sought to benefit from the revivals in the Connecticut Valley of 1735 in the same way they were reaping the harvest of the New Milford revival, they were unsuccessful. "There has been much talk in many parts of the country," Jonathan Edwards reported, "as though the people have symbolized with the Quakers, and the Quakers themselves have been moved with such reports; and came here, once and again, hoping to find good waters to fish in; but without the least success, and seemed to be discouraged and have left off coming."[56] The Connecticut Valley revivalists managed to keep the New Light distinct from the Inner Light.

The revivals between 1712 and 1733 are strikingly similar. The overriding element, of course, is concern with conversion. The reports share an emphasis on sheer numbers of persons joining the church, and on attending the sacraments. Pastoral counseling of awakened sinners is important. A primary concern is moral reformation in general, and in particular suppressing the frivolities of youth. Family worship is urged. The young play a prominent role. The establishment of religious societies for the young, particularly young men, is usual. In some of the revivals the enthusiasm and emotional extremes of the Great Awakening are foreshadowed. While they were localized affairs, the revivals did not transpire in isolation. News of them spread: "The Neighborhood hath rung of it," said Eliphalet Adams about the Windham revival of 1721. Five towns experienced revivals in 1721. New Milford and Woodbury, a dozen miles apart, enjoyed revivals in 1726–1727.

The Earthquake of 1727

Contemporary writings on the local revivals in New England that followed the earthquake in the spring of 1727 provide evidence that New England's ministers shared an orthodox understanding of the mechanisms by which revivals took place, and that they might have found revivals extraordinary because of their magnitude, but neither inexplicable nor unknown.

On 29 October 1727, a Sunday, at about 10:30 P.M., the ground in an area of New England approximately 500 miles in extent trembled violently several minutes. Chimneys crumbled, pewter fell from shelves, wells dried up, and fissures and quagmires appeared in the earth. The next day, throughout New England, people gathered at their meetinghouses to pray for deliverance from any further danger. Aftershocks continued for several weeks, during which time New England's congregations observed days of fasting in order to appease God, and days of thanksgiving in gratitude that he had saved them from harm. For several months after the quake, religious services were well-attended, the people were unusually attentive to the sermons; in many congregations numerous young persons owned the covenant, and surprising numbers of persons offered themselves as candidates for full communion. Probably not a preacher in New England failed to interpret the significance of the earthquake to his congregation. Twenty ministers published some thirty sermons on the earthquake. These sermons reveal the preachers' understanding of the manner in which God operates in bringing about a revival of religion.

Uniformly, the preachers saw the earthquake as evidence of God's anger against the people of New England, a call to repentance and reformation, and a warning of great punishments to come if the people failed to abandon their sins. They saw the earthquake as an awakening providence to bring the people to a sense of their sins and to an acknowledgment of their need to reform. Since the people had long failed to pay attention to God's word from the voices of a faithful, evangelical ministry, God had spoken in a loud voice from the depths of the earth. Christ, said Benjamin Colman, is "striving with us by his awakening SPIRIT, under the *ministry* of his Word, and the *Judgments* of his Providence; And because the common and *ordinary Means* have had so little power and effect on us, GOD has used *One* altogether *new* and extraordinary."[57]

The ministers considered the earthquake to have been a means through which God impressed upon the people a sense of their sins and of their need for grace, and they rejoiced in its good effects.[58] As in the case of all means, however, the earthquake would not produce a lasting change for the better unless an outpouring of the Spirit followed it. In sermon after sermon, the ministers admonished the people not to let the good impressions pass and leave them unreformed and unconverted.[59] "Indeed such *extraordinary* things are not the ordinary means of *conversion* to sinners," wrote Benjamin Colman. "The Impressions of sudden frights and surprizing terrors are apt as suddenly and strangely to wear off." If the earthquake was to make a contribution to a real reformation of morals and revival of religion, the Holy Spirit would have to be poured out to make the ordinary means of grace, the preaching of the gospel, effectual.[60]

The earthquake itself would not produce the desired changes; it could, however, act as a catalyst: "It affords a happy Opportunity for Persons in Places of Authority and Influence, to endeavor a Reformation of Manners, and the Revival of dying Religion."[61] Thomas Foxcroft encouraged preachers to make the most of this occasion. If the Spirit were to be poured out on preachers now, religion would revive.[62]

The ministers suggested that the earthquake was not necessarily a prelude to destruction, but quite probably a prelude to an outpouring of grace. They associated the earthquake not only with the punishment of Sodom and Gomorrah, but also with the pouring out of the Spirit on the Apostles at Pentecost: "When they had prayed the Place was shaken where they assembled together: and they were all filled with the Holy Ghost" (Acts 4:31).[63] The ministers believed that New England, as a people, had grounds to expect that God would save rather than destroy them. Joseph Sewall asked, "May not the People of God in *New England* hope for a Dispensation of the Covenant of Grace relating to them? . . . May we not hope that God will revive His Work among us, and not utterly cast us off?"[64] Foxcroft was confident that "God has come, not to slay us, but only to affright us and call our Sins to Remembrance."[65]

In those towns where reformation and revival were noted, New Englanders believed that God had poured out his Spirit to make the ordinary means of grace effectual. In the aftermath, the Spirit seemed most active in the towns of the Merrimac Valley, where the quake had been strongest. A month after the quake, Thomas Prince, of Boston, recorded that in those towns, Hampton, Newbury, Bradford, and Andover in particular,

> so mightily are many awakened with the Sense of their Danger and the Divine Displeasure, as has produced a wonderful Reformation. Profaneness, Drunkenness and other Vices abandoned; the earnest Pursuits of the World discarded; the Places of Publick Religion vastly thronged; the Worship of GOD set up in Prayerless Families; and great Numbers continually added to the flourishing Churches. Twenty, *Thirty* or *Forty* on a Sabbath have offered either to make or renew their Dedication to the Blessed GOD: In one Assembly above *One Hundred & Fifty* in Three Weeks time; and of these about *Fourscore* in one Day.

Prince identified this as "an happy Effusion of the HOLY SPIRIT!" And he asked for prayers that the same Spirit would be poured out on the Boston churches.[66]

In May 1728, William Williams, Jr., of Weston, reporting that many in his congregation had joined the church since the earthquake, ascribed the revival to the hearing of the Gospel: "The LORD having opened their hearts . . . to give attention unto the *word preached*, which hath become powerful to the further awakening, and I hope conversion of some: This is the ordinary means which he blesses to this end."[67] Nathaniel Gookin published his earthquake sermons when he learned that they had been instrumental for many in bringing the awakening begun by the quake to an apparently successful issue.[68] The ministers, then, did not attribute the revivals to the earthquake. The awakening caused by the quake only prepared the way for the effectiveness of the actual means of grace, the Gospel. The ministers remain convinced that the revival of religion, no matter what the proximate cause, was the responsibility of themselves, who were entrusted with the preaching of the Gospel.

The actual configuration of the earthquake revivals can be followed in John Brown's account of the revival at Haverhill, Massachusetts, on the Merrimac River. The account is in the form of a letter to John Cotton, of Newton, Massachusetts, dated 20 November 1727. Appearing as the appendix to Cotton's *A Holy Fear of God* (Boston, 1727),[69] the letter is the first published revival narrative, a genre that would proliferate during the 1740s.

According to Brown's account, Haverhill held a public fast the Monday following the earthquake, and another on Wednesday. On Thursday, the people met and prayed across the river in Bradford. Each of these services was well attended, and at each the people were "earnest and enlarged in Prayer," and deeply affected. From Thursday on, Brown was kept busy "discoursing People about their Souls---by night & day. . . . *rain* or *shine*; some Days from Morning till 8 a Clock at Night, without so much as time to take any bodily refreshment." Since the day of the earthquake, Brown had admitted eighty-seven persons to full communion and sixty-seven to halfway membership, either through baptism or the renewing of the baptismal covenant. About as many men as women were involved, mostly between fifteen and thirty years of age. Many others were awakened to a sense of their sin. A general moral reformation swept the town. Some of the worst drunkards, blasphemers, and other offenders made seemingly sincere resolutions for change. "Family worship attended with earnestness; Young Men's Meetings multiply, and increase, Family meetings set up, &c."

Carefully examining his records, Brown came to the conclusion that the outpouring of grace had its gradual beginning about a year before the earthquake: "We have for some time past been growing & ripening for this *Harvest*." Of the 109 persons who had owned the covenant since 1719, 54, nearly half, had done so within the last year and a half—more than 20 in the last summer. Of the 130 admitted to full communion between the year 1719 and the time of the earthquake, 30 had been admitted within the year, "which is 3 times so many as in any one Year before, except the first Year of my Ministry [1719], in which I admitted 32."

In Brown's view, then, the revival was not the result of the earthquake. The earthquake was a means of awakening; but the pouring out of the Spirit was an operation separate from the quake.

The Merrimac Valley revivals of 1727 were very similar to the Connecticut Valley revivals of the decade. The major difference was the catalyst. In the Merrimac Valley a thundering earthquake, in the Connecticut Valley a thundering ministry awakened the people to a sense of the danger of remaining unreformed and unconverted. Yet, even this difference is deceiving, for conversionist evangelism played a significant role in the Merrimac Valley awakening. The Sabbath service preceding the earthquake, for instance, Samuel Phillips, of Andover, South Parish, had delivered a sermon steeped in brimstone:

> And Sinner! Thou hangest over this direful Pit, by the single Thread of *thy* Life only; and this is liable to be snapt asunder, every Moment: For,
>
> What art thou? Why a poor Mortal, who cannot stand before a *Worm*, or a *Fly*, if *God* should Commissionate either of them to seize thee, and stop thy Breath.[70]

In April, 1741, at the height of the Great Awakening, Jonathan Edwards would speak very similar words: "Thou art crushed before the moth. A very little thing, a little worm or spider, or some such insect, is able to kill thee. What canst thou do in the hands of God?"[71] At Andover in May 1727 it seemed providential that the earthquake followed that very night to drive home Phillips's frightening message. The next Sabbath, Phillips's church renewed its covenant.

The Middle Colonies

The crucial role of New England's evolving revival tradition in the development of analysis and description of revivals becomes clearer when one contrasts the writings on revivals by New Englanders with those by evangelical clergymen of the Middle Colonies before 1739. In New England, the evangelical revival consisted of the renewed effectiveness of a doctrine long taught. Although ministers talked about the Arminian threat, in the 1730s evangelical conversion and evangelical preaching were still the accepted norms of the society. The evangelical revival in the Middle Colonies, on the other hand, consisted mainly of the reintroduction of evangelical doctrine where it had been replaced by nonevangelical teaching, or its introduction where it had never been taught. In contrast to revivalism in New England, where religious institutions were well established among a homogeneous people, the Great Awakening in the Middle Colonies was largely a response to the churches' difficulties in meeting the needs of a rapidly expanding, ethnically and religiously diverse population.[72] The revival in the Middle Colonies among the Dutch Reformed and the British Presbyterians mirrored the rise of Pietism on the Continent, which had emerged as a reaction against the prevalence of a nonevangelical piety that ignored the conversion experience, emphasized baptismal regeneration, accepted a complacent morality, and looked askance at the display of emotion.[73] During the 1730s, debates over the nature of regeneration and over pastoral styles occupied Reformed clergymen in the Middle Colonies, who speculated little about the nature of the outpouring of the Spirit on communities. Middle Colony revivalists do not seem to have associated the evangelical revival so closely with the sudden effusion of grace on and rapid reformation of individual communities as did New Englanders.

For the Middle Colony revivalists, the hallmark of the evangelical preacher was his differentiating in his sermons among the various spiritual states of his hearers. Hendrik Visscher, translator of Theodorus Jacobus Frelinghuysen's series of sermons giving *A Clear Demonstration of a Righteous and Ungodly Man* (New York, 1731), lamented "how unwillingly do the generality of Professors hear that distinguishing manner of Preaching, and how strange seemeth it to them? As if a Servant of Christ should not any more make a Difference between the *Precious and the Vile*?"[74] Frelinghuysen (1691–c. 1748), Dutch Reformed pastor at Raritan, New Jersey, cautioned against the self-deceit encouraged by teachers who fail to warn the people that outward morality is not enough, "*not rightly dividing the Word*, but (*Armenian like* [*sic*]) throwing out the Promises (in general) to scramble at."[75] Gilbert Tennent, Presbyterian minister at New Brunswick, New Jersey, and later at Philadelphia, attributed his increased zeal as a conversionist to Frelinghuysen's example, "together with a kind letter which he sent me respecting the necessity of dividing the Word aright, and giving to every one his portion in due season."[76] Tennent denounced the "general and undistinguishing way of Preaching," the addressing of the congregation at one moment as if all converted and at the next as if all unconverted.[77]

Like Solomon Stoddard, Tennent believed that one cause of the languishing state of religion was the neglect of preaching conviction of sin.[78] The false hopes of many resulted from "the daubing of unconverted ministers," who were afraid of offending their congregations. In 1735 Tennent related a story about one of Stoddard's spiritual harvests. "After some considerable Time of Coldness and Unsuccessfulness in his publick Labours," Stoddard chose to preach on predestination and

> rais'd this Observation, That one Reason of Men's not receiving the Lord Jesus Christ, was their non-Election. This Sermon . . . (tho' upon so grating and displeasing a Subject) was followed with a very remarkable out-pouring of the Holy Spirit, to the Conviction of many.[79]

The Middle Colony evangelicals advocated passionate address to the affections. "We must brandish the terrors of God's wrath," said Ebenezer Pemberton, of New York, "to alarm the secure and impenitent; and display the wonders of his love, to encourage the weary and heavy-laden sinners." Terror was an important weapon against secure sinners: "We must uncover the mouth of the bottomless pit, and give them affecting views of the amazing agonies & torments which are prepared for the generation of the wicked."[80]

The Middle Colony evangelicals joined their voices to the English evangelicals' call to preach Christ rather than natural religion. "Now a Minister truly preaches Christ," said Tennent,

> when he forms his Discourses upon such Subjects as suit the State of his Audience, so as to have a direct Tendency to convince secure Sinners of their absolute Need of Christ; so as to guide convinc'd Sinners to Christ, and build up converted Persons in him, as the alone Foundation of Justification, Sanctification, and eternal Salvation.[81]

Pemberton urged the Presbyterian clergy to make Christ "the great Subject of our publick Preaching and private Discourses, by inculcating upon our Hearers, the great and distinguishing Articles of the Christian Faith: This indeed is almost grown out of Fashion."[82] Throughout the 1730s the Middle Colony revivalists vehemently denounced nonevangelical preaching. These attacks culminated in Tennent's 1740 Nottingham sermon, *The Danger of an Unconverted Ministry* (Philadelphia, 1740).

The awakening of 1735, Jonathan Edwards notes in *A Faithful Narrative*, was not restricted to New England. "There was no small degree of it in some parts of the Jerseys." While on a visit to New York, he heard from Rev. William Tennent, Jr.,

> of a very great awakening of many in a place called The Mountains, under the ministry of one Mr. Cross; and a very considerable revival of religion in another place under the ministry of his brother, the Rev. Mr. Gilbert Tennent; and also at another place, under the ministry of a very pious young gentleman, a Dutch minister whose name as I remember was Frelinghousa.[83]

News of the revivals in the Middle Colonies reached New England only after the awakening in the Connecticut Valley. On the other hand, the Middle Colony evangelicals made use of news of the Connecticut Valley awakening to stimulate concern for conversion among their auditors.[84]

The revivals in the Middle Colonies before 1735 are poorly documented. Since the best descriptions were not written until 1744, it is difficult to know how closely they resembled the New England revivals of the same period and even whether they were recognized as revivals at the time, or only by hindsight during the later revivals.

Theodorus Jacobus Frelinghuysen is considered the father of Middle Colony revivalism. A Dutch Reformed minister of German birth, Frelinghuysen accepted a call to serve several congregations in the Raritan Valley of New Jersey. Arriving in 1720, he immediately undertook a campaign to introduce evangelical pietism. His violent preaching and "howling prayers," his emotional conversionism, his denunciation of nonevangelical, "carnal" preachers, and his insisting on the right to examine the spiritual state of individuals before admitting them to the sacraments aroused a strong opposition and split the Dutch Reformed in American into two camps.[85]

Revivals have been attributed to Frelinghuysen because of the statement of Jonathan Edwards quoted earlier and statements by itinerating English Methodist George Whitefield and by Gilbert Tennent. Yet, none of the these statements is very revealing. Whitefield recorded in his journal that, when he preached at New Brunswick in 1739:

> Among others who came to hear the Word, were several ministers, whom the Lord has been pleased to honour, in making them instruments of bringing many sons to glory. One was a Dutch Calvinistic minister, named Freling Housen, pastor of a congregation about four miles from New Brunswick. He is a worthy old soldier of Jesus Christ, and was the beginner of the great work which I trust the Lord is carrying on in these parts. He has been strongly opposed by his carnal brethren, but God has appeared for him, in a surprising manner, and made him more than conqueror, through His love. He has long since learnt to fear him only who can destroy both body and soul in hell.[86]

The Christian History (Boston), of 1744, printed the following from a letter by Tennent:

> The Labours of the Reverend Mr. *Frelinghousa* a Dutch Calvinist minister, were much bless'd to the people of *New-Brunswick* and *Places adjacent*, especially about the Time of his coming among them, which was about *twenty-four Years ago*.
>
> When I came *there* which was about *seven Years after*, I had the Pleasure of seeing much of the Fruits of his Ministry: divers of his Hearers with whom I had Opportunity of conversing, appear'd to be converted Persons, by their Soundness in Principle, Christian Experience, and pious Practice: and these Persons declared that the Ministrations of the aforesaid Gentleman were the Means thereof.[87]

Herman Harmelink III argues that:

> the only evidence for an awakening or revival under Frelinghuysen comes from the Tennents, and their reports to Whitefield, Edwards and [Thomas] Prince [Jr., editor of *The Christian History*]. . . . There may have been some kind of "awakening" in Raritan, but the hard evidence points only to a disaffection and division under Frelinghuysen's ministry.[88]

The difficulty lies, in part, in the ambiguity of the term *revival*. Frelinghuysen introduced revivalistic methods and theology, and his ministry resulted in emotional conversion experiences; but whether he presided over distinct periods of intense religious concern recognized by the participants as outpourings of the Spirit is undocumented. He did not write about them as such.

While Frelinghuysen was introducing the evangelical form of ministry to the Dutch Reformed in America, the Tennents were contributing to its revival among the Presbyterians in the Middle Colonies. William Tennent, Sr., the founder of the Tennent dynasty in America, came to America from Ireland in 1717. In 1726 he set up his Log College at Neshaminy, Pennsylvania, where he trained many of the early evangelical leaders of Presbyterianism, including his four sons.

Not until after the revivals of the 1739–1742 period did the Middle Colony revivalists speak of their revivals of the previous decade in terms of special seasons of grace. Compare the two accounts of one of the earliest revivals attributable to the Tennent group. Gilbert Tennent, in his preface to his brother John's sermon on *The Nature of Regeneration* (Boston, 1735), wrote that John, pastor at Freehold, New Jersey, from 1729 to his death in 1732, "was a successful preacher. . . . He gained more poor sinners to Christ in that little compass of time which he had to improve in the ministerial work, which was about three and a half years, than many in the space of twenty, thirty, forty, or fifty years." In 1744 William Tennent, Jr., used the language of the revival narratives to describe his brother's success. Under John's ministry, wrote William:

> the Place of public Worship was usually crouded with People. . . . Many tears were usually shed, when he preached, and sometimes the Body of the Congregation was mov'd or affected. . . . It was no uncommon Thing to see Persons in the Time of Hearing, *sobbing* as if their Hearts would break, but without any public Out-cry; and some have been *carry'd out* of the Assembly (being overcome) as if they had been dead.
>
> Religion was then the general Subject of Discourse. . . . The Terror of GOD fell generally upon the Inhabitants of *this Place*; so that Wickedness as ashamed in a great Measure hid itself; Frolicking, Dancing, Horse-racing, with other profane Meetings were broken up.[89]

The Middle Colony revivalists seem to have expected the phenomena associated with a revival of religion to result from the introduction of evangelical preaching to a community.[90] Gilbert Tennent observed that the majority of Philadelphians "have never had the benefit of a strict religious education," and John Rowland noted that the people of New Providence, Pennsylvania, "before I came into these parts to preach, were but an ignorant sort of people, unacquainted with religion, both as to principles and practice."[91] These revivalists no doubt exaggerated the religious ignorance of their hearers, for a fair proportion of them would have been recent immigrants from Great Britain, many of whom would have shared the revivalists' evangelical orientation. Many of the native-born would have carried on pious traditions through familial religious exercises, despite the long enduring paucity of a settled Presbyterian clergy. In contrast, New Englanders thought of revivals as the product of a special and temporary shower of grace on a community that had lived long under sound preaching. Writing in 1744, Gilbert Tennent described his own ministry during the 1730s. He reported that his preaching at New Brunswick on evangelical themes resulted in "the *Conviction* and *Conversion* of a considerable Number of Persons, at various Times, and in different Places, in that Part of the Country." Significantly, however, at New Brunswick, "I don't remember that there was any great Ingathering of Souls at any one Time." Tennent could remember only one episode similar to the later revivals. When he preached at Staten Island:

> there was about *fifteen* or *sixteen Years ago*, a more *general Concern* about the Affairs of Salvation. . . . Once, in the Time of a Sermon . . . the SPIRIT of GOD was suddenly poured down upon the Assembly; the People were generally affected about the State of their souls; and some to that Degree, that they fell upon their Knees in the Time of the Sermon, in order to pray to GOD for pardoning Mercy: Many went weeping Home from that Sermon; and then the general Inquiry was, *What shall I do to be saved*?[92]

Only in part is my argument here an argument from negative evidence. The absence from their published writings of discussion of a revival of religion as a season of the outpouring of grace for the transformation of a community into a holy people does not necessarily mean that the Middle Colony revivalists did not think in those terms.[93] The main point here is that it was the New England revivalists who publicized this concept. The idea of a communal season of grace can be located without much trouble in the writings of the Connecticut Valley revivalists in the 1720s and 1730s, in the reports of New England's earthquake of 1727, and in a number of the published New England jeremiads since the late seventeenth century. Lacking New England's century of experience, regional identity, sense of divine mission, and jeremiad tradition, the Middle Colony revivalists spoke less about the meaning of the revival of religion for society at large than did their New England counterparts.

In November 1739 George Whitefield wrote that John Cross, of Basking Ridge, New Jersey, had told him, "of many wonderful and sudden conversions that had been wrought by the Lord under his ministry. For some time, eight or nine used to come to him together, in deep distress of soul; and, I think, he said, three hundred of his congregation, which is not a very large one, were brought home to Christ." A year later, when Whitefield asked for a particular account, Cross said, without elaborating, that "it directly answered the account given by Mr. Edwards of the work of God in Northampton."[94] The language used to report Cross's revival was borrowed from Edwards, for 300 is the number of converts mentioned in Edwards's account, and "brought home to Christ" is the phrase Edwards used in this context.[95] It was Edwards, not Cross, who wrote *A Faithful Narrative*. It was Edwards, immersed in the revival tradition of the Connecticut Valley, who conceived of the revival of religion as a specific phenomenon worthy of close analysis.

6

Revivals, History, and Eschatology

The first remarkable pouring out of the Spirit through Christ that ever was . . . was in the days of *Enos*.

JONATHAN EDWARDS, *The History of the Work of Redemption.*

A Faithful Narrative

Solomon Stoddard and Jonathan Edwards perceived most clearly the implications and the potential of communal religious awakenings. In *A Faithful Narrative* Edwards formulated a theory of revivals that had previously been inchoate. By the 1730s the necessary elements, the ideas, and the phenomena, had converged. Following Stoddard's lead, Edwards fitted them together into a coherent whole. Edwards's narrative lay in a tradition of privately circulated clergymen's letters recounting the spiritual condition of their cures. John Brown's brief narrative of the 1727 revival at Haverhill had broken the ground by being published. Edwards's narrative, however, was the first extended account and analysis of a season of religious awakening in a congregation, let alone in a full thirty communities. Before its publication, local revivals had been sporadic, amorphous, and experimental. At the time of the Connecticut Valley awakening, a technical term for these occurrences had not yet emerged. Even though the ministers were seeking to promote a revival of religion, they did not refer to seasons of religious revival as *revivals*. They resorted to descriptive phrases, or to metaphors such as "seasons of grace," and "harvests of souls." In *A Faithful Narrative*, Edwards speaks of "a very great awakening of many," awakening being the first stage in the morphology of conversion, an increased awareness of one's own sinfulness. He used the phrase "general awakening" several times to refer to the renewal of concern for salvation in the community at large. With *A Faithful Narrative*, Edwards not only perfected a new genre of religious literature, the revival narrative, but he also classified the revival as a unique phenomenon whose characteristics were unmistakable. Edwards had intended to report objectively, scientifically, his observation of the revival. In effect, though, by establishing expectations, his narrative became a blueprint for future revivals in America and Great Britain alike.[1]

The progress of Jonathan Edwards's revival in Northampton in 1734–1735 fit the pattern set by the previous Connecticut Valley revivals. English commentators thought it worth noting that

> here was no storm, no earthquake, no inundation of water, no desolation by fire, no pestilence or any other sweeping distemper, nor any cruel invasion by their Indian neighbors, that might force the inhabitants into a serious thoughtfulness, and a religious temper by the fears of approaching death and judgement.[2]

Before the revival, according to Edwards, the young people of Northampton had been guilty of licentiousness, nightwalking, frequenting of taverns, lewd practices, frolics of both sexes through most of the night, and indecent carriage at worship. About the year 1731 they responded to their elders' admonitions "and by degrees left off their frolicking, and grew observedly more decent in their attendance on public worship." In 1733 a sermon against their habit of engaging in mirth and company-keeping on Sunday evenings and after the public lecture resulted in "a thorough reformation." A number of religious conversions were then experienced in a remote village of the town. In the spring of 1734 the deaths of two young people spurred religious concerns among other youths. In the fall, at Edwards's suggestion, the young people began weekly meetings in small groups for "social religion." Their elders soon followed their example. About this time Edwards began a series of sermons on justification by faith alone. "And then it was, in the latter part of December, that the Spirit of God began extraordinarily to set in." Five or six individuals experienced conversion, including "a young woman who had been one of the greatest company-keepers in the whole town." From that hour, religious concern spread quickly through the town. Private religious meetings became frequent and crowded. The careless and immoral noticeably reformed. People seemed to forsake old quarrels, backbiting, and meddling, and were seen to confess wrongs and make up differences. They resorted to the pastor's chamber instead of the tavern and stayed at home except for business or religion. A long-standing political feud lay dormant. After Edwards's uncle killed himself in despondency over his spiritual state in June, the religious ardor gradually cooled. The effects of the revival lingered, however, and the private religious meetings continued.[3]

The Connecticut Valley revival tradition came to a full expression with the revivals in Northampton and some thirty other valley towns in 1734–1735. Edwards recognized the extraordinary character of the awakening. The word *extraordinary* appears thirty times in *A Faithful Narrative*, and *remarkable*, *wonderful*, and *amazing* often. "This seems to have been a very extraordinary dispensation of Providence," Edwards wrote. "God has in many respects gone out of, and much beyond his usual and ordinary way."[4] Edwards found the awakening extraordinary in six respects: it was not restricted to any one category of person; it affected a great number of people; it touched the elderly and the very young, not just young adults; the progress of conversion with many was unusually rapid; the emotional intensity was unusually strong; and it reached so many towns in such a brief span of time.[5] Yet, Edwards argues that the awakening was something neither new nor unaccountable. In fact, in comparison with "former stirrings of this nature," he found it extraordinary only in degree. Comparing the awakening with the spiritual harvests of Solomon Stoddard, hc wrotc:

> The work that has now been wrought on souls is evidently the same that was wrought in my venerable predecessor's days; as I have had abundant opportunity to know, having been in the ministry here two years with him, and so conversed with a considerable number that my grandfather thought to be savingly converted in that time. . . . The work is of the same nature, and has not been attended with any extraordinary circumstances, excepting such as are analogous to the extraordinary degree of it before described.[6]

The title of his narrative, and the word *surprising* in it, are not Edwards's but his English publishers'.[7] Because Puritan religion led to certain recurring types of emotional experiences, precedents for the various kinds of experiences of the subjects of the revivals of 1735 could be found among subjects of earlier conversions. As for the revivals themselves, Edwards could claim to see nothing new in them because, indeed, they had evolved over the previous half-century.

John Howe's *The Prosperous State of the Christian Interest Before the End of Time, By a Plentiful Effusion of the Holy Spirit*, delivered in 1678 but not printed until 1725, intimated the scenario that the British and American evangelical communities expected a communal experience of grace to follow. The revivals of the early eighteenth century in the Connecticut Valley fulfilled these expectations and fleshed out the scenario. The evangelicals believed that, first, God prepares the community by inspiring it to pray for his blessing. God gives ministers zeal for conversion work and increases their skills. As preachers, they awaken the people from spiritual sleep to a view of their danger and show the way to safety through reliance on Christ's righteousness. As pastoral counselors, they guide awakened sinners through the labyrinths of their devious hearts. To be so skilled, the ministers have to be converted men, men who have learned to examined their own hearts. Eventually, grace shows itself. Sinners are convinced of sin. Many are converted. The spiritual experience of saints is enlivened. Love abounds, quarrels cease, and immorality flees. Moral reformation is general, affecting through common grace even those not destined for saving grace. The duration of such a gracious shower varies according to God's secret plan, and those who scoff at the communal transformation are guilty of grieving the Spirit.

With "The Benefit of the Gospel," Solomon Stoddard made a start at a descriptive synthesis of this view of local revivals, but before Jonathan Edwards no one had brought these commonly held assumptions together into a clear exposition. The numerous appeals for the revival of religion in England in the early 1730s were not summonses for local revivals. The English concept of revival was only vaguely related to the experiences of individual congregations; however, in the Connecticut Valley, where a tradition of revivals had developed, local revivals were seen as the basic components of the expected revival of religion. Through *A Faithful Narrative*, Jonathan Edwards exported his perception of the potential of the local revival. English evangelicals accepted the revivals that Edwards reported as the results of the outpouring of the Spirit because they fit their own assumptions about how the Spirit acts. The narrative made the scenario explicit through tangible detail, so that in 1738 Isaac Watts could write that he expected the conversion of the Jews and of the heathen nations before the world's end to be effected by "such a spirit of conversion . . . as appeared in Northampton lately and the towns about it."[8]

Revivals and Church History

Revivals of religion held a central place in a view of history common throughout eighteenth-century British/American evangelical culture. This view of history set evangelicals apart from nonevangelicals.[9] It had three essential parts. First, the course of history could be explained primarily as the product of God's periodic pouring forth of his grace on peoples and then his periodic withdrawing of his grace. Second, in the present period God was withholding his grace generally, religion was on the decline, and the world lay in dire need of a fresh effusion of the Spirit. Third, the millennial kingdom of Christ on earth would be brought about by revivals across the globe through a grand effusion of grace. In 1739, on the very eve of the Great Awakening, Jonathan Edwards synthesized these evangelical ideas into a theory of a revival-driven history of redemption.

The evangelical community's understanding of the scriptural passage Joel 2:28, "I will pour my spirit upon all flesh," summarized their view of the place of religious revivals in the history of the church. They used it to support their emphasis on the operations of the Spirit of God on the heart. It corresponded to their belief in a God who directly interfered with history in order to fulfill his redemptive purposes, and whose special providence was inconstant, appearing at specific times and places. And it confirmed their expectation of a future revival of vital religion. It is significant, then, that a public opponent of the evangelical viewpoint wrote about the book of Joel on the eve of the evangelical revival.

Samuel Chandler, who, as we have seen, engaged John Guyse in pamphlet combat in 1730, published in 1735 *A Paraphrase and Critical Commentary on the Prophecy of Joel.* As early as his "Dedication" he reveals his nonevangelical disposition by defining the end of Christianity as "the virtue and hap[p]iness of mankind" rather than the glory of God. Appended to the *Paraphrase* is a "Dissertation upon the Third Chapter of the Prophecy of Joel." For his purposes, Chandler considers the third chapter to start at Chapter two, verse twenty-eight. The words *I will pour out my Spirit* Chandler interprets "as comprehending the various influences of God upon the minds of men . . . exciting them to a variety of actions, upon particular occasions and emergencies, and conveying to them especially the gift of prophecy." Unlike the evangelicals, he does not consider the words to mean numerous conversions. Also unlike the evangelicals, he believes that the prophecy has already been literally and completely fulfilled. The pouring out of the Spirit has already been accomplished and is not to be expected in the future.[10]

The evangelicals believed that, although Peter had referred to Pentecost as the fulfillment of this prophecy, the prophecy implied further effusions of the Spirit.[11] Chandler disagrees.[12] He interprets Chapter Three of Joel as referring exclusively to the spirit of prophecy that followed Pentecost, the destruction of Jerusalem, and the escape of the Christians from that calamity. Men like Chandler, who exalted reason and reasonableness and distrusted the passions, did not comfortably contemplate a time when God would cause sons and daughters to prophecy, young men to see visions, and old men to dream dreams. They felt more at ease with these events in the distant past than in the near future. Evangelicals, in contrast, eagerly awaited the pouring out of the Spirit for the revival of religion.

Periodic revivals of religion played a major role in the church history of the Puritans.[13] As the Puritans understood it, the chosen people of the Old Testament time after time had abandoned God's commandments and turned to the worship of idols, so that again and again God had to interfere with the course of events to preserve his worship and to restore the holiness of Israel. The New Testament church had also had periods of revival and decay. On the grand scale, the periods of rise and decline had been centuries long. In the age of the Apostles religion had thrived. Gradually, however, antichristian corruption in the church grew. Spiritual declension reigned for centuries until the power of religion revived in the Protestant Reformation. On a more intimate scale, each Christian people had known swings from general piety to general immorality.

By the start of the eighteenth century, Protestants had come to the realization that the Reformation as a period of more than usual activity of God's Spirit had come to an end. Reformed religion had entered a period of decline. In these circumstances, many looked for signs of a renewal of the Spirit's activity. In December 1714, for instance, Robert Wodrow wrote Cotton Mather that he thought the peaceful accession of the Hanoverians might be a good sign for religion.

> We know not if the Lord's controversy with [these] lands be at an end. We long for the inward delivery, for the Lord's returning to the spirits of his people, and to ordinances, when the time of our merciful visitation is lengthened out, and a reviving of a work of conviction and conversion, his repairing our breeches, and taking off the restraint from his Holy Spirit, that the wilderness may be turned to a fruitful field, And that this wonderful turn may extend to all the Churches of Christ.[14]

Wodrow hoped the political providence might herald a spiritual providence. By the 1730s the British/American evangelical community was convinced, more than ever, that the world was ripe for a fresh effusion of grace. The revival of religion in the Reformation had subsided generations before. The Spirit had been withdrawn, and the power of religion had declined. Enemies pressed in on the church from all sides. Doctrinal errors were corrupting the church from within. The evangelicals expected that God would act to preserve the church, that he would grant a new Reformation.

In 1730 Ebenezer Erskine preached a series of sermons on Isaiah 59:19, "When the enemy shall come in like a flood, the Spirit of the Lord shall lift up a standard against him."[15] In November 1730 Robert Bragge used a sermon on the same text to inaugurate the twenty-six Lyme Street lectures by several London ministers.[16] Erskine and Bragge point out the enemy coming in like a flood: Immorality grows more and more prevalent, irreligion, blasphemy, and books ridiculing miracles gain respectability, and the erroneous Arminian and Arian doctrines spread. Erskine and Bragge both assert that the standard lifted up to drive back the flood is Christ displayed in the Gospel. Both preachers offer their text as grounds for hope that the inundation will be stopped. Bragge observes that when Satan is about to be bound for the 1,000 years of Christ's reign on earth, he is to come in like a flood. Erskine says that the enemy shall be driven back and a work of reformation begun "when the Spirit shall be poured out from on high."

The case of the Church of Scotland illustrates the heights to which alarm at the decaying state of religion and expectation of a revival reached during the 1730s. The Church of Scotland had become so corrupt in the eyes of Erskine and his party that they seceded in 1733. For evangelicals who remained within the Church, the Secession was just one more symptom of the malady.

At the General Assembly of the Church in May 1734, John Willison bewailed the withdrawal of the Spirit from Scotland. The remedy, he said was

> to assemble together with one Accord, with Harmony and Love, Unity and Peace. O consider, this is the Way to obtain the Effusion of the Spirit, a Blessing this poor Land stands greatly in Need of; our Mother-Church, our Assemblies, our Congregations, they all need it. It is this that would cement our Breeches, recover our Decays, dispel our Clouds, make Ordinances powerful, awaken the Secure, draw Sinners to Christ, and lead us all in the Way of *Truth and Holiness*.

Mutual concord, according to Willison, means agreement in two areas. First, he calls for agreement on the necessary and essential truths, which he describes as an evangelical Calvinism, emphasizing the necessity of conversion for salvation, which comes through the imputed righteousness of Christ. And second, "*concord* among Ministers lies in their pursuing the *same chief Scope and Way of Management* in their ministerial Labours." The ministerial style for which Willison calls is evangelical conversionism. Were the ministers of the Church of Scotland to follow this advice, then it might be for them as it was for the Apostles: "And when the Day of Pentecost was fully come, they were all with one Accord in one Place. And suddenly there came a Sound from Heaven, as of a mighty rushing Wind . . . (Acts 2:1,2)."[17] Only another Pentecost could save the church.

Revivals and Eschatology

As Pentecost was seen as the chief precedent for and model of a revival, so the millennium was looked to as the ultimate product of the revival of religion. Eighteenth-century evangelicals believed, as Howe and Fleming had argued, that despite periodic setbacks, grace would gradually increase until the whole world would acknowledge God's rule. Not all agreed with Howe's and Fleming's postmillennial, noncatastrophic version of the millennium; but even many who believed that the millennium would be preceded by a cataclysmic interruption of history pictured the coming of the millennium by means of an accelerating revival of religion. In Protestant Britain no orthodox pronouncement on the millennium had been adopted, and a wide latitude in millennial opinions was tolerated. Most had given over the idea of a political dominion of the saints such as the Fifth Monarchy Men had held. Still, many continued to hold premillennial views similar to those of Increase and Cotton Mather, while some, such as Richard Baxter, rejected the entire notion of a future millennium.[18] In the years between the Glorious Revolution and the Great Awakening the notion of a promised postmillennial, non-catastrophic, future period of peace and prosperity for the church on earth gained wider and wider currency. This version of the millennium, known as simple chiliasm and Whitbianism in the eighteenth century, became influential principally through the biblical commentaries of Daniel Whitby (1638–1728), a moderate Anglican; William Lowth (1660–1732), also an Anglican divine; and Moses Lowman (1680–1752), Dissenter and occasional conformist.[19]

These three men interpret the millennium as a period between the fall of Antichrist, which they hold is the papacy, and the Second Coming of Christ. At its commencement the Jews will be converted; then the Gospel will spread throughout the Gentile world. There will be no personal reign of Christ and no reign of the resurrected martyrs. It will be a time of peace, plenty, righteousness, holiness, and pious offspring. This state of affairs will be brought about by a plentiful effusion of the Holy Spirit, "somewhat resembling that which was vouchsafed to the first ages of Christianity,"[20] and as prophesied by such texts as Joel 2:28.[21]

Simple chiliasm took hold more quickly in Britain than in America, where in the first decades of the eighteenth century the complex chiliasm of the Mathers predominated; however, even though the Mathers taught a premillennial Second Coming of Christ to destroy Antichrist and introduce the millennium, they described the kingdom of Christ as spreading across the globe by the preaching of the Gospel and the outpouring of the Spirit in the same way as described by the simple chiliasts. In 1733 Joseph Sewall, of Boston, portrayed the spread of the Gospel to the ends of the earth and the union of Jews and Gentiles in the one fold before the Second Coming of Christ. The Second Coming with its connotations of doomsday is impending, but Sewall hopes that the Gospel will have spread throughout the world before that event.[22] Another Bostonian, Thomas Prince, also spoke of the end of the world in premillennial terms. In his sermon on *Earthquakes the Work of God* in 1727, he affirms that the present state of the world is to end with universal convulsions, and the wicked will be purged from the earth, which will exist in a new state of righteousness. In May 1740 Prince preached *The Endless Increase of Christ's Government* before the annual convention of Massachusetts ministers. Here he offers a chronology that reconciles a gradual spread of the Gospel with a catastrophic introduction of the millennial kingdom. Like the simple chiliasts, Prince outlines the spread of Christianity by means of the outpouring of the Spirit and the preaching of the Gospel from the age of the Apostles to the present and beyond to the millennium. Prince emphasizes the westward trek of Christ's church. The church began in Jerusalem with the gathering of the Apostles and the outpouring of the Spirit at Pentecost. It spread through the Roman Empire, and in the following centuries through Europe. In Prince's day it had reached America and Asia. He expected the Gospel light to continue westward until it returned to Jerusalem, girdling the globe. Unlike the simple chiliasts, who began the millennium at this point, postponing the destruction of the world until the end of 1,000 years, Prince placed the conflagration here. Instead of destroying the earth, the conflagration purifies it. The resurrection of the departed saints and martyrs is literal. He interpreted the 1,000 years prophetically, a year for a day, implying 360,000 years (the biblical year being 360 days). The saints continue to reproduce, and the surplus population is continually translated—as Enoch, Elijah, and Christ had been—into heaven.[23]

Simple chiliasm, then, was not necessary for a belief in the gradual spread of the true church across the globe by means of revivals. Some have argued that Jonathan Edwards, by importing Lowman's postmillennial view, helped transform American evangelicalism into a progressive force that looked forward toward the gradual improvement of man's earthly condition.[24] On the contrary, moving the Second Coming and the cataclysmic interruption of history from the beginning of the millennium to the end did not transform a pessimistic into an optimistic eschatology. Christian millennialism was intrinsically optimistic.[25]

A History of the Work of Redemption

Jonathan Edwards contributed to evangelical historiography by identifying the local revivals of the early eighteenth century with the outpourings of the Spirit throughout biblical and Christian history, and by arguing that these local phenomena were forerunners of the revivals that would characterize the spread of the millennial kingdom. Edwards magnified the revivals experienced by his contemporaries by portraying them as events central to the entire history and future of the relationship between God and mankind. He made the phenomenon of the revival the key element in the drama of redemption. He conceived of revivals as the engines that drive redemption history.[26]

In placing revivals at the center of Christian history, Edwards built on the foundation laid by Solomon Stoddard, his grandfather and predecessor at Northampton. In 1712, in order to illustrate the doctrine that "there are some special Seasons wherein God doth in a remarkable Manner revive Religion among his People," Stoddard did not simply point to the spiritual "harvests" which Northampton was at that moment enjoying. Rather, he turned to Scripture. For all the conclusions about such seasons which he had drawn from his experience at Northampton, he cites biblical examples. He distinguishes between general revivals, "when it is throughout a Country, when in all Parts of a Land there is a turning to God," such as in the first centuries of Christianity, or during the Reformation when whole "Nations broke off from Popery, and embraced the Gospel," and local revivals "when in some particular Towns Religion doth revive and flourish," as in 2 Thessalonians 1:3 the Thessalonians grew in faith and charity, or as religion thrived in the church at Philadelphia while Sardis and Laodicea were declining, (Rev. 3). For the observation that "this reviving is sometimes of longer and sometimes of shorter Continuance," he points to the twenty-nine years of Hezekiah's reign in Israel (2 Chron. 29), and to the year and a half of Paul's residence at Corinth (Acts 18:9–11). Other times when "religion revived" were those of Josiah (2 Kings 23 and 2 Chron. 34, 35) and of John the Baptist (Luke 3:10–14). Stoddard deliberately associates with biblical events the contemporary "special Times, when in particular Towns there is a Noise among the dry Bones, and many are made sensible of their natural Condition, and are laboring after a part in Christ; when one and another are enquiring what they must do to be saved, and some have obtained Mercy." The inferences he draws that ancient revivals paralleled the modern required imagination and a leap of faith.[27]

In 1739 Jonathan Edwards made use of Stoddard's list of biblical revivals in a series of sermons at Northampton, published in 1774 as *A History of the Work of Redemption*. In the manner of Robert Fleming's *Fulfilling of the Scripture* and John Howe's *Prosperous State*, the design of Edwards's *History* is "to comfort the church under her sufferings, and the persecutions of her enemies." Like them, it focuses on

> the constancy and perpetuity of God's mercy and faithfulness towards her, which shall be manifest in the continuance of the fruits of that mercy and faithfulness in continuing to work salvation for her, protecting her against all assaults of her enemies, and carrying her safely through all the changes of the world, and finally crowning her with victory and deliverance.[28]

Like Howe and Fleming, Edwards holds out the promise of the millennium as a ground for hope in times of distress for the church. And like Fleming, Edwards uses the evidence of God's providence toward the church and the fulfilling of scriptural prophecies to confute the deists and to prove the divine authority of the Scriptures.[29]

Adding to and completing Stoddard's enumeration, Edwards discovers revivals of religion in Scripture that he identifies with contemporary revivals.

One significant difference distinguishes Edwards's treatment of revivals from Stoddard's. The latter, who had no interest in millennial speculations, does not place the revivals in an eschatological context. Stoddard emphasizes the arbitrary, sovereign will of God. God revives religion when and where he chooses, without regard to human understanding. It is beyond man to determine God's purpose when he revives religion at any place and time. It might be to prepare the land for great temptation or for great affliction. It might be to lay a foundation for mercy for the next generation, or to gather in his elect before he removes the means of grace and takes away his ordinances.[30] In contrast, Edwards handles revivals within the context of the grand sweep of God's plan for redeeming the elect over time. He identifies the purpose of the revivals of the past and he describes the function of the revivals to come. Every revival fits into a discernable pattern.

In his treatment of revivals and in his eschatological historiography, Edwards was the heir of Stoddard and the New England millennial tradition together. Stoddard's works had played an important part in Edwards's education, and Edwards had assisted Stoddard at Northampton in the final two and a half years of the latter's life. From Stoddard Edwards had learned much about the techniques of evangelism and the morphology of conversion.[31] Reflecting the concerns of other Connecticut Valley revivalists, such as Eliphalet Adams and William Williams, Edwards was intensely interested not only in millennial speculations but also in promoting the advancement of Christ's kingdom toward the millennium.[32] By uniting the millenarian and Stoddardean traditions, Edwards gave the revivals a great historical and eschatological significance. For Stoddard a revival had meant that here in this place and time souls are being saved and the national covenant being kept. For Edwards a revival was an episode in the history of the work of redemption and a step toward its fulfillment.

In *A History of the Work of Redemption* Edwards accepts in general the Puritan periodization of history: The people of God in the Old Testament were continually backsliding, and continually being brought back to the service of the true God; after Christ's Ascension, the original purity and holiness of the church gradually diminished, corruption began, the antichristian papacy emerged, and true religion declined more and more, until revived by the Reformation. Edwards's objective, however, was not merely to demonstrate the operation of the Spirit of God in the affairs of men. His principal intention was to outline God's plan for the redemption of the elect as it has unfolded from Adam's sin to the present, and as it will unfold until completed at the end of time. Therefore, his emphasis was on the continual, inevitable, irresistible building up of the Kingdom of Christ.

The doctrine the *History* presents is that "the Work of Redemption is a work that God carries on from the fall of man to the end of the world."[33] Edwards began by explaining that here he is not concerned with the work of redemption as it pertains to individual souls, whose salvation God effects through all ages by repeating the same work in each of the elect.[34] Rather, Edwards dealt with the work of redemption "with respect to the grand design in general." In this respect, God accomplishes redemption "not merely by the repeating and renewing the same effect on the different subjects of it, but by many successive works and dispensations of God, all tending to one great end and effect, all united as the several parts of a scheme, and altogether making up one great work." The work of redemption is one united work "brought about by various steps, one step in each age and another in another."[35] Edwards compared the work of redemption to the building of a house or temple and carried this conceit through the entire *History*. With the final judgment, the building is complete. "And now the whole Work of Redemption is finished. . . . Now the topstone of the building is laid."[36]

Edwards divided the history of the work of redemption into three periods. First, from the fall to the incarnation, Christ's coming was prepared for. Second, from Christ's incarnation to his resurrection, redemption was purchased. Third, from Christ's resurrection to the end of the world the effects of the purchase are accomplished. Throughout the whole space of time from the fall to the end of the world, all the works of providence are united as one work, with one end, redemption. Principal among the works of providence contributing to this one end are those that insure the perpetual continuance of the church. Throughout the Old Testament and the New, God has seen to it that the church has never been totally destroyed either by its enemies from without, or by corruption from within. In both testaments, God has revived and sustained the church by fresh effusions of his Spirit.

According to Edwards, the first religious revival in human history occurred shortly after the expulsion of Adam and Eve from the garden, in the days of their grandson Enos. "The first remarkable pouring out of the Spirit through Christ that ever was . . . was in the days of *Enos*." Edwards based this remarkable assertion on Genesis 4:26, "Then began men to call upon the name of the Lord." This, Edwards argued, shows that "there was something new in the visible church of God with respect to the duty of prayer, or calling on the name of the [Lord], that there was a great addition to the performance of this duty, and that in some respect or other it was carried far beyond whatever it had been before, which must be the consequence of a remarkable pouring out of the Spirit of God." The use of the word *began* in this scriptural passage implies that "this was the first remarkable season of this nature that ever was; it was the beginning or the first of such a kind of work of God." Edwards justified deriving these conclusions from the brief statement, "Then began men to call upon the name of the Lord," by arguing that an increase in prayer is associated with all the revivals recorded in Scripture, with the prophecies of the outpouring of the Spirit in the last days, and also with all the revivals contemporaries know by experience. "We see by experience that a remarkable pouring out of the Spirit of God is always attended with . . . a great increase of the performance of the duty of prayer." He thus identified the Northampton revivals and harvests with the very first revival. The revival at the time of Enos, by Edwards's description, was of the same nature as all subsequent revivals: "There had [been] a saving work of God in the hearts of some before, but now God was pleased to grant a more large effusion of his Spirit for the bringing in an harvest of souls to Christ."[37] The revival in the days of Enos was but the first of a long series of revivals. "Time after time, when religion seemed to be almost gone and it was come to the last extremity, then God granted a revival and sent some angel or prophet or raised up some eminent person to be an instrument of their reformation."[38] Whenever the Israelites decreased in piety, God intervened to sustain his church.

> Idolatry seemed to be ready totally to swallow all up, yet God kept the lamp alive. And [God] was pleased several times after exceeding great degeneracy, and things seemed to come to extremity, and religion seemed to be come to its last gasp, then God granted blessed revivals by remarkable outpourings of his Spirit, particularly in Hezekiah's and Josiah's times.[39]

Edwards noted how, at every new establishment of the visible church, God poured out his Spirit in order to strengthen it. He did so when the Jews first set up regular ordinances in Canaan in the time of Joshua. He did so again when the Israelites returned from the Babylonian Captivity. And he did so once more at the beginning of the Christian church.[40]

Edwards described the revival after the Babylonian Captivity in the terms used to describe eighteenth-century revivals. The people listened attentively to the preaching and expounding of the law by Ezra and the other priests. The people "were greatly affected [and] wept when they heard the words of the law." "There appeared a very general and great mourning of the congregation of Israel for their sins, which was accompanied with a solemn covenant that the people entered into with God, and this was followed with a great and general reformation."[41]

Down to the coming of Christ the worship of the true God had been sustained, the covenant remembered, and the Coming of the Messiah prepared for. One more pre-Christian revival prepared immediately for Christ's public ministry, the "very remarkable outpouring of the Spirit of God" that attended the preaching of John the Baptist.[42] With Christ's resurrection begins the gradual setting up of his kingdom. The kingdom will be accomplished through grace, by the gradual advancement to the millennium, and then in glory, with the Second Coming, the Last Judgment, and eternity.[43] The outpouring of the Spirit at Pentecost began the process of Christ's coming in a spiritual manner to set up his kingdom. After the Spirit was poured out on the Jews in Jerusalem, it was poured out on the Samaritans, and then, accompanying Paul's preaching, on the Gentiles. There was a remarkable outpouring in the city of Ephesus, and at Corinth.[44] And so the Gospel has been spreading across the earth, and will continue so until the millennium. The kingdom of Christ on earth will be accompanied by the overthrow of the kingdom of Satan, and by the most extensive outpouring of the Spirit in all of history. The introduction of the millennium will be swift, but not all at once: "This is a work that will be accomplished by means, by the preaching of the gospel, and the use of the ordinary means of grace, and so shall be gradually brought to pass." Edwards's millennium comes about much as John Howe's and Robert Fleming's:

> Some shall be converted and be the means of others' conversion; God's Spirit shall be poured out, first to raise up instruments, and then those instruments shall be improved and succeeded. And doubtless one nation shall be enlightened and converted after another, one false religion and false way of worship exploded after another.[45]

The pouring out of the Spirit will facilitate the propagation of religion across the globe. The Gospel will be preached in every language and multitudes from every nation will be converted. Satan, seeing the success of the Gospel, will put up a mighty resistance with his antichristian, Mohametan, and heathen forces. "We know not particularly in what manner this opposition shall be made. 'Tis represented as a battle," Edwards wrote. In any case, "Christ and his church shall in this battle obtain a complete and entire victory over their enemies." Satan will be bound for 1,000 years and the church will enjoy peace, holiness, and temporal prosperity. At the end of the 1,000 years Satan will be loosed, and great apostasy will threaten the church. At this point Christ will come to judge the world. The dead will rise from the grave. All who ever lived will be divided into the saved and the damned. The saved will be taken to heaven and the damned cast into hell for all eternity. History, and revivals, will end. The work of redemption will be fulfilled.[46]

With *A Faithful Narrative* Edwards established models for local and regional revivals of religion. With *The History of the Work of Redemption* he defined the central place of revivals in God's plan for his elect. In both works, Edwards carried forward the tradition we have here traced from the era of the Restoration. He built on the foundation laid by his predecessors who helped transform a strain of the Puritan legacy into Reformed evangelicalism. During the height of the Evangelical Revival in America and Great Britain, the publication of dozens of revival narratives confirmed Edwards's version of the dynamics of the local revival. Assessments of the significance of the revivals subsequent to Great Awakening would ensconce in American and British evangelical culture Edwards's vision of the pivotal role of revivals in God's grand scheme for mankind.

II

The Evangelical Revival

7

Catalyst

By the 1730s, intellectual and political developments had renewed a sense of crisis for Protestant Christianity among evangelicals and an expectation of impending divine intervention to preserve and strengthen it. When the appearance on the scene of itinerant preachers presenting the doctrine of the New Birth in dramatic fashion fed on and appeared as fulfillment of those expectations, the Evangelical Revival broke out in power in Great Britain and America. News of the itinerants' successes created new expectations, which fed new successes. An international network for exchanging revival news sustained the impression that the Holy Spirit was unusually active among congregations throughout Protestantism. Printed accounts of the revivals prepared the way for the itinerants by informing their potential audiences of appropriate kinds of behavior in a revival. As preachers, itinerants had advantages over settle pastors in creating excitement: novelty and the opportunity to stress the most compelling doctrines of salvation without the need or responsibility of teaching the full scope of Christian belief and practice. The itinerants spread examples of successful techniques rapidly as they traveled, and local revivalists learned through imitation.[1]

The intellectual developments, namely, the spread of rationalism and reactions to it, that underlay the sense of crisis within Protestantism were long term but had a cumulative effect, especially on pious scholars who came of age during the first third of the eighteenth century, such as Jonathan Edwards and John Wesley, both born in 1703. A clear sense of crisis, an urgent defense of orthodox doctrine against the rationalist assault, is unmistakable in the writings of these evangelicals. The sense of crisis was shared by Calvinist revivalist in New England and Arminian Methodist in England and provided the same imperative tone and impassioned spirit to the preaching of revivalists of both theological persuasions. Flee from the wrath that is to come![2] Evangelical Pietism's immediate experience of God's presence—John Wesley's "I felt my heart strangely warmed"[3]—supplied an unanswerable counter to rationalism's logic. Pious intellectuals who had faced down the rationalist enemy in solving their own spiritual crises confidently took the offensive in behalf of evangelical truth.[4]

The peculiar abilities of the early leaders of the Revival, such as Whitefield's dramatic talent and John Wesley's organizational skills, facilitated their assuming leadership of the movement; however, their ability to evoke emotional responses from large numbers of followers derived from another quality, the congruity of their own personal religious crises, and their solutions, with the intellectual and spiritual crises of the Revival's constituency.[5] Men like Edwards, Whitefield, and the Wesleys were better able to make the familiar doctrines that underlie the New Birth fresh, relevant, and compelling to contemporaries because their own spiritual crises had been vivid and profound. Solomon Stoddard was right: those preachers who had experienced for themselves the doubts, humiliations, false starts, ultimate surrender to God, and joy in forgiveness and justification were the most successful conversionists. Having made the journey themselves, they could recognize and point out guideposts to the lost and the anxious. They could develop skill in leading men and women in self-examination through the deceptive labyrinth of the soul. They knew the searching questions to ask from the pulpit or in the pastor's chamber to awaken concern, to shake false security, and to drive home the necessity of total reliance on God's grace. It was not the introduction of a new theology that set the Revival in motion but a new and powerful presentation of some basic teachings of the Reformed tradition.

As W. R. Ward has cogently argued, political developments in Europe created fears for the future of Protestantism and fed the anxieties that fueled the Revival in Britain and America. Protestants in Great Britain and America recognized a Protestant Crisis on the Continent. By the early decades of the eighteenth century, a number of Protestant rulers had converted to Rome, and the Reformed house in the Palatinate had died out and been succeeded by a Catholic branch. Within the Holy Roman Empire, large numbers of Protestants were subject to persecution by their Catholic rulers. Where oppressed, Protestant populations developed means of maintaining their faith. Domestic piety and informal class meetings replaced public worship; clergy, deprived of their livings, became itinerants; the people "turned to inwardness of faith, a patient steadfast trust in God, a certain tenderness of piety." The piety of these groups came to the public's attention as a result of notorious acts of religious persecution such as the expulsion of Salzburg's Protestants in 1731.[6]

Events in Central Europe nourished the British and American revivalists' millennial expectations that were entwined in their understanding of the Revival. The revivalists saw the remarkable evidence of faith given by the

Salzburgers and other oppressed Pietists of Central Europe as tokens of the inevitable triumph of Christ over the whole earth. The contest between Protestantism and Catholicism in Central Europe provided an international context for the urgent call for revival, and the power of Pietism revealed in the experience of the oppressed Protestants in Central Europe provided grounds for hope of revival and an indication of the means by which it would be accomplished. The Protestant Crisis confirmed expectations shared by many Britons and Americans that God was about to do marvelous things for the faith and confirmed men like the Wesleys and George Whitefield in their calling to evangelize in unorthodox fashion.

The most concrete, direct, and immediate catalysts of the Revival in Great Britain and America were the men who introduced the new style of preaching conversionist doctrine with vivid imagery and dramatic flair. The most influential of these preachers was George Whitefield. Although John Wesley would have a more permanent and extensive role in the formation of Methodism, during the first few years of the Revival the public at large identified Whitefield as the principal leader of the movement, and the great majority of early anti-Methodist tracts were aimed at him, not Wesley. Through his extensive travels, Whitefield was able to demonstrate his style in England, America, and Scotland and thus to provide an example for would-be imitators. Interested ministers took pains to learn Whitefield's methods. In New England, Whitefield's preaching tour was the immediate precipitant of and signal for the revivals.

Beginnings of Methodism

Evangelicalism in the Church of England was the product of a merger of High Church piety with Puritan practical divinity.[7] Most early Anglican evangelicals came not out of the Puritan, but out of the High Church tradition. High Church piety supplied an intense Christian perfectionism and a powerful impulse toward benevolent activity. In the case of many early Anglican evangelicals, perfectionism created an intense awareness of sin, which, when coupled with inability to live up to the demands of their ideal and failure to find inward satisfaction in external duties, led to emotional spiritual experiences of the type known to Puritans as conversion. In the 1730s a number of Anglicans found in the practical divinity of the Puritan writers both explanation and justification of these experiences. The Thirty-nine Articles of the Church were ambiguous enough to allow for the teaching of either baptismal regeneration or the necessity of conversion. The evangelical movement in the Church of England began in the 1730s when a number of Anglican ministers denounced the rationalist and legal divinity of the age, asserted the necessity of an instantaneous, internally perceptible conversion experience, stressed the emotional life of the soul, and developed an affective preaching style.

During the two decades following the Glorious Revolution, the Church of England experienced a burst of religious activity, which included private societies for prayer and conference, societies for missionary work at home and abroad, and charity schools. These activities formed an important part of the background of the Anglican evangelical movement.[8] The piety cultivated in the charity schools and taught in the numerous publications of the Society for the Propagation of Christian Knowledge (S.P.C.K.) was of a High Church type.[9] It focused on the development of personal holiness through frequent reception of the Lord's Supper, and through penitential discipline, with an emphasis on fasting.[10] The High Church tradition provided a counterweight to the rationalist and latitudinarian tendencies in the Church, by emphasizing the life of the soul rather that the life of the mind, by insisting on inward piety as well as external morality, and by holding a high ethical ideal.

The activities of Griffith Jones (1684–1761) in Wales in the second decade of the eighteenth century, foreshadowed the evangelical movement in the Church of England. Jones, ordained an Anglican priest in 1709 and settled in 1711 at Llandilo Abercowyn, Wales, united in his person Anglican piety and practical Puritan divinity. He taught a Calvinistic doctrine and preached in Welsh with an affective, conversionist manner. By 1713, much like the Erskines in Scotland, he was preaching to audiences of 500 to 1,000 in the open air. In 1714 he was tried by his bishop for preaching in the open air in other men's parishes and for departing from the established prayers. In 1715 he expressed in a letter to his bishop distress at the wickedness of the clergy and the ignorance and sinfulness of the laity. While he was getting into trouble for irregularities, Jones was also involved with approved pious activities. About 1712 he considered going as a missionary to the East Indies for the S.P.C.K., but decided he could do more good in Wales. In 1713 he became a corresponding member of the S.P.C.K., whose charity school movement was spreading through Wales. In 1716 Jones was promoted to the cure of Llanddower, where, for the next fifteen years he seems to have restricted his evangelism to his own parish. Few details of Jones's life between 1716 and 1731 survive. In 1731 he began establishing Welsh catechistic schools. The S.P.C.K. had nearly 100 charity schools in Wales, but they taught in English. Jones felt that if the Christian religion were to be their own, the people must be taught to read in their own language. The sole object of his schools was to teach piety. Hence, they taught only reading, no writing or arithmetic. Catechisms and devotional literature were the textbooks. By 1737 Jones had thirty-seven schools in operation. In that year the schools began circulating. Schoolmasters would travel in circuits, teaching perhaps for six-week stretches in each parish. The instructors were allowed to catechize, read from the Bible, read prayers from the Prayer Book, and exhort privately. They were to give no grounds for accusations of supervising public worship, however. Soon the schools were objected to for spreading Methodism and dissent. More than twenty of the early teachers are known to have been Methodists. In 1741 Jones split with the Welsh Methodists. He dismissed the Methodist teachers, prohibited lay exhorting, and placed control of the schools in the hands of the local priests.[11]

A conversionist zeal, a desire for the revival of religion, and a millennial vision inspired Jones's reading schools. He argued that ever since human reasoning had replaced discussion of the Holy Scriptures in public discourses, conversions had decreased in number. This, he thought, justified teaching the poor to read their Bibles. He thought the most useful method to revive religion was "explanatory catechizing, together with prayer, psalmody, conference, and brotherly exhortations." Christian knowledge was a prerequisite for moral reformation. Finally, in the days of the Apostles, the Gospel was preached to the poor and spread all over the earth. Is it not likely, he asked, that God would choose the same method to restore the Gospel to its primitive force? He hoped "that more glorious times than the present are not very far off."[12] Jones's preaching influenced several of the leaders of the early Welsh evangelical movement, including Howell Harris (1717–1773) and Daniel Rowlands (1713–1790).

The nonjurer William Law (1686–1761) was the bridge between High Church piety and revivalism in the Church. Law is known as the father of the Methodist revival because of the influence of his *Practical Treatise upon Christian Perfection* (London, 1726), and his *Serious Call to a Devout and Holy Life* (London, 1729). Law was neither revivalist nor evangelical preacher. Nor did he approve of sudden conversions, or the Methodist version of the New Birth. He is considered father of the Methodist revival because his insisting on a deep personal religious experience and his uncompromising demand for a sinless life were primary influences on a great many of the early Anglican evangelical leaders. Furthermore, the inability of these men to live up to the ideal of the Christian life set by Law brought them to seize on an evangelical version of salvation by faith alone through the New Birth. As an early Anglican evangelical punned, "Law came before the Gospel."[13] The Methodists continued to embrace Law's high ethical ideal while they rejected his mysticism and version of regeneration.

Law's message of imitation of Christ resembles, superficially, the legal preaching of the age. Both stressed moral obligation, and both shared a faith in the ability of man to comply with that obligation. The kind of moral behavior expected by Law, however, differed greatly from that expected by the latitudinarian preachers. Tillotson and his followers taught that the Christian life was easy, natural, and conducive to temporal happiness. In contrast, Law demanded a complete perfection, an entire imitation of Christ. In *Christian Perfection* he taught that it is "an entire change of Temper, that makes us true Christians." "The Corruption of our Nature," he wrote, "makes Mortification, Self-denial, and the Death of our Bodies necessary. Because human Nature must be thus unmade, Flesh and Blood must be thus changed, before it can enter into the Kingdom of Heaven." Like the evangelicals, Law believed that "all the Precepts and Doctrines of the Gospel are founded on these two great Truths, the deplorable Corruption of human Nature, and its new Birth in Christ Jesus." Yet, he taught a High Church version of the New Birth. The Christian's regeneration begins with baptism, and progresses as the Christian grows toward perfection through reception of God's Spirit in the sacraments. One was to prepare for the reception of the Spirit by loving God and obeying his commandments. Christianity carried with it obligations to constant purity and holiness. It required renunciation of the world. And it called men to a state of self-denial, mortification, and constant prayer and devotion.[14]

Oxford University in John Wesley's time was strongly High Church. The theological view taught there was closer to that of the Caroline divines than to that of the latitudinarians, and patristic studies were encouraged. The Oxford Holy Club (1729-ca. 1736) sought to develop a meaningful High Church piety by consciously imitating the Religious Societies and by deliberately attempting to put into practice William Law's Christian perfectionism. Following the practice of the Religious Societies, the members met frequently for prayer and pious reading. They made constant use of such guides to practical holiness as Thomas a Kempis's *Imitation of Christ*, Jeremy Taylor's *Holy Living* and *Holy Dying*, and William Law's *Christian Perfection* and *Serious Call*. By frugality, frequent fasts, giving to the poor, and visiting prisoners, they sought to put into practice the precepts of these books.[15]

The spiritual journeys from this kind of High Church piety to evangelical piety made by the Oxford Methodists and by many of the other leaders of the Anglican evangelical movement were much like John Wesley's. Before entering Oxford University, Wesley had expected to be saved by keeping all the commandments of God, or at least by being better than most and fulfilling the external duties of prayer, public worship, and Scripture reading. Upon entering Oxford (1720) he continued employing the external duties, but still had no notion of inward holiness. Between 1725 and 1729 his reading of Taylor, Thomas a Kempis, and Law convinced him of the need not only for outward conformity to the divine will but inward as well. He joined the Holy Club and undertook to implement the demands of Christian perfectionism. Between

1732 and 1734 he visited William Law several times. Law recommended he read the Christian mystics.[16]

Wesley's failure to live up to Law's ideal brought him, about 1735, to begin questioning Law's position. By May 1738, with the help of the Moravians, he had rejected Law and had become an evangelical. In 1735 Wesley sailed to Georgia as minister to Savannah, with hopes of preaching as well to Indians. The white settlers resented his imperious manner, and the warring natives had no interest in Christianity. These setbacks aggravated Wesley's distress over his inability to live up to the standard set by Law. In the meantime, Wesley had come into contact with Moravians. They spoke to him of the need for personal assurance in Christ as one's own savior. Wesley returned to England in February 1738 in a state of spiritual turmoil. His Georgia experience left him with a profound sense of personal failure, and he feared he still was not saved. In London and Oxford he held long conversations with the Moravian Peter Bohler, who spoke of salvation by faith alone and the need for conversion. These conversations convinced Wesley that instantaneous conversions were the norm. In March he began following the advice of Bohler to "preach faith until you have it." At this time he broke with Law, accusing him of failing to preach the necessity of saving faith in Christ and total reliance on Christ's righteousness. On 14 May 1738, Wesley experienced his conversion at the Moravian meeting at Aldersgate Street, London, when he felt his heart "strangely warmed," and believed himself "saved from the law of sin and death."[17]

In the period Wesley was making the transition to evangelicalism, changing from belief in salvation by faith and works to belief in salvation by faith alone, other Anglicans were making the same transition by similar paths. Some were influenced by Law's *Serious Call*, others by Puritan works such as John Owen's *On Justification*, Thomas Shepherd's *Sincere Convert*, and John Bunyan's *Grace Abounding*. Many were influenced by personal contact with those who had already made the transition.[18]

George Whitefield's conversion to evangelicalism typifies the experience of early Anglican evangelicals perhaps better than Wesley's does, for Wesley was influenced by the Moravians more directly than most and continued to affirm Law's perfectionism when most expected sin to persist in the believer. Whitefield's case unites the influence of Law, of the Oxford Methodists, of Puritan authors, and of the conversion experience. When Whitefield joined the Oxford Methodists in 1733, they were still reading books on devotion, self-discipline, and good works. Whitefield, who was undergoing deep spiritual distress, embraced the ascetic life of the Holy Club with a passion. But self-denial did not sooth fears for his soul. His reading of Law's works and Henry Scougal's *Life of God in the Soul of Man* convinced him of the necessity of the rebirth. Later he would write of this period that "God showed me that I must be born again, or be damned! I learned that a man may go to church, say his prayers, receive the sacrament, and yet not be a Christian." Finally, in the spring of 1735, in a state of exhaustion, he threw himself down with the words "I thirst." At that moment he felt assurance that Christ had died for his sins. He then began reading works which explained the nature of the New Birth from the Puritan perspective: Joseph Allein's *Alarm to the Unconverted* (London, 1672), Richard Baxter's *Call to the Unconverted* (London, 1657), and James Janeway's *Life* of John Janeway (London, 1673). Gradually he came to see the importance of the doctrine of salvation by faith alone. Not until his return from Georgia in 1738, however, when he found his friends John and Charles Wesley preaching salvation by faith alone as the essence of Christianity, did this doctrine assume the central place in his preaching.[19]

While the Wesleys and Whitefield were turning to salvation by faith alone, the New Birth, and conversionist, evangelical preaching, a similar movement was developing independently in Wales. Individuals who trod similar spiritual paths around 1735 made contact and began to form their own evangelical movement.[20]

The Work God Has Begun

Leaders of the Church of England in the late seventeenth and early eighteenth centuries believed that vital religion was languishing. The Anglican religious societies were thought to be an "effective means for restoring our decaying Christianity to its Primitive Life and Vigour."[21] Seeking to revive religion, several bishops promoted the religious and reformation societies, the S.P.C.K. and the charity schools, and vigorously worked to improve the pastoral care available in their churches.[22] The reforming bishops, of course, were not looking for a revival of evangelical piety. Nevertheless, the Methodists considered their movement just such a revival of religion as was needed to transform England into a truly Christian nation.

The Methodists believed their movement was more than a restoration of the official doctrines of the Church of England concerning justification and regeneration. It was a work of God for the spiritual redemption and moral reformation of England—probably for Europe as well, and perhaps for the world. In 1739, Whitefield wrote, "I believe the Lord will work a great work upon the earth." Again, "That a great work has begun is evident and that it will be carried on I doubt not." In the autumn, from New York, he wrote to Jonathan Edwards, "The Journal sent with this will shew you what the Lord is about to do in Europe." In London in April 1742 he wrote, "I believe there is such a work begun, as neither we nor our fathers have heard of. The beginnings are amazing; how unspeakably glorious will the end be!" Whitefield's words mirror those of Wesley's preface to the third extract of his journal (Bristol, 1742): "Such a work this hath been, in many respects, as neither we nor our fathers had known. . . . These extraordinary circumstances seem to have been designed by God for the further manifestation of His work, to cause His power to be known, and to awaken the attention of a drowsy world." *The Work* is generally the term by which Wesley and Whitefield referred to the Methodist movement. Wesley also considered the Work to be characterized in part as an *awakening*. On 22 November 1738 he wrote from Oxford, "After a long sleep, there seems now to be a great awakening in this place also." On 12 April 1740 he wrote, "I am just come from Wales, where there is indeed a great awakening." Whitefield considered the goal of the movement to be "a revival of true and undefiled religion in all sects whatsoever." In Wales, the term revival was used to describe the movement. In January 1739, for example, Howell Harris wrote, "There is a revival in Cardiganshire, through Mr. D. Rowlands. . . . And in this county where I am now the revival prospers."[23]

On 9 October 1738, as he walked from London to Oxford, John Wesley read Jonathan Edwards's *Faithful Narrative*, spending three or four hours reflecting on it. "Surely," he exclaimed, " 'this is the Lord's doing, and it is marvelous in our eyes.' "[24] Edwards's account may have played a role in revealing to the evangelicals in the Church of England the potential of mass conversions.[25] Still, the revival of religion implied to the Methodists a general movement, such as the awakening in the Connecticut Valley—or in Cardiganshire—as a whole, and held few connotations of an experience of grace of a particular community, such as a revival at Northampton, Massachusetts—or, at Bristol, England.[26]

The Methodists did not use the terms *the Work*, *a great awakening*, and *the revival* to describe the experience of grace of an individual congregation or a single gathering of worshippers. Nevertheless, the Methodists did believe that God occasionally manifested himself in a special way to such gatherings by an immediate effusion of his grace. Describing a love feast of about sixty members of the Methodist Society at Fetter Lane, London, on 1 January 1739, Wesley reported,

> About three in the morning, as we were continuing instant in prayer, the power of God came mightily upon us, insomuch that many cried out for exceeding joy, and many fell to the ground. As soon as we were recovered a little from that awe and amazement at the presence of His majesty we broke out with one voice, "We praise Thee, O God; we acknowledge Thee to be the Lord."

Five months later the Fetter Lane Society met

> to humble ourselves before God, and own He had justly withdrawn His Spirit from us for our manifold unfaithfulness. . . . In that hour we found God with us as at the first. Some fell prostrate upon the ground; others burst out, as with one consent, into loud praise and thanksgiving. And many openly testified there had been no such day as this since January the first preceding.[27]

These divine visits occurred more often at the private meetings of the Methodist societies than at the huge gatherings of crowds to hear the field-preachers.[28]

Unlike the revivalists in New England, few of the early Methodists had settled congregations. They were itinerants, preaching to crowds, great and small, and gathering those they had awakened into religious societies. They did not intend the religious societies to be new congregations in competition with the established churches, but to be ancillary bodies. Neither the field meeting nor the religious societies represented the community at large on which the effects of an outpouring of grace might be described. Whitefield portrayed the effects of his preaching on his auditors, but he did not write about an outpouring of the Spirit on Bristol.

Between 1735 and 1740 the Methodists emerged and developed the doctrines and practices that would link them with the evangelical movements in Scotland and America. In these years they adopted the evangelical message: the sinfulness of man and his need of salvation; the atonement for sin by Christ and justification of sinners by the imputation to them of Christ's righteousness; and regeneration by the operation of the Holy Spirit on the soul. They taught a conversion experience that began with a sense of personal sinfulness, and ended with assurance that Christ had died for one's own sins. They preached the law in order to awaken men to a sense of their sins, and they preached the Gospel to show them the way to be saved. They advocated a strict, puritanical morality and associated revival of religion with moral reformation. By their stress on sin, hell, the atonement, and salvation by faith alone, they set themselves off from their fellow Anglican ministers, and they aroused opposition by repeating the Dissenters' criticisms of the "dead morality" of Anglican preaching. In the same years, they adopted and innovated with evangelical methods: itinerancy, private pastoral counseling, use of small groups for social religion, and affective preaching.

The Methodists, particularly Whitefield, were willing to cooperate with fellow evangelicals across denominational lines. By 1739 they had made contact with the evangelical movements in every part of the British world, conceiving each, despite denominational differences, as part of one divine work to revive experiential religion.

Whitefield in New England

The sources of the Evangelical Revival in both Great Britain and America were not solely indigenous. The revival tradition in the Connecticut Valley set the stage in New England. Nevertheless, it was news of the Methodist movement across the ocean that notified New England that the drama was about to begin. England's George Whitefield filled the leading role. The transatlantic character of the Evangelical Revival is most evident in the case of Scotland. There, the accounts of the American revivals, including Edwards's *Faithful Narrative* and Whitefield's journals and sermons, taught the players their parts. Whitefield's Scottish visit, after his successful tour of the colonies, served as a lesson in performing style. Only after the Scots had carefully studied the American and English methods did they produce their own revivals. Both in America and Scotland the revivals of 1739–1742 combined features of the Connecticut Valley and of the Methodist experience.

George Whitefield is the central figure of the British/American Evangelical Revival not only because he linked the various revival movements through his transatlantic crossings, but also because he sought consciously to forge a transatlantic evangelical connection. By 1739 Whitefield had become convinced that God was beginning a great work that would be carried on to a general revival of religion throughout the British world. He believed that the evangelical movements in England, Wales, Scotland, and America were parts of one divine work.[29] By the end of 1739 he had corresponded or met with evangelical leaders in each of these areas.[30] He hoped that belief in the New Birth, in salvation by faith alone, and in predestination would provide the doctrinal basis for interdenominational cooperation.[31]

American and Scottish evangelicals considered Whitefield to be the leader of the Methodists, a group of fervent evangelical preachers raised up out of the Church of England, the bulwark of dry morality, to preach a living faith and revive orthodox Reformed doctrine. They believed a revival of religion had begun and that Whitefield was one of God's principal instruments in that revival. And they were determined that the revival would not pass them by.

In his account of the Great Awakening in Boston, printed in *The Christian History* in early 1745, Thomas Prince attributed the beginning of the revival to the visit of George Whitefield. He readily admitted the function that Whitefield's reputation had in producing great expectations from his ministrations. "It was in the year 1738," wrote Prince, "we were first surpriz'd with the News of Mr. Whitefield as a young Minister of the *Church of England* of flaming Piety, and Zeal for the Power of Godliness."[32] Since his first sermon after his ordination as deacon in June 1736, which had reportedly driven fifteen persons mad, Whitefield had made news. By December 1737 his name had appeared in American newspapers.[33]

Whitefield spent most of 1738 in Georgia. When he returned to England in December to receive priest's orders and to raise funds for an orphanage for Georgia, he found himself excluded from most Anglican pulpits. His *Journals* of his voyage to Gibraltar and America, and nine of his sermons had been published by then. They had made him vulnerable to criticism for spiritual pride, reliance on impulses, and enthusiasm, and had offended many ministers of the Church of England by their implication they neglected the doctrine of regeneration. Anglicans now filled the press with pieces attacking Methodist enthusiasm and the Methodist version of the New Birth. The majority of these pieces in 1739 were directed against Whitefield in particular. Believing he had a divine mission, Whitefield decided to preach in the open air. The newspapers began reporting the huge audiences he attracted. Whitefield's preaching during 1739 became more evangelical than it had been previously. From the Wesleys he borrowed an emphasis on salvation by faith alone. By the time he returned to America in October, he had become confirmed in his Calvinism.[34]

Whitefield's success as an evangelical preacher held the attention of American evangelical preachers. In 1739, according to Thomas Prince,

> we were yet more surpriz'd to hear of his Preaching the *Doctrines* of the *Martyrs* and other *Reformers*, which were the same our *Fore-Fathers* brought over hither: Particularly the great Doctrines of *Original Sin*, of *Regeneration by the* DIVINE SPIRIT, *Justification by Faith only*, etc. and this with amazing Assiduity Power and Success: which extraordinary Appearance, especially in the *Church* of *England*, together with the vast Multitudes of People that flock'd to hear him, drew our Attention to every Thing that was published concerning him.[35]

By the autumn, New England ministers were following Whitefield's career in America.

Whitefield's intention at this point was to serve the parish at Savannah for a year, and then to resign in order to be free to tour America, "if God ever call me to such a work." Evidently, he heard the call sooner than he had expected to, for, instead of returning directly to Georgia, Whitefield sailed first for Pennsylvania. In November he arrived at Philadelphia. The elder William Tennent came to meet him. After ten days of preaching in Philadelphia, Whitefield set out for New York. At New Brunswick, New Jersey, he met Gilbert Tennent. Tennent accompanied Whitefield to New York and stayed with him until 23 November. On 3 December Whitefield started south for a journey through the southern colonies to Georgia. Whitefield considered his visit to the Middle Colonies a success. His preaching in the churches and out of doors had drawn great crowds and had evoked evident emotional responses. Whitefield spent the winter in the South, where he supervised the building of his orphanage. In Charleston, South Carolina, he preached with visible results in the meetinghouses of the Presbyterian minister Josiah Smith and of the Baptist minister Isaac Chanler.[36]

The reports of Whitefield's success in the Middle Colonies and in Charleston raised the expectation of the evangelical ministers in Boston. Benjamin Colman, William Cooper, Thomas Foxcroft, Thomas Prince, and others decided to do all they could so that New England might have a share in the revival underway elsewhere. The first step was to see that Whitefield came to New England. While Whitefield was in New York late in 1739, Colman wrote his friend the Presbyterian minister Ebenezer Pemberton in reference to Whitefield. Pemberton showed the letter to Whitefield, who wrote directly to Colman. Colman replied on 3 December, telling Whitefield how much pleasure his sermons and *Journals* had given. He informed Whitefield that he would be welcome by many in Boston and offered the use of his own meetinghouse. Colman's associate, William Cooper, wrote Whitefield also.[37] Whitefield announced his intention to visit New England in the summer, before returning to England.

Colman and Foxcroft each wrote Isaac Watts for his impression of Whitefield, having learned from the latter's *Journals* that the two men were acquainted. In his letter of 28 January 1740, Foxcroft confessed that he was "very much prejudiced in his Favour, and . . . pleas'd very highly with the Prospect of his visiting Boston next Summer."[38] Colman reported to Watts on 16 January the good results of Whitefield's sojourn through the Middle Colonies. "And our town and country," wrote Colman, "stand ready to receive him as an angel of God."[39] Replying to Colman, Watts expressed ambivalent feelings. He thought that Whitefield was rash and relied too heavily on inward impulses. Watts also felt "that the complements to Mr. Whitefield in your newspapers are something extravagant." On the other hand, Whitefield was "a man of serious piety and uncommon zeal for the Gospel of Christ." He preached sound, evangelical doctrine. And his ability to labor in the ministry so constantly without fatigue seemed to the invalid Watts evidence of an extraordinary call. Since Whitefield had done real service in the awakening of many, and, Watts hoped, the conversion of several, Watts "could not but say, Go on and prosper."[40] Watts's ambivalence did not dampen Boston's interest in Whitefield's visit.

The Boston ministers conceived of the revival of religion led by Whitefield as an extension of the English Dissenters' campaign to promote "preaching Christ." Whitefield vociferously called for the preaching of Christ, while he vigorously denounced mere moralists. After having spent ten days in the company of Gilbert Tennent, Whitefield grew even more strident in his denunciations of carnal preachers. At Savannah he characterized the majority of clergymen as "slothful shepherds and dumb dogs," and accused the author of *The Whole Duty of Man* of sending thousands to hell. In March Benjamin Franklin published at Philadelphia *Three Letters* from Whitefield justifying his assertion that Archbishop Tillotson had known less than Mahomet of true Christianity.[41] In the spring of 1740 Benjamin Colman had John Jennings's *Two Discourses* on evangelical preaching reprinted at Boston.[42] In May, before the annual ministers' convention, Thomas Prince, referring specifically to Jennings's discourses, denounced those who failed to preach Christ. His phrases echoed the English debate. "How astonishing to see them treat even his own religion without a reference to him," he said, "that one would think they had only studied Seneca and Plato; and had never read . . . the New Testament."[43] Once Whitefield had preached in Boston, the evangelicals would interpret his visit in terms of preaching Christ. On 23 October Thomas Foxcroft offered *Some Seasonable Thoughts on Evangelical Preaching*, "Occasion'd by the late Visit, and uncommon Labours, in daily and powerful Preaching, of the Rev. Mr. Whitefield." The title page quotation, taken from Isaac Watts's *Humble Attempt*, called for looking to the prophets and the Apostles rather than to Plato, Seneca, Euclid, Locke, or Newton for authorities in one's preaching. In the body and notes of the sermon, Foxcroft referred to both Watts's *Humble Attempt* and Jennings's discourses. Foxcroft repeated the standard elements of the call to preach Christ: Christ, in his person and in his offices, is the special subject of preaching; morality is to be consistently related to Christ; one is to preach on the nature and necessity of conversion; and one is to preach justification through imputed righteousness without the law. Foxcroft echoed Whitefield's censure of Tillotson for neglecting "the things concerning Christ." Foxcroft summed up his advice by urging preachers to follow Paul's example. He finished by comparing Whitefield to Paul:

> We have in a fresh Instance seen this Pauline Spirit and Doctrine remarkably exemplify'd among us. We have seen *a Preacher of Righteousness, fervent in Spirit, teaching diligently the Things of the Lord, ceasing not* even *daily to preach the Kingdom of God*, and *the Things concerning Christ*; and this *with all Confidence*. May I not say, that the *Gospel* he preach'd, *came not unto us in Word only, but in Power, and in the Holy Ghost, and in much Assurance?* . . . And shall we not now strive to imitate![44]

Finally, the Governor of Massachusetts called Whitefield nothing less than the Apostle Paul.[45]

Through the spring and summer preceding Whitefield's visit, anticipation continued to grow. "In April 1740," according to Prince, "we had an Account of his Arrival at Philadelphia . . . of his setting out for *New York*, and his intending to return from thence to Philadelphia, and thence to Georgia, before he came to Boston; where he designed God willing to be in *July or August*."[46]

Two sermons from Charleston, South Carolina, received at Boston during the summer confirmed the ministers' perception of Whitefield as an evangelical preacher raised up to revive religion: *The Character, Preaching, etc. of the Reverend Mr. Geo. Whitefield*, by Josiah Smith; and *New Converts Exhorted to Cleave to the Lord*, by Isaac Chanler. Both were published in Boston, Smith's in June with a preface by Colman and Cooper, and Chanler's during Whitefield's visit at the end of September with a preface by Cooper. Chanler describes Whitefield as a "*sincere, true* and *faithful* Servant of the living God, sent forth to preach the everlasting Gospel to poor Sinners, in its *primitive Purity* and *Power*." Whitefield exemplified Paul's advice to ministers. Who was more successful than he, Chanler asks, in "giving *new Life* to Religion in so many Parts of the World, both in *Europe* and *America*?"[47] Smith shared Chanler's opinion of Whitefield, whose emergence, he thought, augured well for a general revival of religion. "And now behold God seems to have reviv'd the ancient Spirit and Doctrines. He is raising up of our young Men, with Zeal and Courage, to stem the Torrent. . . . It looks as if some happy Period were opening, to bless the World with *another* Reformation. Some great Things seem to be upon the Anvil, some big Prophecy *at the Birth*."[48] "By such Accounts as these," wrote Prince, "many Ministers and People were excited to desire [Whitefield's] Assistance in carrying on that Revival of Religion, which some Years before was begun in some Parts of our Land, and were prepared to embrace him."[49]

In their preface to Smith's sermon, Colman and Cooper recapitulate the maxim that "when God is about to carry on Salvation Work with any Remarkable Success, he will raise up suitable Instruments to work by. He will form and spirit Men for great and extraordinary Undertakings, when he has any great and extraordinary Purposes to serve." He raised up Joshua for the building of the second temple. He endowed with power twelve poor, illiterate fishermen for the establishment of the Gospel Church. And he brought forth Zwingli, Luther and others for the Reformation.

> And when it pleases God to *renew the Face of Religion*; when primitive Christianity, & the Power of Godliness, shall be reviv'd in the *Reforming Churches*; . . . when these expected *Times of Reformation* shall come on, is it not reasonable to suppose God will raise up those to effect it, whom he will furnish with a good Measure of the *primitive apostolic Spirit*?[50]

In 1739 Jonathan Edwards had observed how, whenever in the history of God's people religion seemed on the verge of extinction, "God granted a revival and sent some angel or prophet or raised up some eminent person to be an instrument of their reformation."[51] To Boston's evangelicals, Whitefield was such a prophet. They would write to Smith that Whitefield "has been received here as an Angel of God."[52]

Whitefield arrived in New England on 14 September, was met and led into Boston by a delegation of gentlemen on the eighteenth, preached in Boston and vicinity for twenty-six days, and toured New England until 29 October. The familiar story of his New England visit and the subsequent year and a half of local revivals need not be retold here; however, the English contribution to the beginning of the Great Awakening in New England should be underscored. The New England ministers considered the awakening to be a part of a general British/American revival of religion. They accepted the English Methodist Whitefield as God's principal instrument in that revival. And they identified their own concerns about evangelical preaching with the concerns of Dissenters in England about preaching Christ.

From America, the awakening spread to Scotland. The Scottish revivals illustrate clearly the transatlantic nature of the awakening.

The Cam'slang Wark

In Scotland, as in America, a group of evangelical ministers perceived the beginnings of a revival of religion, first in England and then in America, and made up their minds that, God willing, they would have a share in it.

The Scots began to hear of Whitefield about the same time the Americans did. Much of the early news of Whitefield was unfavorable. The *Scots Magazine* carried several anti-Whitefield and anti-Methodist pieces from the London *Weekly Miscellany* beginning in 1739.[53] On the other hand, the Scottish evangelicals read with approval Whitefield's sermons and *Journals*. Once Whitefield began his American tour in November 1739, they began to receive firsthand accounts from their American correspondents. For two years before Whitefield came to Scotland, revival news from England and America nourished expectations for his visit.

Aside from Whitefield's *Journals*, the principal medium through which the Scots received revival news was a weekly Methodist journal begun in London in August 1740 under the name *The Christian's Amusement*.[54] John Lewis, who described himself as "printer to the religious societies," founded the magazine as a vehicle to promote the Calvinist cause among the Methodists. On his return to England in the spring of 1741, Whitefield sanctioned the magazine, which, given a new title, *The Weekly History*,[55] became the official organ for Whitefield and Calvinist Methodism.[56] The letters concerning the activities of Whitefield and his fellow laborers of which it was primarily composed consistently attributed the revival of religion in America to the labors of Whitefield. In this way the Scots were led to see Whitefield as the chief instrument of the revival.

In July 1741 Whitefield arrived in Scotland, when accounts of the New England revivals following his visit there began filling the pages of the *Weekly History*. At the same time, American evangelicals were writing their Scottish correspondents of the American revivals and encouraging them to expect the same soon in Scotland. Robert Abercrombie, a graduate of Edinburgh University recently arrived in New England, on 11 June wrote "a friend in Edinburgh . . . how refreshing would it be unto me to hear of a Revival in my native Land." John Moorhead, born near Belfast, Ireland, educated at Edinburgh University, and minister of Arlington Street Church, Boston, since 1730, sent two letters to Scotland on 19 June. To a "friend in Glasgow," he described how the preaching of Whitefield and subsequently of Gilbert Tennent had awakened many in New England, and how the "harvest daily increased after their departure." "My soul," he continued, "longs to hear of God's Spirit being poured out upon you." To John Willison, of Dundee, after reporting that, in the judgment of charity, thousands had been converted in New England since the visits of Whitefield and Tennent, he exclaimed, "O that I could hear of the like blessed effusion with you, and all the reformed World. I trust the Almighty is about to do great Things for his Church."[57]

Landing at Leith on 30 July 1741, Whitefield went directly to Dunfermline to meet several members of the Secession's Associate Presbytery, from whom he had received one of several invitations to Scotland.[58] The Seceders insisted that Whitefield preach only from their pulpits, to avoid giving countenance to the corrupt churches, and that he allow them to enlighten him in the matter of church government. Whitefield replied that he would preach to anyone who would listen, and that he was too busy preaching the essentials of salvation to study the Solemn League and Covenant. By 6 August the Associate Presbytery had rejected Whitefield, and he had denounced them for building a Babel.[59]

During the next twelve weeks Whitefield preached in about thirty towns, including Edinburgh, Glasgow, Aberdeen, and Dundee. The moderate clergy were cool toward him. The "warmer Brethren," they believed, "wanted to use him as a Tool for establishing their enthusiastic Notions, and to bear down some of their Brethren who (in their Phrase) are termed MORAL PREACHERS."[60] The evangelical party within the Church heartily received the evangelist, however.[61] Whitefield favorably impressed the Scottish evangelicals, among whom he made many new friends, in particular William McCulloch, John Maclaurin, James Ogilvie, Alexander Webster, and John Willison. Willison looked "upon this Youth, as raised up of God for special Service, and spirited for making new and singular Attempts, for promoting true Christianity in the World, and for reviving it where it is decayed." He prayed that God would use Whitefield's coming to Scotland to revive God's work there.[62]

Like the Seceders, the Church ministers hoped to bring Whitefield over to Presbyterianism. In the middle of his Scottish tour, Whitefield wrote Willison: "I wish you would not trouble yourself or me by writing about the corruptions of the Church of England. I believe there is no Church perfect under heaven; but as God is pleased to send me forth simply to preach the gospel to all, I think there is no need of casting myself out."[63] Unlike the Seceders, the Church evangelicals were willing to accept Whitefield as an imperfect ally. They quickly learned to use his catholic rhetoric. Willison wrote on 8 October, for example, that though Whitefield

> is ordained a minister of the Church of England, he has always conformed to us in other points. God, by owning him so wonderfully, is pleased to give a rebuke to our intemperate bigotry and party zeal, and to tell us that neither circumcision nor uncircumcision availeth anything, but the new creature.[64]

The evangelicals did not give up trying to win him over to Presbyterianism. The next summer, Whitefield felt obliged to chide Willison once more: "Your letter gave me some concern. I thought it breathed a sectarian spirit, to which I hoped dear Mr. Willison was quite averse. You seem not satisfied, unless I declare myself a Presbyterian and openly renounce the Church of England."[65]

In the months following Whitefield's return to England, the Scottish evangelicals noted renewed interest in religion but nothing they could recognize as an extraordinary outpouring of the Spirit.[66] Intent on duplicating the American experience, the Scots sought to have Gilbert Tennent visit Scotland, as he had New England, after Whitefield. Whitefield passed the Scottish suggestion to Tennent, but Tennent declined.[67]

Willison's twelve sermons preached during the autumn and published in January 1742 under the title *The Balm of Gilead for Healing a Diseased Land* underscore the urgency of those who sought an outpouring of the Holy Spirit. The series began with a sermon on Jeremiah 8:20, "The Harvest is past, the Summer is ended, and we are not saved." Willison warned Scotland not to let pass the opportune moment. "It is sad when likely Seasons for saving a People from temporal Enemies and Grievances are lost," he wrote, "but it is yet sadder for them to lose hopeful and promising Seasons for saving of their Souls." Several sermons describe the spiritual diseases besetting Scotland, making it barren of conversions. God had withdrawn his Spirit because of the formality and hypocrisy of the people, because the workings of the Holy Spirit were "reproached with the odious names of Enthusiasm, whimsical Notions, and melancholy Imaginations," and because the ministers had swerved from Reformation principles. Yet, there were hopeful signs of a season of grace—in particular, the revivals in America.

> It makes the Season somewhat promising, that we hear of Christ's goings as a Conqueror in other parts of the world, and many bowing down at his feet. Now if the King of Zion be rising up to make his circuit thro' his Churches, to display his glorious power and grace among them; may it not raise our hopes, and encourage us to look out for his marching towards us?

Willison's prescription for taking advantage of this season of healing was to apply the balm, the blood of Christ, through the preaching of the Word. A number of the subjects of the subsequent revivals would remark how influential these sermons were in their awakening.[68]

Efforts to bring about a visible outpouring of the Holy Spirit finally showed results beginning in February 1742 in Cambuslang, a suburb of Glasgow. No spontaneous release of spiritual energy, the "Cam'slang Wark" was the product of the calculated labors of its minister, William McCulloch, with the cooperation of a group of interested laymen.

Cambuslang was a rural parish five miles southeast of Glasgow on the river Clyde in the Presbytery of Hamilton, in Lanarkshire. In 1742, when the revival began there, the land was held by eleven heritors, with two-thirds belonging to the nonresident Duke of Hamilton. The population numbered less than 1,000, divided into some 200 families, most of whom made their living as tenant farmers. Several families were colliers, legally bound for life, like serfs, to specific coal pits. Several spun or wove linen. Improvement had not yet touched the land: The new crops and agricultural methods would not be introduced until the 1760s; for now, farming was still done under the "run-rig" system of planting the traditional "oats, pease, beans, and barley," as the nursery song goes, in ridges in a common field, and tenants were still obliged to render their landlords various services, such as carrying coals and harvesting. The parish's economy, however, was becoming tied into the expanding linen industry. Weaving of fine linen, introduced in the 1730s, by now employed several looms. At this early stage, the weavers maintained control of the processes from buying the yarn, through weaving, bleaching, and bringing it to market. Only later, about mid-century, would they obtain the yarn from dealers in Glasgow, to whom they returned it woven into webs.[69] Weavers, among other craftsmen such as shoemakers, would be prominent among the lay leaders of the Cambuslang revival, and they and their wives and children would constitute a significant percentage of the revival's converts.[70]

For seven years before the settlement of McCulloch in 1731 the manse at Cambuslang had remained empty because of an impasse between the principal patron, the Duke of Hamilton, and the parish over choice of minister. Not until the Duke's nominee was forced on another parish did he consent to McCulloch. The parish had a history of evangelical activism. Robert Fleming, who would chronicle the revivals of the 1620s and 1630s in western Scotland in his *The Fulfilling of the Scripture* from his exile in Rotterdam,[71] ministered to Cambuslang from 1653 until expelled in 1662. Over the subsequent decades of persecution, several leading parishioners were fined or imprisoned for attending conventicles or for refusing to assist conforming ministers in administering discipline. Several had the king's soldiers billeted on them from time to time. McCulloch's immediate predecessor read from the pulpit his reasons for refusing the oath of abjuration in 1712. In 1720, parishioner Hugh Cumin presented the kirk session a testimony against defections from the Word of God and covenant engagements, that is to say, against the Union with England. In June 1740, the presbytery deposed the majority of Cambuslang's elders, including Cumin, for refusing to exercise their office until they could be made to see "that Patronage is agreeable to the word of God." The elders' statement of grievances echoes those of the Seceders. That 19 of the 106 Cambuslang converts whose accounts McCulloch later recorded mention hearing Secession preachers indicates local sympathy for the Secession. McCulloch had strong reasons to work for a religious revival to demonstrate the validity of his ministry. Given its recent experience with patronage, evangelical past, and current leanings toward the Secession, Cambuslang was as likely a candidate for a revival as any Scottish parish.[72]

McCulloch closely followed the activities of Whitefield and studied his methods. Early in 1741 he began to preach regularly on the nature and necessity of the New Birth. In September he observed Whitefield preach in Glasgow. McCulloch was an undistinguished orator, known as a yill [ale] minister, the term applied to speakers whose delivery was so dry that when their turn came to preach at outdoor communion gatherings, many resorted to the ale barrels for refreshment.[73] After his return from Glasgow, his parishioners found he "preached much better than he used to Do."[74] Particularly memorable were his warnings of the immediate danger of dropping into hell. His words of 2 February 1742 so strongly impressed one auditor that months later she could repeat an entire passage from memory:

> The very dust under your feet, the very seats you sit on, nay, The Devil himself, may go as soon to heaven, as you can go there, Except ye be born again: and how can ye allow yourselves quietly to eat or sleep when ye know Nothing of the new birth, and when there is nothing but the frail thread of life between you and everlasting burnings, and if that should be snapt asunder, as you do not know but it may this night you instantly drop doun into the pit of hell.[75]

Obviously an apt pupil of Whitefield, McCulloch may have been modeling his new sermon style after Jonathan Edwards as well. While there is nothing unique about the imagery in this passage, it is strikingly similar to the imagery in Edwards' *Sinners in the Hands of an Angry God*:

> You hang by a slender thread [over a] great furnace of wrath. . . . It would be a wonder if some that are now present should not be in hell in a very short time. . . . And it would be no wonder if some persons, that now sit here . . . in health, quiet and secure, should be there before to-morrow morning.[76]

Delivered in July 1741 and published in Boston within a few months, Edwards' sermon could easily have been circulating in Glasgow by February 1742.

About the time of Whitefield's Scottish tour, McCulloch began after every service to read to the congregation of Whitefield's successes in America as reported in the London *Weekly History*. In December he began editing his own *Glasgow-Weekly-History*, which consisted mainly of letters printed first in the London *Weekly History*.[77]

News of the Great Awakening in America deeply affected the pious of Cambuslang. Janet Jackson, young unmarried servant woman, daughter of elder James Jackson, told of eagerly reading Whitefield's sermons and the *Weekly History* after hearing him at Glasgow. Her sister Elizabeth, who had also heard Whitefield, reported that, hearing McCulloch "read some papers relating to the Success of the Gospel abroad; I was greatly affected at the thought that so many were getting good, and I was getting None." Margaret Richie, a twenty-year old spinner, even considered going to America:

> When I heard Mr Edwards Narrative of the Surprizing work of God at Northampton read; I was very glad to hear that there was such a work of conversion in these far distant places; and I thought that if I were there, I might perhaps get a cast of Grace among others, and I was busy from time to time contriving methods how I might get there.

Reading in Whitefield's *Journals*, before the itinerant's visit to Scotland, how many were benefiting from his preaching, she "thought that if I might hear him I might get good also" and in the winter of 1741–1742 "was much and oft taken up in praying for a Revival of Religion."[78]

Those who had heard Whitefield at Glasgow acted as leaven within the private religious societies at Cambuslang, raising the level of religious emotion. Ingram More, a shoemaker, was especially active in arousing the convictions of others by the lively description of his own feelings. In January, together with Robert Bowman, a weaver, he gathered the signatures of ninety heads of families to a petition for a weekly lecture.[79] McCulloch complied, reporting this to Whitefield as a hopeful sign: "I know not what God is about to do, among this poor People under my Charge; but wou'd gladly hope, that he is on his Way to come and do mighty Works of Grace among us."[80]

The stage for the revival in Cambuslang was now set. Through the winter, the spiritual concern of many had been building. The pastor had provided descriptions of the appropriate behavior in a revival. The people were expectant and awaiting their cue. On Sunday, 14 February, Catherine Jackson, another of Elder Jackson's daughters, showed signs of deep distress during the service. Afterward, while McCulloch was counseling her at the manse, she related evidence of conversion and peace with God. McCulloch wrote out an account of her experience, which he read the next day at a general meeting of the religious societies. On Tuesday the religious societies again met together, in concert with societies elsewhere, to pray for the success of the Gospel. They held yet a third meeting on Wednesday. Thursday was the weekly lecture.[81] In the course of his extemporaneous prayer after the lecture, McCulloch asked "where is the Fruit of my poor Labours among this People?" At this several persons cried out. A critic describes what followed:

> Some of the People thus prepared began, in the same Manner they had heard described, to cry out in the most publick Manner, of *their lost and undone Condition*, saying, *They now saw Hell open for them, and heard the Shrieks of the Damned!* and express'd their Agony not only in Words, but by clapping their Hands, beating their Breasts, terrible Shakings, frequent Faintings and Convulsions; the Minister often calling out to them, *Not to stifle or smother the Convictions, but encourage them.*[82]

After the sermon approximately fifty persons under convictions came to the minister's house. Thereafter, McCulloch preached almost every day. Convictions and conversions continued. Bowman and More, the men who had promoted the petition for the weekly lecture, continued their work in fostering the revival. Whenever anyone showed signs of distress during the revival services, they would encourage the affected to express their feelings. The two artisans took the awakened in hand, organizing them into prayer meetings, coordinating their conferences with McCulloch in the manse, or arranging for counseling by some experienced Christian. Within a few weeks the people had started a dozen new societies for prayer.[83]

Reports of the revival spread rapidly, attracting the curious, the skeptical, and the hopeful. Within weeks, sympathetic pastors, including Maclaurin from nearby Glasgow, and Willison from distant Dundee, arrived to assist. By the middle of March, Whitefield had been informed that "the work" had begun in Scotland.[84] By the middle of April, Boston knew of the revival.[85] James Robe, of nearby Kilsyth, and many of his parishioners came to observe and by the middle of May had brought the revival home. Now the revival began to spread to many parishes in western Scotland.

Whitefield returned to Scotland at the beginning of June and remained five months, dividing his time between Edinburgh—where, in the park at Heriot's Hospital, 2,000 seats were erected for his auditors—and the western towns where the revivals were occurring. At mid-July Cambuslang hosted a great Communion service. Approximately 20,000 persons attended, and 1,700 received Communion. Whitefield had never seen anything to equal the fervor. On his first day there,

> such a commotion was surely never heard of, especially about eleven o'clock at night. It far outdid all that ever I saw in America. For about an hour and a half there was such weeping and so many falling into deep distress, and manifesting it in various ways, that description is impossible. The people seemed to be smitten by scores. They were carried off and brought into the house like wounded soldiers taken from a field of battle.[86]

The usual practice was to hold but one Communion a year in each parish. At the suggestion of Whitefield and Alexander Webster, however, a second communion at Cambuslang was called for August, at which about 30,000 attended and 13 pastors administered to 3,000 communicants. Again, the emotion was visible. "Some of both sexes, and all ages, from the stoutest man to the tenderest child," wrote Webster, "shake and tremble, and a few fall down as dead."[87]

The second Communion of 1742 was the high point of the Cambuslang revival. From Cambuslang, the revival spread to neighboring parishes in the Presbytery of Hamilton and nearby parishes in Glasgow Presbytery, including several in Glasgow itself. From there, the Awakening worked its way northward to the Presbytery of Auchterarder. North of the River Tay, aside from some evidence in Dundee and Aberdeen, awakenings were not noticeable, until in 1743 and 1744 the revival movement appeared in the northern counties of Ross and Sutherland. After 1744, there were no notable new awakenings, but at least into the next decade the Awakening's effects could be measured by a significant increase in participation at yearly Communion services in Glasgow, as well as in several of the awakened rural parishes.[88]

The accounts of the revivals in New England provided the evangelical ministers of Scotland the information they needed to become successful revivalists. The accounts revealed the doctrines and the style of preaching that proved effective. Whitefield's first Scottish tour served as a practical demonstration. In addition, the Scottish pastors learned what behavior in their flocks to look for as signs of a revival and were enabled to communicate their expectations to their congregations. Scottish revivalists exploited the expectations aroused by the Great Awakening in America to bring about a revival of religion they devoutly desired. Like the Great Awakening in America, the Scottish revival drew on the people's deeply rooted, traditional piety.

Revival Literature as Catalyst

Two cases, at the far corners of the British provinces, Golspie, in the north of Scotland, and Durham, New Hampshire, illustrate the processes by which revival accounts served as training manuals as well as stimuli for revivals.

Before the Glorious Revolution, Golspie, under the protection of the Sutherland family, had been a sanctuary for persons persecuted for refusal to conform to the Episcopal establishment. When the Reverend John Sutherland came to the parish in 1731, he found a number of devout Christians, descendants of those refugees. Until 1744, however, the spiritually awakened were so few, and the behavior of others so scandalous, that Sutherland felt his labors wasted. When printed accounts of the Awakening in America and in the southwest of Scotland reached him, he took pains to inform his congregation. Then, he reported to them what he witnessed in person when visiting Cambuslang, Kilsyth, and Muthil on his way to the General Assembly of 1743, "if by these means, I might provoke the people to emulation." In August 1743, while assisting at a Communion service at Nigg, he learned from its minister the value of societies for prayer in stimulating an awakening and "resolved to essay the like means in imitation of his successful example." He initiated three prayer groups, which met weekly, encouraging them to pray for an outpouring of the Holy Spirit. Finally, about a year later, the awakened began calling on him. After some spoke of being reluctant to disclose their spiritual turmoil, Sutherland explained to the congregation the duty of awakened sinners to discuss their state with ministers and experienced Christians. By August 1744, about seventy persons had come to him "under various exercises of soul."[89] Golspie's experience suggests that a clergyman could not create a revival simply by mastering revival techniques, but that his congregation had to be ready to respond to revivalism. A revivalist's labors brought forth a quicker and fuller harvest when, as in the case of Cambuslang, it corresponded with lay initiative.

Despite his somewhat remote location in Durham, New Hampshire, frequent contacts with Boston gave the Reverend Nicholas Gilman access to the latest revival news. Committed to conversionism, Gilman read in 1740 and 1741, besides seventeenth-century standard Puritan works on conversion, a number of works on the Methodist movement, including the life of Whitefield, Whitefield's *Journal*, *Nine Sermons*, *Sermon on Religious Societies*, and *Answer to the Bishop of London's Pastoral Letter*, in addition to *An Account of the Rise of the Methodists at Oxford*. He also read John Jennings's two discourses on *Preaching Christ*, and *Evangelical Preaching*, and accounts of revivals printed in the *Boston Gazette*. In mid-November 1741, the day after receiving a copy of Jonathan Edwards's *Sinners in the Hands of an Angry*

God, he began reading aloud from Edwards's sermons and *Faithful Narrative*. Crowds gathered at his house to hear, and he continued to read from these works daily. By early December many had been awakened and expressed their strong feelings openly: A revival had begun. During the revival, Gilman continued his practice of reading evangelical literature aloud and included works by Benjamin Colman, Ebenezer and Ralph Erskine, and Gilbert Tennent, as well as narratives of the Scottish revivals. In a process remarkably like the coming of the revival to Cambuslang, revival literature brought home to minister and people of Durham not only the doctrines of regeneration, with which Scots and New Englanders were already familiar, but also their own expected roles in an awakening.[90]

By 1742 the British/American evangelical community had before it two models that suggested a specific form for the long-awaited revival of religion. Edwards's *Faithful Narrative* embodied one model. Edwards portrays the spiritual awakening of an entire community led by its settled pastor. There was more to the awakening than a spate of new converts and the rededication of saints to their calling. The people perceived the revival as a communal event and joined with one another to reform their community. Parents united to discipline children. Youths joined in societies for prayer and pious reading. It seemed love between neighbors might be attainable and contentiousness might be overcome. Whitefield's *Journals* mirrored the second model, the Communion at Shotts in 1625. Whitefield was an evangelist with an extraordinary call to preach to all who would listen. Instead of leading to religious awakening and moral reformation a specific community over which he exercised recognized spiritual authority, he traveled from place to place, seeking to transform the entire nation. The phenomena associated with the Methodist revival were the itinerant preacher, large outdoor religious gatherings, and the organization of the awakened into religious societies, not so much to help reform the community as to witness to a rejection of secular values. Anticipating Whitefield's visit, the New England clergy had both forms of activity in mind. Writing to Whitefield in December 1739, Benjamin Colman juxtaposed the New England and the Methodist experience:

> When I read your Journals [I am convinced that God is with you]; the evident Impressions of the Word upon the Minds and Hearts of the Multitudes who hear you [must give you inward comfort]. I write not this to you, my dear Brother, because it has been with me as it is with you, nor because I have seen the like in our *New-England* Churches, altho' there have been at Times, uncommon Operations of the holy Spirit on the Souls of many under the Ministry of the Word; as in our Country of *Hampshire* of late; the Narrative of which Mr. *Edwards*, I suppose you may have seen; but I do indeed account the Impressions from God upon the *Methodists* (our dear, Brethren) to be something very extraordinary.[91]

The Great Awakening in New England would partake of elements from both models, for, while the itinerant was important in stimulating spiritual concern, the revivals often broke out only after the evangelist had continued on his journey, or in towns not even visited, leaving the settled pastor to direct the revival.[92]

The revivals in Scotland also borrowed from both models. The beginnings of the revivals in Cambuslang and Kilsyth directly imitated New England forms, as expressed in the revival narrative; the great Communion services in the summer of 1742 were quintessentially Whitefieldian and reminiscent of the revival at Shotts.

8

The British/American Evangelical Connection

During the revivals of the 1740s revivalists developed a transatlantic network of connections for promoting and defending the revivals. Ministers in the centers of that network, Boston, Edinburgh, Glasgow, and London, maintained contacts with ministers in the other centers through the frequent exchange of letters. They shared these letters with fellow clergy, who read them to gatherings of their people. From each of the centers promoters published a weekly magazine that printed news about the revivals throughout the empire. The ministers received from each other the newest published accounts and justifications of the revivals, and reprinted many of them locally. Revival literature spread understanding of techniques, confirmed the validity of the awakenings, and fostered a sense of community among widely separated evangelicals. Revivalists relied on one another for help in refuting their critics. Indeed, from their own transatlantic connections, the critics, themselves, borrowed ammunition for use against the Revival. Both promoters and opponents recognized the transoceanic character of the movement.[1]

Transatlantic Antirevival Connections

Both those who opposed and those who defended the Scottish Awakening called on America for support. Opposition came from either end of the spectrum: from the Seceders and the Cameronians, and from the liberals.

The Seceders had reason to oppose the revivals, for the revivals undermined their position. If the revivals were genuine, then God was owning churches they would not, and the standing churches were not so corrupt as to justify secession. James Fisher, Secession minister at Glasgow, accused the supporters of Whitefield of using him as a tool to break the Secession. If Whitefield was correct in teaching that particular forms of church government were not fundamentals of Christianity, then the protest of the Seceders would be pointless.[2] The revivalists did not hesitate to use the Awakening as an argument against the Secession. Alexander Webster suggested that the Awakening might teach the Seceders that God still dwelt in the established church and make them "more cautious in separating from those whom the great Master of Assemblies condescends to countenance so remarkably with his Presence."[3] It was said that even some liberals who disliked the revivals were willing to embrace them as a way to destroy the Secession.[4]

Between 1733 and 1740 a wall of enmity had been raised between the Seceding ministers and the church evangelicals. The Seceders denounced the evangelicals for failing to witness adequately against the corruptions of the Church. The evangelicals resented the refusal of the Seceders to respond to their efforts to rectify the corruptions during the General Assemblies of 1734–1739. John Willison blamed the Seceders' stubbornness for stopping a national reformation.[5] The two groups now competed for the loyalty of the populace.

The revival proved particularly influential among persons who had been inclined toward the Secession. One man wrote to James Robe how, ever since he had begun to repent his evil ways, he had had "the desire to the Associate brethren; because I thought they were contending for the truth; . . . but while I thought on these things, the news of a surprising work at Cambuslang . . . was told me." The revival convinced him to remain in the church. A tailor, a subject of the Cambuslang work, compared the sermons of a revivalist and of a Seceder on the same text, and found the former as solidly evangelical as the latter. Another Cambuslang convert had been perplexed about joining the Secession or staying in a corrupt church until a pamphlet by John Willison convinced him that God could be present in an imperfect church.[6]

The Seceders commenced their condemnation of the revival by denouncing George Whitefield and continued by denouncing Jonathan Edwards. Adam Gib, Secession minister at Edinburgh, began the offensive in early June 1742 with *A Warning Against Countenancing the Ministrations of Mr. George Whitefield*, in which he demonstrated "that Mr. Whitefield is no Minister of Jesus Christ; that his Practice is disorderly and fertile of Disorder; that his whole Doctrine is, and his Success must be, diabolical." The idolization of Whitefield in America and Scotland appalled Gib. His main objection to Whitefield was his catholic scheme, which made men skeptical about the divinely ordained presbyterian discipline and government.[7] The Associate Presbytery considered Whitefield and the Awakening to be divine punishments for Scotland's sins. They appointed 4 August for a public fast to ask

God's deliverance from the revival, which they viewed as the grossest enthusiasm and the work of the devil.[8]

Like the Seceders, the Cameronians rejected the Awakening as diabolical because of Whitefield's hand in it. In August 1742, calling themselves "the suffering Remnant of the Anti-Popish, Anti-Lutheran, Anti-Prelatick, Anti-Whitefieldian, Anti-Erastian, Anti-Sectarian, true Presbyterian Church of Christ in Scotland," they lamented that ministers calling themselves Presbyterian should invite into their pulpits "an abjured prelatick Hireling, of as lax Toleration-Principles as any that ever set up for the advancing of the Kingdom of Satan."[9]

The Scottish revivalists paid little attention to the Cameronians, who were insignificant in numbers and no threat to the Church of Scotland. The Seceders, on the other hand, they took seriously. James Robe called the Associate Presbytery's act for a fast "the most heaven daring paper, that hath been published by any set of men in Britain these hundred years past," for to denounce the revivals was to denounce the work of God.[10] Any objections against the revivals, thought Robe, had been fully answered by Jonathan Edwards's *Distinguishing Marks of A Work of the Spirit of God.*[11] James Fisher replied to Robe with an attack on Edwards's work. "It would require a Treatise by itself," said Fisher,

> to follow the Chain of *Error*, both in *Philosophy* and Divinity, that runs thro' the *whole* of this *Performance*; only because it is cry'd up, as the *Standard-Piece* of this *Work*: I shall endeavour to make it appear, that, whatever was the Intention of the *Author*, yet, the manifest *Design of his Work*, is to *overthrow the very Foundation of Faith, and all practical Godliness*, and to *establish mere Enthusiasm, and strong Delusion*, in the *Room* of the *true Religion*, revealed and required in the Word.

Fisher argued that Edwards's view "that we cannot think upon any Thing invisible or spiritual, without some Degree of imagination" led to the worship of mental images rather than of the true Christ. He concluded that the horrible representations on the imaginations of the subjects of the revivals were "impulses of Satan."[12] Robe and Willison came to Edwards's defense. In their replies to Fisher they argued that the imaginary ideas that attended one's meditations of God were not the same as imaginary ideas of God, and, therefore, not idolatrous, and that the Holy Spirit certainly could make use of a person's imagination in the conversion process.[13] Over the next couple of years Ralph Erskine continued the Seceders' attack on Edwards, culminating with the lengthy *Faith no Fancy: or A Treatise of Mental Images* (Edinburgh, 1745). The Seceders formed connections with America's radical New Lights, such as Andrew Croswell, whose doctrine that the essence of faith is assurance matched the views of the Marrowmen.[14]

Revivalism in Scotland lost a natural ally in the radicals. Whereas the radical evangelical movement in America embraced revivalism, out of which it had been born, the Scottish Secession rejected the revivals, which had injured it. Seceders repudiated the demonstrative emotionalism of the revivals and this form of public piety did not come to characterize Scottish Presbyterian Dissent as it grew to significant dimensions in the latter half of the eighteenth century.[15]

The liberal opponents of the Scottish Awakening relied on direct aid from the American critics of the Great Awakening, as one Scot acknowledged: "*A Tide of Enthusiasm* that had almost over-whelm'd your Churches, has been flowing in fast *among* us also: But one of the most effectual means of stemming it with us, was the seasonable Information we got of what you had suffered by it."[16] Until August 1742, asidc from the Associate Presbytery's fulminations, only one piece had been published against the Scottish revivals, and that had come out but three weeks after the first cryings-out at Cambuslang.[17] Until August, as well, none of the favorable accounts of the revivals in America, contained in the *Weekly History* and in letters from Boston circulated by Maclaurin, had been contradicted. Then there appeared from the Glasgow press a letter, dated 24 May 1742, from a gentleman in Boston, signing himself "A. M.," showing the American Awakening to be mere enthusiasm. "A. M." denounced Whitefield as "a bold and importunate beggar . . . insatiable as the Grave." He called Boston's prorevivalist ministers "flaming Zealots for certain favorite Opinions and Tenets." Gilbert Tennent was "an awkward and ridiculous Ape of Whitefield . . . roaring out and bellowing, *Hell*, *Damnation*, *Devils*. . . . Every one that was not exactly of his Mind, he damned without Mercy." "A. M." enumerated the disorders—the itinerants, the lay exhorters, the visions, trances, and convulsions, and Davenport's madness—and denied there had been any moral reformation in Boston: Contrary to William Cooper's preface to Edwards's *Distinguishing Marks*, the taverns and dancing schools were as frequented as ever. His appendix consisted of excerpts of writings by American opponents of the revival—Charles Chauncy, John Caldwell and Samuel Mather—as well as excerpts cautioning against disorders by the revival's promoters—Ebenezer Turell, Jonathan Parsons, and Benjamin Colman.[18] The *State of Religion in New England* sold quickly. Soon after, New England antirevival pieces appeared in Scottish editions, including Charles Chauncy's *New Creature* (Boston, 1741; Edinburgh, 1742), John Caldwell's *Impartial Trial of the Spirit* (Boston, 1742; Glasgow, 1742), and the anonymous *Wonderful Narrative . . . of the French Prophets* (Boston, 1742; Glasgow, 1742).

With the publication of these writings, Scots no longer needed to assume the validity of the American Awakening, and if the American revivals produced little good and much disorder, then what was to be expected from their exact parallel in Scotland? "Tho' nothing had been directly published against the Work at Cambuslang," wrote a liberal to his friend in New England, "yet it was generally understood to be attack'd indirectly by the Printing of the *New-England* Papers." He said that "the finishing Stroke to that Credulity wherein-to we had been betrayed by the first Accounts we had from New England" was a letter from Charles Chauncy to George Wishart, one of Edinburgh's liberal ministers, regarded for fine preaching, and brother of the principal of the University of Edinburgh. The letter was published anonymously in Edinburgh in November, a few days after Whitefield had left for England.[19]

Alexander Malcolm, Anglican minister at Marblehead, Massachusetts—probably the author ("A. M.") of the *State of Religion in New England*—and William Hooper, pastor of Boston's West Church, had brought Chauncy to the attention of Scottish liberals in the summer of 1741. Hooper had been born in Scotland and was graduated at Edinburgh University in 1723. In 1741 he headed a campaign to obtain for Chauncy an honorary doctor of divinity degree from the university, thinking, as Malcolm explained, "it would be of use to the cause of reason and religion, if in the present situation of things such a mark of distinction were putt upon a man of worth that dares to oppose such a tide of nonsense and madness" as the revivals were. Malcolm wrote Charles Mackie, professor of history at Edinburgh University, that Chauncy "is a very valuable gentleman, both as a scholar and clergyman; an example and patron of good sense, virtue and true religion, in opposition to a spirit of enthusiasm, which Whitefield has kindled in this country to the great prejudice of religion." Hooper wrote to others, including George Wishart.[20] The university awarded Chauncy the degree in March 1742.[21]

Chauncy's letter to Wishart was similar to "A. M." 's letter, but avoided attacking characters. Chauncy ridiculed the treatment of Whitefield as an angel of God. He accused Whitefield of taking advantage of the susceptibilities of women and the young. The violent and frightful language of the revival sermons, the extemporaneous preaching, the illiterate exhorters, and the fits, swooning, screams, visions, trances, and loud laughing of the revival meetings he denounced as dangerous enthusiasm.

The liberal element in the Church of Scotland could occasionally bring effective pressure to bear against clerical supporters of the revivals. For instance, during the autumn 1742 Communion season, no ministers in Edinburgh or Glasgow invited Whitefield to assist. "Some members of the University and some Elders of Rank and Distinction" in Glasgow prevailed with Maclaurin to spare them the offense.[22] Willison was unwilling to preface the Scottish edition of Edwards's *Thoughts on the Revival* because of the abuse he had received for prefacing the *Distinguishing Marks* and could find no one else to accept the duty.[23] The Edinburgh edition of Edwards's *Thoughts* appeared in 1743 without a Scottish preface.

The lengths to which the Scottish opponents of the revivals went to expose the falseness of the American Awakening measures the strength of the Awakening's influence in the Scottish Revival.

The Transatlantic Prorevival Connection

In their defense of the Revival, the Scottish evangelicals relied heavily on American materials. Edwards's *Distinguishing Marks* took preeminence. John Willison wrote the epistle to the reader of the Scottish edition on 23 June 1742. He thought Edwards's treatise should remove all doubts about the divinity of the work at Cambuslang as it did for the similar work in America. In *Divine Influence the True Spring of the Extraordinary Work at Cambuslang*, Alexander Webster at the end of August referred to Edwards's *Distinguishing Marks* and repeated several of its arguments. John Maclaurin answered the *State of Religion in New England* with a long letter to the *Glasgow-Weekly-History*. He showed that the writings of Colman, Parsons, and Turell used by "A. M." to prove the enthusiasm of the New England Awakening actually attested to a marvelous work of God's grace in many parts of the land.[24] Maclaurin, Robe, and Willison each cited the numerous attestations and sermons in America, many reprinted in London, Glasgow, and Edinburgh, confirming the reality of the Awakening.

During the Evangelical Revival in New England and Scotland, personal associations between evangelicals across the Atlantic multiplied. Correspondents informed each other of all new developments. They exchanged encouraging words and ideas about the revivals. The principal figures in this transatlantic correspondence during 1742–1743, as evidenced from the items published in the revival magazines, were New Englanders and Scots: on the American side, Robert Abercrombie, Benjamin Colman, William Cooper, John Moorhead, Thomas Prince, all of Boston, and, by 1743, Jonathan Edwards; on the Scottish side: John Hamilton, William McCulloch, John Maclaurin, James Robe, Alexander Webster, and John Willison. Several of these associations had formed before the revivals. Abercrombie and Moorhead had come from Scotland. Colman with Willison, and Cooper with Maclaurin, had been corresponding since the early 1730s. Other associations resulted from the revivals. Webster contacted Colman in January 1743 to join with Willison in sending revival news. Hamilton was corresponding with Prince by the summer of 1742. McCulloch, Maclaurin, Robe, and Willison began writing to Prince in 1743, apparently because of his guiding role in the Boston

Christian History edited by his son. Perhaps through an introduction by Cooper, Edwards and Maclaurin began corresponding sometime in 1742. Through Maclaurin's services, Robe and McCulloch entered into a correspondence with Edwards in the spring of 1743. Of the few transatlantic letters from the Middle Colonies printed in the revival magazines most are those of the Tennents to Whitefield.

The Scottish revivalists had informed their American friends as soon as the Cambuslang work began and they kept the Americans informed of subsequent developments as they occurred. Two narratives of the revival at Cambuslang, published in Scotland in the spring of 1742, were reprinted at Boston and at Philadelphia before the close of the year.[25] In September John Maclaurin sent William Cooper the "Kilsyth Narrative" by James Robe, as well as copies of the *Glasgow-Weekly-History* containing Robe's case histories of individual converts.[26] The "Kilsyth Narrative" appeared in Boston in March 1743 in the first several issues of Thomas Prince's revival magazine, *The Christian History*. Through 1743 and 1744 *The Christian History* printed letters from Scottish revivalists describing the prosperous state of religion in Scotland.

The Scottish revivals did not produce the ecclesiastical disorders that came out of the Great Awakening in New England: unsolicited itinerant preachers did not invade the parishes of settled ministers; no James Davenport identified unconverted ministers by name or burned books; and the revivals acted to keep people within the established church rather than to create separations. Hence, American antirevivalists did not point to the example of Scotland to sully the American Awakening. They did, however, seek to show that the Scottish revivals shared in the enthusiasm of the American. Antirevival pieces published in Scotland were available in Boston by early 1743.[27] In the spring of that year *A Letter from a Gentleman in Scotland to his Friend in New England* was published in Boston. The letter was expressly in repayment for the letters of "A. M." and Chauncy, and served a similar purpose, to correct the false impression given by the favorable reports of the Scottish revivals.

In 1743 the New England clergy formally assessed the Great Awakening. In May the opponents of the Awakening in Massachusetts issued their *Testimony* against the errors and disorders of the revivals.[28] In July, the proponents answered with their *Testimony and Advice*, attesting to the reality of a late remarkable effusion of the Spirit.[29] In 1743 Jonathan Edwards published *Some Thoughts Concerning the Present Revival of Religion in New-England*, and Charles Chauncy countered with *Seasonable Thoughts on the State of Religion in New-England*. During their debate, the New Englanders, preoccupied with their own affairs, gave some thought to the impact of their conflict on Britain. Even though Boston's William Cooper wrote to Glasgow in June that he and others believed the May *Testimony* was a scheme laid "more with a view to

Scotland than to New England" to give an unfavorable view of the revival,[30] the proponents of the revival in America were somewhat tardy in sending their defense, the *Testimony and Advice*, to Scotland. The Scots first received copies through a London edition put out by Isaac Watts, who had been unwilling to send his own only remaining copy to Scots eager to see it.[31] John Maclaurin called it "the much longed-for ATTESTATION."[32] In the middle of 1744 Thomas Prince sent William McCulloch a large number of the first volume of *The Christian History*, which, in numbers twenty and twenty-one, contain the text of the *Testimony and Advice*.[33]

Through Whitefield, the Americans and Scots had constant contact with the Calvinist wing of the English Methodist movement. John Syms, Whitefield's traveling companion, secretary, and confidential friend until he joined the Moravians in August 1743, was Thomas Prince's contact with London Methodism. The Welsh Methodists had few contacts in America, but their leader, Howell Harris, corresponded frequently with William McCulloch in Scotland,[34] as well as with Whitefield, Syms, and the Wesleys in England. During these years, the Wesleyan Methodists concentrated their efforts within England. Their Arminianism divided them from the evangelical movements in Wales, Scotland, and America. In 1751, when John Wesley made his first visit to Scotland, Whitefield made his fifth.

English Dissent and the Transatlantic Awakening

Whereas the Revival in England began in the late 1730s through the efforts of the new evangelical movement within the established church, evangelicals among the Dissenters lagged well behind. Although English evangelical Dissenters shared with the American and Scottish revivalists a commitment to evangelical doctrine ("preaching Christ") and to affective religious experience, their reading of the *Faithful Narrative* did not turn them into successful revivalists. The Dissenters' consciousness of their minority status (just over 6 percent of the population of England and Wales)[35] inhibited the ministers from countenancing in their own congregations any extraordinary emotional behavior that would have exposed the Dissenters to ridicule and contempt. While they supported the American and Scottish revivals, evangelical Dissenting ministers on the whole remained aloof from the English revival movement during the 1730s and 1740s, for they dreaded the taint of enthusiasm and were jealous of the success of the Methodists. Michael Watts suggests other reasons why the Revival began among the Methodists of the Church of England rather than among the Dissenting churches. Restricted to licensed places of worship, Dissenters could not legally preach in the fields. The

Dissenters had a small percentage of the populace as an audience. The independency of their congregations limited cooperation—the widespread evangelizing efforts of Richard Davis in the 1690s had provoked the resentment of other Dissenting ministers. And fifty years of toleration had reinforced the impulse toward separation from, rather than conversion of, the world.[36]

The key to the actions of the evangelical Dissenting ministers in granting or withholding support of the Revival was their status as Dissenters. This position explains why they endorsed the American and Scottish revivals as genuine works of God but remained cautious about the Methodist revivals in England. The Methodists carried with them the taint of cant and enthusiasm. Whitefield was inexperienced, rash, incautious, spiritually proud, and overly confident of internal impressions. Wesley was Arminian, and his Moravian associates were fond of odd notions. The Dissenters lacked the self-confidence being part of an establishment provided. The evangelical Dissenters were happy to find in the Church of England men committed to the evangelical New Birth, but thought of them as a blessing for the Church of England, for Dissent already possessed the evangelical truth. The Methodists were Church of England men. Their success showed up the Dissenters, as well as lured away members. In his correspondence with Colman, Watts treated the Methodist revivals in England as an affair of the Church of England alone. In the midst of the Methodist revivals he continued to lament the decline of vital piety and persisted in looking for revivals in England. "I am glad to hear of such a spirit of religion raised and spread thro your country," he wrote in November 1741. "I wish we could find it amongst us."[37] In April of 1742 he wrote: "We pray for the same revival, and as we rejoice in your mercys we wish the same blessings were diffused amongst our churches."[38]

Whitefield had sought out Watts's acquaintance early in 1739, and Watts quickly adopted the role of semiofficial spokesman of evangelical Dissent's view of Whitefield. Watts admired Whitefield's piety and zeal but found himself continually embarrassed by Whitefield's indiscretions. "I cannot but say I have a great love for the man," Watts wrote in the summer of 1740, yet by his letters defending "such an unadvised sentence as that Archbishop Tillotson knew no more of Christianity than Mahomet, . . . he had done himself and his ministry unspeakable hurt. . . . This conduct, I say, must fall under the charge of great imprudence."[39] When Edmund Gibson, Archbishop of London, issued his *Pastoral Letter . . . By Way of Caution, Against Lukewarmness on one hand and Enthusiasm on the other* (London, 1739), directed against Whitefield, Watts replied privately in a letter to the bishop. He agreed with the bishop that Whitefield lacked convincing evidence of many of the extraordinary influences to which he laid claim, and that he walked dangerously close to delusion. Watts, however, believed that Whitefield's censure of the clergy of the established church bore a measure of truth. The bishop, thought Watts, could do nothing more effectual "to make all the Whitefields less regarded and less dangerous to the church, as to induce the ministers under your care to preach and converse among their people with . . . evangelical spirit, . . . zeal for the honor of God and success of the gospel, and with . . . compassion for the souls of men."[40]

Benjamin Colman assumed that the unity of interests between the English Dissenters and the American evangelicals should lead to a common attitude toward Whitefield. In Colman's first letter to Whitefield, he expressed an interest in the treatment Whitefield had received from Watts and others in London, and speculated on why the Dissenters had not invited him into—or he had not accepted—their pulpits.[41] Shortly thereafter, Colman wrote to Watts directly.[42] As soon as Whitefield had departed Boston, Colman wrote to Watts, John Guyse, William Harris, and Daniel Neal as representatives of the Dissenters, to report that Whitefield had avoided the Church of England services but had preached and worshipped in the Congregational and Presbyterian meetinghouses. He wrote in order to give "the most early Notice I can of these Occurances among us, that when Mr. Whitefield shal return to London, you may the better prepare your tho'ts how to receive him, should he go further from the Established Chh, or that cast him out, wch I plainly see he expects, and I must own I do likewise."[43] In other words, he thought that to win over Whitefield would be a coup for Dissent.

Watts told Colman he doubted that Whitefield would join the Dissenters. In later letters, Watts made it clear he hoped Whitefield would not do so. Whitefield, Watts thought, "seems to have been raised up, like Luther in the Reformation from Popery, to rouse the generality of the Church of England from its formalitys in religion and from some of the growing errors of the times." The Church, not Dissent, needed Whitefield. On 19 August 1741, Watts wrote Colman that Whitefield "is travelling to several parts of the nation, preaching in the same manner as amongst you, but intirely as a minister of the Church of England, of which I am very glad; for tho I think God has succeeded his ministry greatly, yet his joining himself to us would be no advantage to the success of his work." On 16 April 1742 Watts wrote Colman that Whitefield "has not yet made any motion towards joining with the Dissenters, which we are much best pleased with, for we fear it would hinder his usefulness among the Church of E[ngland]." Four days later Watts explained to Colman why the Dissenting ministers were Whitefield's "cooler friends":

> His narrow zeal for the Church of England as a party, and some imprudencys, made him less accepted here in the beginning of his publick preaching. Besides that I think his services in America have been much more conspicuous and large, and God has now honor'd him to a much greater degree, and therefore

> several of us honor him more now. But it is not fit to discover it too much, lest we should seem to invite him amongst us which we think will attain no good end. I must confess also there are several of us who rather despise than honor him.

Whitefield was orthodox enough to be a friend but too eccentric to be welcome as a member of the family. In the Church, he was an ally; within Dissent, he would be an embarrassment. Watts added that he had not published Colman's account of the beginning of the Awakening in New England because "I well knew the first appearance of the representation of this great work of God in London would be more offensive to many, and less pleasing to most, by dating the spring of it from Mr. Whitefield's name."[44]

The case of Philip Doddridge reveals the impediments to cooperation between Dissenters and Methodists. Doddridge became acquainted with Whitefield in the spring of 1739. Like Watts, he thought Whitefield sincere but rash.[45] Out of a catholic spirit broader than most of his associates' he cooperated more closely with the Methodists and Moravians than was comfortable for other prominent Dissenting ministers. Doddridge was a respected minister and the director of a famous academy. His disgrace would reflect on the rest of evangelical Dissent. When his taking part in services at Whitefield's Tabernacle in London in 1743 was severely criticized, he received advice from friends.[46] John Guyse warned Doddridge to avoid prejudicing belief in all influences of the Holy Spirit by endorsing superficial appearances. Watts said that he, himself, could easily have come into Doddridge's predicament had he not exercised greater caution. "There is a medium of prudence with regard to this sort of connection and acquaintance which," he suggested, "is hard to hit exactly. . . . But at present, I think it is best to keep ourselves as Dissenters, entirely as a separate group."[47]

The evangelical Dissenters, then, considered the English revivals as an affair of the Church of England. They preferred to wait for a more acceptable revival, in the words of John Guyse, "for a plentiful effusion of the divine Spirit, that shall approve itself by substantial and abiding effects to be of God, and shall put a credit upon the Redeemer's name and grace."[48]

Although the leading evangelical Dissenting ministers gave little credence to the Methodist movement in England, Watts, for one, gave full backing to the American and Scottish revivals. Throughout the period of the revivals, Watts believed them to be genuine works of God. He sent narratives of the Scottish revivals to America, and American revival literature to Scotland.[49] He handled the London publication of Jonathan Edwards's works. When reports of the disorders resulting from the Awakening appeared, Watts continued to give the revivals his approval. When, in the autumn of 1742, Thomas Clap, rector of Yale College, asked Watts for his reaction to several pamphlets that detailed the irregularities, Watts referred Clap to Benjamin Colman, a discriminating but firm supporter of the revivals.[50] In February 1743 Watts wrote: "I hope it will not be in the power of all the evill agents, either in this or the infernal world, to stop the work of God. I am glad to hear any peaceful tydings concerning this work among you. I hope it encreases also in Scotland."[51] On 25 May 1743, the annual ministers' convention in Boston adopted the *Testimony* against the errors and disorders of the revivals. The next day the following excerpts of a letter from Watts to Rev. William Shurtliff of Portsmouth, New Hampshire, endorsing the Awakening, appeared in the prorevival *Boston News-Letter*:

> As it is the Labour and Desire of my Life to see the Kingdom of Christ make its Progress among Men, so I am heartily pleased in the latter End of Life, to see the Grace of God break out afresh in so powerful a manner in so many Places of our Plantations in *America*. . . . I am persuaded 'tis the Work of our Lord Jesus Christ making some steps towards his glorious Kingdom, so I cannot but encourage it with all good Words and Prayers. I am very sensible, That in the midst of such Effusions of the Spirit there will be many humane Weaknesses and Imperfections [therefore, discretion is needed].[52]

When the prorevival *Testimony and Advice* of July reached England, Watts had it reprinted in order to meet the British demand for copies. The number of these reprints Watts distributed in Scotland may have amounted to the tens of thousands.[53] In the preface that Watts added, he reaffirmed his belief that, despite the disorders, there had been an extraordinary outpouring of the Spirit of God in America.[54]

In the summer of 1743 Colman sent Watts a copy of Edwards's *Thoughts on the Revival*, to which he had attached a newspaper clipping advertising Charles Chauncy's intention of publishing his *Seasonable Thoughts* in reply. Watts wrote back that he had never received from Colman's hand anything that had made him "so uneasy as that little scrip of the Boston newspaper." He conceded "that these bad things ought not to be allowed, but we need not oppose nor discourage, nor darken the work of grace because such bad things . . . mingle with them." Cooper, Edwards, Dickinson, and Webster had all demonstrated convincingly that the awakenings included "Some uncommon, almighty, converting work." Watts, for his part, said, "I know not what the work of God is on the hearts of men, and I am a stranger to the great business of the pastoral office if God and His grace are not found glorious in these historys which I hear from Scotland and New England."[55] When Colman showed Chauncy one of Watts's letters, apparently the one just referred to, Chauncy sent Watts his *Seasonable Thoughts* and a long letter. Watts wrote Chauncy "a very long answer" to both the letter and the book.[56]

The revivalists perceived the Evangelical Revival as a unified work of God for the revival of religion throughout the British Empire. Hence, they cooperated in whatever ways they thought would promote the movement as a whole. They assumed that opposition to the Revival in one quarter was detrimental to all. Even when denominational rivalry prevented him from openly aiding the movement locally, Isaac Watts went to lengths to support and defend the Revival at large. The attempts of Thomas Clap and Charles Chauncy to change Watts's mind indicate the influence they attributed to his opinion in America.

9

The Morphology of Religious Revival

During the early 1740s supporters of the religious awakenings employed the phrase *revival of religion* in two ways. First, the phrase referred to all the revivals together as a single phenomenon, what Methodists called "the Work." Second, it referred to individual local awakenings. Writers for the Boston *Christian History* spoke of "the late revival of religion" in America, and "particular instances of the Revival of Religion"; "the remarkable revival in this Country," and the "more extensive REVIVALS" of religion; "this remarkable Revival in the Land," and "the late Revival in this Town." By 1743 the phrase had become so common that, without ambiguity, authors could write simply of "the revival," dropping the explanatory "of religion." One Scot called heaven itself "the eternal revival."[1] The dual use of the term indicates the solidification of the concept of a local revival as a particular phenomenon. In describing the individual local revivals, observers created a model by which participants in subsequent revivals could identify each stage of progress through what became a standard morphology.

Revival and Covenant

The British/American concept of a revival of religion had its roots in the notion of the covenant. Conversion of a large number of unassociated individuals did not constitute a revival of religion. Revival meant the transformation by grace of a community, a group of people bound together as a single moral entity, be it social unit, church, or nation, by a covenant, implied or explicit, with God, he to be their God and they to be his people. In New England, two basic kinds of rituals expressed the people's sense of themselves as a community needing revival: Covenant renewal and days of public prayer and fasting. The former focused on the requirement that the people reform themselves, the latter on the people's need for God's help.

In the decades immediately following the Stuart Restoration, British and American clergymen told their people that the grace that God had withdrawn from the community because of its sins would not be restored unless the community repented and reformed. In the 1670s the New England clergy instituted covenant renewal as a mechanism to bring about repentance and reformation in preparation for an outpouring of grace on the people. Increase Mather proclaimed "it is a known Principle . . . *That renewal of covenant is the way to attain Church Reformation*."[2] By the early eighteenth century, however, the clergy had concluded that no attempt at reformation would succeed until there was an outpouring of grace. Revival must precede reformation. In 1705 when religion revived at Taunton, in thanksgiving the people renewed their covenant of reformation made during King Philip's War. Yet,after religion revived at Taunton in 1741, the covenant was not renewed. Only one account of the New England revivals of the 1740s mentions renewal of covenant. At Northampton, in March 1742, Jonathan Edwards "led the People into a solemn publick RENEWAL of their COVENANT with GOD." The covenant, which Edwards had drafted, implied that it was the revival that made reformation possible:

> Acknowledging GOD's great goodness to us . . . in the *very late* spiritual *Revival*. . . . We do *this Day* present our Selves before the LORD, to renounce our evil Ways, and put away our Abominations from before GOD's Eyes, and with one Accord, to *Renew our Engagements* to seek and serve GOD.[3]

Nathaniel Leonard, of Plymouth, agreed that revival must precede reformation. Disturbed by the "Growth of *Impiety, Prophaneness, Sabbath-breaking, Gaming, Tavern-haunting, Intemperance*, and other Evils," his congregation had kept a day of fasting and prayer for an outpouring of God's Spirit year after year. The town authorities tried to stop the growing intemperance by closing the taverns at 9:00 P.M. and by punishing the disorderly persons who frequented them, "but all the Methods used one Way and the other, proved of little Effect." After the outpouring of grace during the Great Awakening, however, for several months scarcely anyone but travelers could be found at Plymouth's taverns. Leonard, if not the tavern owners, drew the lesson: "Nothing but a Stream of Grace from that Fountain where all Fulness dwells, can maintain and carry on a Work of Reformation against the Devices of the Devil, the Snares of the World, and the Opposition of Mens Hearts."[4]

During the decades following the turn of the eighteenth century, prayer for revival gradually took the place of covenant renewal as the leading communal device for promoting revival of religion among the Reformed churches of New England. Perry Miller believed, on the contrary, that the revivals of the Great Awakening represented the culmination of the tradition of covenant renewal.[5] At the end of the summer of 1722 the churches of Boston agreed to keep in each successively "a Day of Prayer and Fasting to ask of GOD the Effusion of his HOLY SPIRIT." In the summer of 1734 the Boston churches observed a similar course of days of prayer and fasting for the same purpose. And in January 1742 most of the Boston churches held yet another course of observances to thank God for spiritual blessings already received and to request "the more plentiful Effusion of the HOLY SPIRIT." Inspired by Edwards's *Faithful Narrative*, John White, of Gloucester, Massachusetts, persuaded his church to "set apart a *Day of Fasting and Prayer*, to wait upon GOD for this Blessing; *viz*. that the Dews and Showers of the HOLY GHOST might fall upon us." White reported, "GOD did speedily and plentifully answer our Prayers." The outbreak of the revivals in 1740 moved many congregations to set aside days to pray that the shower of grace might touch them. After the visits of Whitefield and Tennent, Portsmouth, New Hampshire, and the neighboring towns agreed to a monthly fast to seek revivals. Days of prayer and fasting preceded the revivals at Gloucester, Halifax, and Middleborough, Massachusetts. The first evidences of the revivals at Portsmouth, New Hampshire, and at Wrentham, Massachusetts, appeared during fast day services.[6]

Opponents of the revivals in New England agreed with supporters that reformation must come as a product of an outpouring of grace. They denied, however, that any real reformation had occurred, and they believed that the immoral effects of the revivals far outweighed any good that had come from them.

In Scotland, where a group of evangelicals had reason to oppose the revivals, the case was different. Staunch traditionalists, Seceders retained the notion that the church must reform before the Spirit would return. The question of the sequence of reformation and revival became a debating point between pro-revivalists and Seceders.

In contrast to New England's evangelicals, Scotland's Seceders re-emphasized the ritual of covenant renewal. Having alerted the people weeks or months in advance, the minister would open the ceremony with a sermon, the names of those subscribing would be announced, and then the covenant and an acknowledgement of sins and engagement to duties would be read. Prayers of confession, administration of oath, an exhortation, and signing of the covenant followed, accompanied by the singing of psalms. In these ceremonies, the Seceders renewed their commitment to moral reform, Knoxian Presbyterianism, and opposition to religious innovation. They frowned on the kinds of emotional displays found in the revivals in the non-Secession churches.[7]

In 1742, James Fisher, one of the Secession's principal spokesmen, attacked the revivals as bogus because they had not been preceded by a recommitment to the Solemn League and Covenant and a national reformation. God would not bless a church, he asserted, that was still corrupt and unfaithful.[8] John Willison answered the Seceders in his preface to Edwards's *Distinguishing Marks*:

> When some alledge, God will not be pacified, nor return to us, till we thorowly reform ourselves; it is to say, We must first chuse him before he chuse us; which is impossible, and contrary to the method of God's preventing free Grace: For still God pities and turns to us, before we really and unfeignedly turn to him. He must pour out his Spirit to cause us *mourn and reform*, before we can do either.[9]

In March of 1743 Willison publicly addressed Fisher. "D. Sir." he wrote,

> Do you ever expect to see National Reformation till you see Personal? And do you ever expect true personal Reformation until the Spirit concur with the Word for convincing and converting Sinners to God? . . . Tho' you and others in this divided State should attempt to renew our Covenants, that will not bring about a *National Reformation*, until there be a pouring out of the Spirit upon all Ranks.[10]

In April, when Fisher told two tradesmen, converts of the revivals, that "God will never return to a Church going on in backsliding till they return to him," one of the tradesmen replied,

> Ye are far out Sir: . . . God returned to the Church of the Jews before they returned to him. . . . Does not the Lord say by the Prophet (Isa. 57. 17, 18, 19) for the iniquity of his covetousness was I wroth I smote him and hid my face from him, but he went on frowardly in the ways of his own heart. I have seen his ways and will heal him.[11]

This lay supporter of the revivals had learned his lessons well: Revivalism rested on an emphasis, in the paradoxical logic of the covenant, not on man's faithfulness to the covenant's conditions, but on God's promises to his people to pour out his Spirit on them.

The Revival Narrative

A revival meant the outpouring of the Spirit not only on numerous individuals but also on entire communities. As much as the revivalists were interested in the individual—and in mapping the conversion process in all its variations—

they were also excited by the transformation of their communities. It was in the revival narrative that the revivalists expressed their concern with the town as a moral unit. Just as the conversion narrative traced the rise and progress of grace in the soul, so the revival narrative traced the rise and progress of grace changing a town from sin to holiness. The focus of Edwards when writing his *Faithful Narrative* had been the transformation that had taken place in Northampton as an entity. In the letter to Benjamin Colman of 30 May 1735 that preceded his full *Narrative*, he wrote:

> Then a concern about the great things of religion began . . . to prevail abundantly in the town, till in a very little time it became universal throughout the town. . . . This town never was so full of love, nor so full of joy, nor so full of distress as it has lately been. . . . There is an alteration made in the town in a few months that strangers can scarcely conceive of. . . . The town seems to be full of the presence of God.[12]

The final text of *A Faithful Narrative* gives a fuller account of the history of the spiritual conditions of Northampton before the revival and of the changes effected by the revival. At the end of the narrative Edwards writes: "We still remain a reformed people, and God has evidently made us a new people. . . . And we are evidently a people blessed of the Lord!"[13]

Twenty-five American revival narratives were written and printed between 1741 and 1745. Not counting the two that were continuations or further accounts (see Appendix 2), there were twenty-three separate narratives: in New England, thirteen from Massachusetts, three from Connecticut, one each from New Hampshire and Rhode Island; in the Middle Colonies, two from New Jersey, one from Pennsylvania, and two concerning revivals in both New Jersey and Pennsylvania. All but one were printed in *The Christian History* (Boston, 1743–1745). John Rowland's narrative, although intended for the magazine, was finished too late to make the last issue. Samuel Blair's on the revival at New Londonderry, Pennsylvania, was printed in *The Christian History* and also separately at Philadelphia. The majority seem to have been modeled on Edwards's *Faithful Narrative*. Two of the narratives were written well before *The Christian History* was even contemplated. Some were addressed to William Cooper, who in his preface to Edwards's *Distinguishing Marks*, 20 November 1741, had suggested that revival accounts be collected and compiled into a narrative on the model of the *Faithful Narrative*. He thought such a work "would be one of the most useful pieces of Church History the people of God are blessed with. Perhaps it would come the nearest to the Acts of the Apostles of any thing extant."[14] Most were written in response to the call for such accounts by the first issue of *The Christian History*, 5 March 1743. The narratives were relatively brief, ranging between two pages and forty-four, and averaging thirteen. Each narrative, of course, bears the character of its author; however, when they are considered together, a pattern emerges.

The typical revival narrative had most or all of the following elements, all possessed by Edwards's *Faithful Narrative:*

- A brief history of the town concerned—when it was founded, its population, its spiritual state on the arrival of the present pastor;
- A statement that, although the people are generally sober and honest, and had kept up external religion, the power of godliness had for a long time been in little evidence; religious formality had set in; the prevailing vices, usually frolics and mischief among the young, tavern haunting and contentiousness, might be mentioned;
- A description of the beginning of the revival—what the awakening influences were, and how the awakening spread through the town; the frequent sermons, and the crowded religious conferences;
- A description of the moral reformation in the town—the replacement of frolics and tavern haunting with religious meetings; the ending of feuds and the increase of family religious observance;
- A brief analysis of the kinds of persons awakened—age, sex, previous religious and moral condition;
- Descriptions of the various experiences of the awakened—the various degrees and duration of terror, the various causes of joy, and so on;
- Comments on the extent and prevalence of extraordinary bodily effects;
- Comments on disorders in practice and errors in doctrine, and the measures taken to correct them;
- Accounts of the experiences of individual converts;
- Mention of the extent to which new convictions continue or have ceased to appear;
- An attestation to the evidence that the revival is a genuine work of God.

The narrative writers used the revivals as laboratories for observing the morphology of conversion. They recorded every nuance in the experiences of the awakened, and identified elements that appeared to be universals. The revivalists seem to have been eager to learn about the conversion process from the experiences of the awakened.

Just as the revivals were laboratories for observing the morphology of conversion, so the Awakening was a laboratory for observing the morphology of the revival. The evangelicals saw the local revival as a process. It had a beginning, built to a climax, had a finite duration, and subsided. In a number of ways, the progress of revivals in congregations and towns paralleled the morphology of conversion. As James West Davidson notes, for instance, just as before conversion the individual had to be brought to see the depth of his sinfulness, the justice in his condemnation to eternal punishment, and his inability to save himself, so too, communities were revived only after times of spiritual declension and deadness.[15] Concerning the beginning of local awakenings, the narrative writers recorded the apparent mediate causes and made conclusions concerning the relative usefulness of various means of promoting revivals, including the news of revivals elsewhere, the visits of itinerant preachers, and the example of others. Revivalists seized on the usefulness of the news of revivals in spreading the awakening to justify printing the narratives. The observation that the same means that sometimes produced no results would at other times produce extraordinary results confirmed the belief that the awakening was brought about only when the influences of the Spirit accompanied the means of grace. The first signs of a revival were often an uncommon attentiveness to sermons and a decision by the young people to lay aside "their customary Tavern-haunting, Frolicks, and other youthful Extravagancies" in favor of societies for prayer and religious discussion. The narrators would describe the evidence of building emotion and the transformation of the towns into little Zions. The towns appeared most clearly as holy communities in two ways. First, they resembled holy communities in terms of external morality: "We are almost in every Respect a reformed People," wrote Samuel Allis, of Somers, Massachusetts; "Quarrels and Contentions between Neighbour and Neighbour which have subsisted for many Years, and no Means could effect a Reconciliation, are now at an End";[16] "The very Face of the Town," said Thomas Prince about Boston, "seem'd to be strangely altered."[17] Second, they resembled holy communities in the gathering together as a people of God for worship: "To see them at these Seasons," thought John Porter, of Bridgewater, Massachusetts, "chanting forth the Praises of our glorious REDEEMER: They do in some low Degree resemble the humble Worshippers of Heaven." "O," he exclaimed, "it looked probable now that the whole society . . . would have . . . received a whole Christ with a whole soul."[18] The narrators identified the sources of contaminations to the revivals, the disorders and doctrinal errors. And they noted the time when the appearance of new convictions began to fall off and the causes of the decline of general concern.

In Scotland, two revivalists, William McCulloch and James Robe, collected materials for the study of the morphology of conversion and the morphology of revival more deliberately and systematically than anyone had. Edwards's *Faithful Narrative* influenced both men.

From the start of the revival at Cambuslang, McCulloch adopted a careful procedure:

> After Sermon is over, the Minister retires to his own House, attended with as many of those that are under Convictions, as the House can contain, where he takes down in Writ the Names of the new ones, with their Designations, Place of Abode, Time and Manner of their being seiz'd.[19]

McCulloch probably undertook this practice in order more easily to keep track of the spiritual progress of each of his numerous awakened parishioners. By 1743, however, he had begun transcribing in their own words the conversion accounts of the subjects of the revival, with the intention of editing and publishing them. In all, there are narratives of 106 individuals. Most of the accounts were recorded in 1743 and 1744, although one is dated as late as 1749. Four ministers, Thomas Gillespie, James Ogilvie, Alexander Webster, and John Willison, gave McCulloch editorial assistance. McCulloch never realized his ambition of publishing these accounts, and they remain in manuscript.[20]

At Kilsyth, James Robe also kept a case book that summarized the progress of several of the distressed to whom he ministered in private. "I had a prevailing Inclination from the Beginning," he explained,

> with all the Exactness I was capable of, to observe every Thing that past; and with the most scrupulous Niceness, to examine every uncommon Circumstance, and to take down Notes of what appeared to me most material. . . . This hath issued in a JOURNAL of what was most observable in the Case of many in this Congregation; who have applied to me from Time to Time, for Instruction and Direction under their spiritual Distress.[21]

In McCulloch's *Glasgow-Weekly-History*, Robe found at hand a vehicle for publishing his "Journals." The first appeared in early June 1742. By the end of the summer, ten had been printed.

Edwards's accounts of Abigail Hutchison and Phoebe Bartlett in the *Faithful Narrative* may have been the stimulus for Robe's Journals and McCulloch's conversion accounts. Robe clearly modeled his *Faithful Narrative of the Extraordinary Work of the Spirit of God, at Kilsyth*, on Edwards's *A Faithful Narrative of the Surprizing Work of God*.

Several short revival narratives had already appeared in Scotland (see Appendix 2). Shortly after an antirevival narrative, *A Short Account of the Remarkable Conversions at Cambuslang*, was published in early March 1742, two narratives came out in defense of the Cambuslang revival. The first, *A True Account of the Wonderful Conversions at Cambuslang*, published at Glasgow in early April, did no more than assert that the revival was genuine. The second, *A Short Narrative of the Extraordinary Work at Cambuslang*, published at Glasgow in early May, was a more carefully considered defense. It had the attestations of McCulloch, seven other ministers, and five elders. This narrative itself is traditionally attributed to James Robe, but according to the author of *A Letter from a Gentleman in Scotland to His Friend in New England*, it was prepared by McCulloch and John Maclaurin with the advice of several others.[22]

In contrast to these short pieces, written in a brief amount of time, in answer to a specific attack, Robe's "Kilsyth Narrative" was book-length, and published in segments at intervals between 1742 and 1745, with a final installment, evaluating the lasting effects of the revivals, published in 1751. The contents and the organization follow those of Edwards's narrative.

The narratives by Edwards and Robe begin with a description of the town and of its inhabitants. Compare, Edwards: "The people of the county, in general, I suppose, are as sober, and orderly, and good sort of people, as in any part of New England"; Robe: "The people of the said parish . . . are, for the most part, of a discreet and towardly disposition." Edwards: "Take the town in general, and so far as I can judge, they are as rational and understanding a people as most I have been acquainted with: many of them have been noted for religion, and particularly have been remarkable for their distinct knowledge in things that relate to heart religion and Christian experience"; Robe: "The most of them . . . have attained such a measure of knowledge of the principles of religion as renders them inferior to few of their station and education." Each author devotes several paragraphs to a history of the spiritual conditions in his town, Edwards beginning with the ministry of Stoddard, and Robe with the start of his own ministry in 1713. Edwards tells of Stoddard's "harvests" and how "after the last of these came a far more degenerate time (at least among the young people)." Robe reports "for several years" his parishioners "appeared to profit under the gospel ordinances, by the blessing of the Lord upon them," but "after this the state of religion declined and grew every year worse with us. . . . The younger set attained indeed to knowledge, . . . but I could observe little of the power of godliness in their lives." In Northampton "licentiousness for some years greatly prevailed among the youth of the town; they were many of them very much addicted to night-walking, and frequenting the tavern, and lewd practices." In Kilsyth, even those who professed grace "seemed sensibly to degenerate into a negligence and indifferency about spiritual things, and some of them into drunkenness and other vices."[23] Next each author describes the circumstances leading to the awakening. In the year 1734 Edwards began preaching on justification by faith alone. In the year 1740 Robe "began to preach upon the doctrine of regeneration." In both towns, the meeting together of the young people for prayer and conference marked the beginning of the awakening. News of the awakening first among the youths and then among the adults of Northampton helped spread the revival through the other towns of the Connecticut Valley. Knowledge of the revival in Cambuslang helped bring the revival to Kilsyth. Both authors described the transformation of the town into a more godly community, and the spread of the revival to other towns.[24]

The remainder of Edwards's narrative is devoted to a lengthy analysis of the experiences of the awakened, a description of the experiences of two exemplary converts, and concluding remarks on the end of the revival, and its lasting effects. The continuation of Robe's narrative follows a similar pattern.[25] Robe designed his own categories, but he followed Edwards in dissecting the varieties of conversion experiences. Indeed, both authors were fond of the word *variety*. Whereas Edwards offered the models of two exemplary converts, Robe drew from his journals to provide illustrations for each of his categories of converts.

In their private letters, reformed ministers throughout the British world had become practiced in describing the state of their spiritual charges. Edwards brought that expertise to fruition in the *Faithful Narrative*, analyzing the morphology of the transformation of a community by God's grace, much as the conversion narrative describes the progress of grace in the individual. In Scotland, before the Evangelical Revival, accounts of revivals, such as those in western Scotland in the 1620s and 1630s, had appeared in either national religious histories, such as Robert Flemings's *The Fulfilling of the Scripture,* or lives of exemplary ministers,[26] neither of which genre focuses on the local community as a moral entity in the way the revival narrative does. Robe justified publication of the Kilsyth narrative on the basis of the complaints he had heard of the "omission of our worthy Forefathers to transmit to Posterity, a full and circumstantial Account" of the Communion of Shotts in 1630 and the other revivals in the west of Scotland of that earlier era.[27] With his *Faithful Narrative*, Edwards established a new religious genre, creating the model of American revival narratives for a century to come. The Scots imported the genre along with the awakening.

An American plant that flourished in its native New England soil where the strong sense of communal identity of the towns provided an appropriate climate, the revival narrative successfully transplanted to Scotland. It did not take root in south Britain, however. The genre was inappropriate for itinerant evangelists such as George Whitefield. They were seldom among one community long enough to observe the entire process from start to finish. Instead, they wrote journals of their travels. The journal and the autobiography of the itinerant evangelist were the typical narrative forms of the Methodist movement.[28] The Methodists did write descriptive narrations of local revivals in letters, such as "A Letter from Mr. Joseph Humphreys, to the Religious Society belonging to Deptford and Greenwich; containing an Account of the Work of God there,"[29] which were read at the monthly "letter days" of the Religious Societies, but these narrations tended to be episodic and to lack the unity of scheme of most American narratives. In America the revivals often occurred within congregations that were nearly synonymous with a precinct or town, whereas in England the revivals usually took place among religious societies called out from and opposed by the community at large. Hence, the English Methodists were more likely to write about their experiences as a despised minority than about the moral transformation of a community.[30] Whitefield and Wesley both claimed the motto, "The world is my parish." Between the religious society and the nation at large, the Methodists had no intermediate community to treat as a moral unit.

The Spirit's Presence with the Community

In theory, a revival was more than a conjunction of numerous conversion experiences. It was the outpouring of grace for the reformation of the community. Boston's Charles Chauncy could accept such a definition of a revival. He had awaited the outpouring of the Spirit for the revival of religion. "O let us not be insensible of the sad low state of religion among us!" he had pleaded in 1737. "Let that be our ardent prayer to God, the God of all grace. O Lord, revive thy work in the midst of the years."[31] On 13 May 1742, when Boston's First Church observed a day of prayer "to ask of God the effusion of his SPIRIT," Chauncy, the associate minister, preached on *The Outpouring of the Holy Ghost* and called for prayer "that there might be a revival of true primitive Christianity."[32] For Chauncy, however, a revival was recognizable in moral reformation. He turned against the awakening when he concluded that it consisted more of heated passion than of moral transformation.[33] "Han't the Talk of a *Revival of Religion*," he asked, "arisen *more* from the *general* Appearance of some Extraordinaries, (which there may be where there is not the *Power of Godliness*) than from such Things as are *sure Evidences* of a *real Work of God in Men's Hearts?*"[34] He found in scripture the identifying marks of an outpouring of the Spirit:

> The outpouring of the SPIRIT is again spoken of, where GOD promises to give his People a *new Heart*, and to *put his* SPIRIT *within them*. And what is the Effect, the *visible appearance*? It follows in the next Words, *Ye shall walk in my Statutes, and ye shall keep my Judgments and do them.*[35]

While conversion was a change in the "inward frame of mind," it could be judged only by a change in "the outward course and manner of Life." Just so, said Chauncy, an outpouring of the Spirit could be judged only by the visible sanctification of the community.[36]

For the revivalists, in contrast, the outpouring of the Spirit was more directly experiential than simply the presence of an abundance of grace making possible the moral improvement of the community. They thought of the outpouring of the Spirit as being immediately sensed by the gathered people during religious functions. "God is verily with us in our religious Meetings," wrote John Porter, pastor at Bridgewater, Massachusetts. "Tis frequent on *Lecture-Days* and on *Lord's Days*, while we are supplicating the *Divine Majesty*, singing the high Praises of GOD, hearing his Word, celebrating the holy Supper, that we see some of the . . . Influences [of God's Spirit.]"[37] On 21 May 1742, James Robe, of Kilsyth, Scotland, wrote: "Sabbath last we were surpriz'd with a great and uncommon out-pouring of the Spirit from on high; a numerous Congregation were brought to a deep Concern, were all in Tears."[38] For these evangelicals, the outpouring of the Spirit was a moment of intense, shared emotion. That moment could be pinpointed, as when George Griswold described the revival at New London, Connecticut, North Parish, December 1741:

> I preached a *Lecture* in the *Evening* of the *same Day*; and there seem'd a very great *Pouring out* of the SPIRIT; Many were in great Distress. . . . On *Monday* I preached again at the Meeting-House; and there seemed to be a great *Pouring out* of the SPIRIT of GOD; and many in Distress. . . . Within the Space of about two or three *Minutes* after the *Blessing* was given, there seemed to be a wonderful *Outpouring* of the SPIRIT: many Souls in great Distress.[39]

At Gloucester, Massachusetts,

> the first most visible and powerful Effusion of The Spirit was on the last Sabbath in January [1742] and especially as I [John White] was preaching in the Afternoon and on the Evening in two religious Societies in the Harbour; many were impressed both with Distress and with Joy above Measure.[40]

White spoke of the outpouring as taking place at specific moments, and as being witnessed almost as clearly as the tongues of fire that appeared over the Apostles at Pentecost.

Sometimes the metaphor was changed to the image of the wind that accompanied the tongues of fire in the account of Pentecost in Acts 2:2. A witness of an evening lecture at Campsie, Scotland, given by Reverend James Burnside, of Kirkintilloch (a suburb of Glasgow), wrote that

> a little after he began, the Spirit of the Lord, like a mighty rushing, Wind, fill'd the House in such a Manner, that almost the whole Congregation was in a Flood of Tears, accompany'd with bitter Outcries, by several immediately awakened.[41]

The comparison of these "special Seasons of divine Influence" to Pentecost was a constant. Jonathan Parsons, of Lyme, Connecticut, said: "I can't pass over our Penticost, on the 11th Day of the following October [1741] . . . especially when the Lord's Supper was administered. GOD pour'd out his Spirit in a wonderful Measure."[42]

Jonathan Edwards noted that the revival of 1740–1742 in Northampton differed from the revival of 1734–1735 in that the conversions were frequently wrought more sensibly and visibly, so much so that a bystander could often almost observe, step by step, the progress of conviction. He noted further that "these apparent or visible Conversions (if I may so call them) were more frequent in the Presence of others, at religious Meetings."[43] The people sensed the revivalists' expectation of communal expressions of religious feelings. Whitefield made it clear that he wanted his auditors to react in an observable manner to his preaching. When they failed to do so, they disappointed him. Neshaminy, New Jersey, 22 November 1739: "I began to speak. At first the people seemed unaffected, but in the midst of my discourse, the hearers began to be melted down, and cried much." Greenwich, New Jersey, 21 April 1740: "At first I thought I was speaking to stocks and stones; but before I had done, a gracious melting was visible in most that heard." Revivalists and supporters believed that if the Spirit were active in an assembly, he would manifest himself. Whiteclay Creek, Pennsylvania, 2 December 1739: "I preached a second time. . . . God magnified His strength, and caused His power to be known in the congregation. Many souls were melted down." 30 April 1740, New York: "Towards the conclusion of my discourse, God's Spirit came upon the preacher and the people, so that they were melted down exceedingly."[44] In Scotland, John Maclaurin believed he could observe the presence of the Holy Spirit in Whitefield's assemblies. On 8 October 1741 he wrote, "I have my self been witness to the Holy Ghost falling upon him and his Hearers ofter then once, I don't say in a miraculous, tho' observable Manner."[45]

Jonathan Edwards shared the belief that the special presence of the Spirit could be discerned within an assembly. "How common a thing has it been for great part of a congregation to be at once moved, by a mighty invisible power?" he asked.[46] Edwards promoted this attitude. In *Thoughts on the Revival* he defended ministers who have been

> blamed for making so much of outcries, faintings, and other bodily effects; speaking of them as tokens of the presence of God, and arguments of the success of preaching; seeming to strive to their utmost to bring a congregation to that pass, and seeming to rejoice in it . . . when they see these effects.

Edwards answers these criticisms directly: "For speaking of such effects as probable tokens of God's presence, and arguments of the success of preaching, it seems to me they are not to be blamed; because I think they are so indeed." By experience and investigation, Edwards had found that those who cried out in such circumstances usually gave good evidence of being under the influences of God's Spirit. Furthermore, such effects were inevitable when God visited a gathered people:

> Though the degree of the influence of the Spirit of God on particular persons, is by no means to be judged of by the degree of external appearances, because of the different constitution, tempers and circumstances of men; yet, if there be a very powerful influence of the Spirit of God on a mixed multitude, it will cause . . . visible commotion.[47]

Edwards's defense gave color to Chauncy's accusation in his reply that "so high were the People in their Opinion" of the commotion in some of the houses of worship

> as a *Sign* of the *extraordinary* Presence of the SPIRIT with them, that if you talk'd with them to shew them the *Indecency* of such Carryings on, they would only pity you, and speak of you, as poor *carnal Sinners*, destitute of the SPIRIT, and in the *broad Way to Hell*.[48]

In May 1743, *The Testimony of the Pastors of the Churches in Massachusetts* expressly condemned the notion that the demonstration of strong emotion within an assembly indicated the special presence of the Spirit.[49]

Before the Great Awakening, theorists had contended that an outpouring of the Holy Spirit would produce several desirable effects. Saints would grow in holiness, many sinners would be converted, and the unconverted, through the influence of common grace, would improve in the outward performance of duty. The growth in grace of individuals would become visible in the moral improvement of the community as a whole. Not the least important result of this reformation would be a renewed respect and even love for the ministers of the Gospel. In its initial stages in New England, the Great Awakening seemed to be fulfilling all these expectations. Revivals were measured by the number of new converts. Quarreling neighbors patched up their differences. Factions in towns appeared to dissolve. The people gave up their "frolics" and "merry meetings" in favor of meetings for prayer and pious reading. The pastor's chamber replaced the tavern as place of common resort. The ministers were suddenly subject to demands for frequent public and private lectures, and for private and group conferences. The minister's importance as spiritual leader was once again recognized by his flock. He was loved: "As I was turning from the Crowd," reported Peter Thacher, of Middleborough, Massachusetts, "many whisper'd thus in my Ears, '*O my dear Minister* I never loved you before, but now I do.'"[50] Benefiting directly from them, at first most settled ministers in New England supported the revivals.[51]

As the Connecticut Valley revivalists of the 1720s had argued must be the case, the renewed respect for the minister was based not on his office, but on his charisma: the presence of the Spirit with him, that is, his ability as a revivalist. The Spirit was expected to be poured out on ministers, making them able preachers and pastoral counselors, giving them gifts to reach the hearts of the people. Herein lay a problem, for the revival did not raise the prestige of the clergy as a group, but only as individuals. Only those ministers who appeared to have been given by the Holy Spirit the power to preach for conviction and to guide the awakened to conversion won the peoples' affection. Ministers who shied away from the more vigorous styles of revivalist preaching, who failed to evoke an emotional response, or who frowned on commotion in the meetinghouse found themselves excluded from the affection of the people and even denounced as carnal legalists. Discord followed, with the preacher the *causa belli*. Congregations divided over loyalty to their pastors. People left their own preachers in order to hear men from whom they expected to get more good. Itinerant evangelists invaded the parishes of suspected carnal ministers. Even laymen were taking upon themselves the minister's function of expounding the Word, claiming extraordinary gifts. If charisma was the necessary quality of a good preacher, what use was human learning? Several quickly concluded that such disorders could not be God's doing: The revivals were delusions produced by enthusiasts appealing to the animal passions of the people, the "conversions" mainly hysteria, and the reports of moral reformation exaggerated. They concluded that the Great Awakening was not an outpouring of the Spirit because it did not produce the expected results. It undermined, rather than enhanced, the authority of the clergy. As a communal experience, it promoted not harmony but discord.[52]

An Evangelical Understanding of Revivals

The progress of the Great Awakening seemed to fulfill John Howe's 1678 vision of *The Prosperous State of the Christian Interest Before the End of Time, By a Plentiful Effusion of the Holy Spirit*, spreading holiness outward from one transformed place to whole regions and countries, like ripples around a stone dropped in a pond. In Howe's vision of the millennium, first comes surprise at the changes in men in some locality, a whole city for instance, turning from sin to holiness. News of the event travels abroad. Others investigate, approve of the transformation, and apprehend God's hand in it. This leads to favorable inclination toward religion. Serious religion comes into credit, and the revival of religious concern spreads.[53] Fifty more years of revivalist speculation and experimentation after Howe's work informed the evangelicals' understanding of the religious revivals of the Great Awakening in Great Britain and America. During that period, two developments, in particular, supported the evangelicals' understanding of the process of revival: one, justification of the expression of pious emotion in public worship and of impassioned preaching and soul-stirring singing for that purpose; and two, the accentuating of the doctrine that the community must experience grace before it can reform, leading to the precedence of prayers for the outpouring of the Holy Spirit over covenant renewal. These two developments placed emphasis on the local community, the locus of preaching, worship, and prayer, led by its minister. Setting the local community in the center of the process of the revival of religion led, naturally, to the revival narrative.

In the first half of the eighteenth century the phrase *revival of religion* referred both to the general return through the society to the full complement of phenomena associated with vital piety—widespread interest in religion, frequent conversions, a high level of private and public morality, active missionary enterprise, and so on—and to a period of intense personal concern for salvation within a community—what Stoddard called a spiritual "harvest." Before the Evangelical Revival, evangelicals often discussed a revival of religion in the former sense, without reference to local revivals; by the early 1740s, the revival of religion and local revivals had become, among the evangelicals, intimately associated.

10

Laity, Community, and Revivals: The Scottish Experience Understood from a New England Perspective

In contrast to clergymen's revival narratives, which trace the progress of grace in awakened communities, lay persons' accounts of the revivals are primarily conversion narratives that trace the progress of grace in their own souls. Nevertheless, through lay accounts it is possible to discover the roles of community in the spiritual life of the awakened and to deduce something of the meaning of a revival to the evangelical laity. There is little to suggest that the laity's understanding of a revival was fundamentally different from that of their pastors.

The significant collection of lay persons' accounts of their experiences in the revivals of the 1740s contained in the McCulloch Manuscripts allows an intimate and detailed study of the laity's point of view. A reading of these accounts in the light of the revivals in New England provides a better understanding of the Scottish experience and of the transatlantic movement as a whole.

Conversion and Community

Conversion was a deeply personal, individual experience, but a revival was in its essence communal. The congregation gathered to hear a sermon, the group conference of the awakened after the revival service, the society of young people for prayer and pious reading, and the company of travelers singing psalms in the streets were focal points of the movement. These groups both stimulated and provided an outlet to express shared emotions.

Two remarkably similar dreams or visions, one by a Connecticut farmer and one by a Scottish youth, both converts of the Revival, underscore the importance of communal revival exercises in the personal religious experiences of the subjects of the revivals. At Cambuslang, Mary Colquhoun, an eighteen-year-old daughter of a tenant farmer, reported to McCulloch,

> When I was lying in my bed, I was awaked with the most delightful sound, as of a great company of people singing Psalms: and lay listning to with great pleasure after I awaked for about the space of an hour. And was sorry when it ceased. But tho' I thought the Sound was if the Company singing had been round about me in the place where I was, and did not stop as we usually do, when we sing Psalms, but went on in one continued Song of praise, yet I could not understand any of their words.[1]

Nathan Cole, a farmer of Kensington, Connecticut, converted under the influence of Whitefield's preaching, recorded that,

> One morning I . . . dreamed that I was going to a meeting of the Saints, and that when I came near the house I heard such melodious singing that it ravished my Soul; . . . there was a multitude of people Circled in with great glory all their faces one way hundreds or thousands and they sang so gloriously that no tongue can any way express it to man; their voice was with great power and Strength: but not louder than common; their voice was as strong at the beginning and end of a note as it was in the middle of the note—I could hear every voice Clear and distinct at the same time; and yet hear them all at once and no hindrance at all, every tongue begins the note as one and Ends as one; for there was no stop or pause to take breath every one had a cheerful voice and countenance; Great Glory and joy sat upon their brow; . . . and whilst I was trying to learn their song I awaked and remembered their last words which were these— And in the name of Jesus Christ Jerusalem shall stand for ever, Stand for ever, stand for ever.[2]

Differences between the Scottish youth's and the Connecticut farmer's experiences mirror disparities between their musical cultures. The saints' musicality in Cole's dream would delight any modern choirmaster and reflects the revolution in performance of church music that had taken place in New England. That the company in Colquhoun's account makes "delightful sound," sings psalms, and generates words that are unintelligible, in contrast to the "melodious music," hymns, and impeccable enunciation of Cole's saints, suggests the traditional, folklike mode of Scottish Presbyterian church music. Nevertheless, the similarities between the two descriptions are more striking than the differences. Whether of dream, imagination, hallucination, or vision, the descriptions suggest the psychological effect of actual participation in or witnessing of group performance of religious song. "The singing of Dr. WATT'S *Hymns*," wrote John White, of Gloucester, Massachusetts, "is the chief Recreation of Christians when they convene."[3] The Cambuslang conversion narratives are replete with evidence of the importance of psalm singing in the spiritual experiences of the awakened there. Union in prayer, especially sung praises, reinforced the feeling of belonging to a select group of saints and augmented the convert's joy. Jonathan Edwards defended the prevalence of group singing from its usefulness in stimulating pious emotions. He thought that those who objected to so much singing in religious meetings were those who did not believe that the revivals were the work of God and who thus doubted that the singers were expressing any real love of and joy in God. In the community singing God's praises Edwards saw a foreshadowing of the millennium and an image of the communion of saints in heaven.[4] Both Cole's and Colquhoun's descriptions intimate similar attitudes. Each underscores by implication or directly the joy of the singers. By contrasting the unbroken continuity of the praise in the dreams/visions with the discontinuity of earth-bound performances—"did not stop as we usually do, when we sing Psalms, but went on in one continued Song of praise," "there was no stop or pause to take breath"—the speakers suggest the heavenly host's eternal adoration. And the final words of the hymn remembered by Cole, "in the name of Jesus Christ Jerusalem shall stand for ever, Stand for ever, stand for ever," allude to both heaven and the millennial kingdom and underscore the significance of the communion of saints.

Along with a private experience of personal reconciliation with and joy in God, conversion brought a sense of unity with others who shared similar experiences. The significance of this sense can be illustrated by one who experienced it unusually powerfully, Connecticut farmer Nathan Cole. Cole laid claim to a sort of spiritual telepathy with other saints. "I am no stranger to the saints having Communion one with another when they are some miles apart," he wrote. Once, he related,

> I seemed to feel that a number of Saints some miles from me were distresd and in the dark about some things, and wanted to converse with other certain Saints; and while I was plowing in the field my mind ran so upon them that at unawares I broke out into a prayer with a loud voice not thinking but that I was in the midst of them, and I believe they were then praying that I might be sent to them. I finished my days work, and went and found them together: I had not been there long before there came an other from nine or ten miles of[f] on the same Errand. And said that the saints in that town felt the Distress of the saints in this Town: and sent him to see how it was; and I believe that Saints have been fetched from town to town many a time by the strength of prayer.[5]

Cole recounted a number of other similar occurrences. Because those who had gone through the New Birth of the Evangelical Revival felt that they possessed a new sense of things such that only others reborn could understand—"Saints have meat to eat that the world knows not of. . . . The Saints have that joy that sinners meddle not with"[6]—they longed for communion with fellow saints. The revivalists spoke of the exclusivity of the experiences of converts and of their need of fellowship with one another to share the feelings emanating from those experiences. Samuel Finley explained:

> Now, 'tis from this Sameness of Experiences that their Knowledge of one another proceeds. When they tell each other what GOD has done for their Souls, they understand one another sweetly; and the one sees himself in the other's Heart when it is told him.[7]

It was natural, and accorded with the established practice of fellowship meetings in both New England and Scotland, that during the revivals the awakened would seek counsel concerning their spiritual concerns not only from their pastors but also from "experienced Christians." Uncertain about the validity of his conversion experience, Nathan Cole, for example, first consulted a licensed candidate for the ministry and then studied Thomas Hooker's *Poor Doubting Christian Drawn to Christ*, without resolving his doubts. In Kensington, Connecticut, where Cole lived, conference meetings consisted of gatherings of church members with their pastor where questions of Scripture were propounded and answered.[8] Cole was praying in such a meeting, when he decided to consult an "experienced Christian" who was there. To Cole's relief, this man not only had known a difficulty like Cole's, but also could cite the specific Scripture by which to resolve it.[9] Many of the Cambuslang conversion narratives show the awakened there consulting private Christians known for their piety and wisdom. There is no evidence that the revivalist clergy on either side of the Atlantic found this objectionable. Gilbert Tennent told believers they were obliged to help bring others through the New Birth by answering their questions and applying words to convict or to comfort according to the needs of the individual.[10]

The clergy expected religious revival to transform towns into holy places where people dwelt in harmony. Indeed, the New Birth could produce such a new outlook that old animosities were obliterated. "Now I had for some years a bitter prejudice against three scornful men that had wronged me, but now all that was gone away Clear, and my Soul longed for them and loved them," wrote Cole of the aftermath of his conversion, in a passage that confirms statements by the ministers concerning the effects of the revivals.[11] For the individual, conversion might mean integration into local society. James Walsh sees such a scenario occurring in Woodbury, Connecticut, during the Great Awakening, arguing that at the heart of the revival there was family reconciliation. By converting, the youth of the town were fulfilling the wishes of their parents and becoming responsible and respectable members of church and community.[12] For others, however, conversion might just as well involve separation from local society and its values and entry into membership in a select group. In the revival at Norton, Massachusetts, for instance, John Bumsted believed he found generational conflict. Young men, frustrated in their desire for land, showed up their elders by having spectacular public conversions.[13] In New England perhaps both processes, of integration into local society and of alienation and creation of a new community, were going on, depending on local circumstances. For those who separated from the standing churches to form New Light Separate or Baptist churches, the holy community was not the town but the separate group of saints. Such was the case with Nathan Cole. Cole reports that following his conversion experience he "had in mind the form of A Gospel Church, and the place where it was settled, and Angels hovering over it, saying, the Glory of the town, the Glory of the town, and strangers that came pressing by had the same to say." The town to whose glory he refers is not Kensington, as becomes evident after he leaves the Kensington church in 1747, but the local Separate communion of which he became a lay leader.[14] At the time of his conversion, Cole reconciled with three men who had wronged him. After he joined the Separates, however, many former friends forsook him and he grew contemptuous of the "hireling ministers," of the tax collectors who forced him to support the established church, and of the community leaders and the rich men who sought to see the Separate movement fail.[15]

The meaning of community for lay people in the Scottish revivals was as ambiguous as it was in New England's Great Awakening. The revivals confirmed for many the legitimacy of their society by demonstrating that God still worked through its institutions and that the established church was not corrupt enough to justify secession. Ned Landsman believes that the revivals helped individuals in the emerging commercial crafts, confused by the contrast between commercial values and those of traditional peasant society, by reaffirming their association with their traditional church without challenging the new values required for success in the market place. In these ways, the Scottish revivals meant affirmation of individuals' places in society. Yet, in other ways, according to Landsman, participation in the revivals entailed for the same individuals alienation from traditional society, the forging of new identity, and confirmation of membership in a distinct and separate community.[16]

Lay and Clerical Control of the Revivals

Because Ned Landsman makes a forceful argument for the existence in Scotland of a lay understanding of the revivals dissimilar to that of the evangelical clergy and of a dynamic evangelical lay leadership with objectives different from those of the preachers, it is worth our while to examine his argument and evidence closely.

Observing that the Scottish revivals of the 1740s were concentrated in those geographical areas that had closest contact with the Anglo-American commercial world, Landsman proposes that the changes in the Scottish linen industry, by bringing weavers and spinners into more intimate contact with the commercial market, challenged them to reorient their system of values. Leaders of the newly self-conscious and self-confident community of weavers, he suggests, used the revivals to promote personal characteristics it valued, such as self-esteem, assertiveness, sobriety, gravity, diligence in one's calling, and identification with a select group separate from society at large.[17]

That at Cambuslang we are witnessing the budding of an alicnation from traditional agricultural communal life is as plausible to believe as that we are seeing its full flowering. This process that Landsman believes he discovers operating in the Cambuslang revival others find fully developed only later in the century, when the agricultural and industrial revolutions had made a greater impact on society. In the latter part of the century, dissenting Presbyterian and Methodist churches found their strength among occupational groups that were not tied to landed society and among agricultural laborers who had become alienated from the landowning class.[18]

With the Union of Parliaments in 1707, Glasgow began her rise to prominence as entrepot of the American tobacco trade, and, encouraged by the Commissioners and Trustees for Improving Fisheries and Manufactures in Scotland, set up in 1727, the linen industry began to modernize its technology and rationalize its organization and marketing, particularly in the vicinity of Glasgow. There, on the outskirts of Glasgow, the Scottish revivals would have their beginning and most active center. Still, during the 1730s, the volume of Scottish linen production grew only slightly.[19] By the beginning of the next decade, the gradually rising demand for their skills does seem to have given weavers a growing self-confidence, which they began to exercise in local religious affairs. In the year 1740, for example, in Inveresk, just east of Edinburgh, a body of weavers, said to number eighty, petitioned against the transfer of a minister to their parish because they thought his reasons for desiring the transfer were material rather than spiritual.[20] In contrast to weaving, a valued skill practiced by a self-conscious community, spinning was common, practiced by 80 percent of Scottish women, and an occupation of low esteem. Even so, the rise of the linen industry increased the income of spinners' households.[21] Participation in the commercial linen industry by weaving or spinning lessened families' vulnerability to the wrath of landlords for religious behavior on which they frowned, another reason craftsmen may have been more willing to participate in revivals than were agricultural workers. Landowners, many of whom disapproved of popular religious activism,[22] could punish their dependents for participating in revivals: "The Gentle Man in whose ground I lived," twenty-one-year-old John Wier reported, "sent his officer for me and another Lad. He discharged us to go to Cambuslang: threatening that if we did he would arrest our crop, and turn us out of his Land." Wier's landlord made good on his threat.[23]

Landsman points to the central place of psalm singing in the spiritual life of the subjects of the Cambuslang Work as evidence of the alienation of the converts from the folkways of agriculture society. Several Cambuslang converts forsook the secular harvest song in favor of psalm singing, and a number spurned the "carnality" of group activities of the harvest, sheep shearing, and fairs. In rejecting these traditional activities of agricultural society and replacing them with participation in religious, self-help, and mutual aid societies, the artisan community, Landsman argues, was asserting its new, separate identity and values.[24] In line with these speculations that the psalms were substitutes for harvest songs, folklorist David Buchan argues that as the agricultural and industrial revolutions of the later eighteenth century made society less cooperative and communal in spirit, more individualistic and competitive, daily life became less conducive to the story-telling of balladry. Although the weaver community continued to have its trade songs, by the end of the century they were no longer traditional but were literary productions.[25] It is possible to make too much of the rejection of secular harvest songs, known for their bawdry,[26] as evidence of the alienation of commercial craftsmen from agricultural folkways, for the replacement of secular by religious song was more a Reformed and evangelical phenomenon than it was an artisanal or industrial one. The seventeenth-century Puritan migrants to New England replaced ballads with psalm singing,[27] and converts of America's Great Awakening regularly forswore lewd songs and substituted religious conference for merry meetings.

To buttress his argument that revivalism found fruit at Cambuslang because the population there was adjusting to its sudden entry into the international commercial network, Landsman asserts that soon after 1740 the cloth industry involved "almost half the inhabitants" of Cambuslang. This is most likely an overstatement. According to the *Statistical Account*, in 1791, when the number of inhabitants was 1,288, having grown by 200 since 1783 primarily because of an increase of weavers and colliers, and when the proportion of cloth workers was certainly greater than it had been a half century earlier, the number of cloth workers, (weavers, including journeymen, and cotton spinners, including boys and girls) stood at 170, 13 percent of the total population.[28]

Landsman is without doubt right in observing that the artisan community participated in the Cambuslang revival in numbers far beyond their proportion in the population; however, he seems to be inflating the proportion of weaving families among the Cambuslang converts whose accounts appear in McCulloch's compilation. Landsman can identify the "social origins" of sixty-two of the narrators. He says that, of those, two-thirds, or forty-two, came from the artisan community and half of that group, or twenty-one, from weaving families. In a footnote, however, he identifies weaving families as "weavers *and spinners* or their families" (emphasis added).[29] The only way to approach his numbers of weaving families is to count every female whose narrative mentions her spinning; however, since the great majority of Scottish women included spinning among their other chores, we cannot confidently identify those particular women as primarily associated with the clothmaking industry. In the analysis of the occupational background of the narrators (see Appendix 1) thirty-seven from artisan families were found and of these only ten from families of weavers. If Landsman is correct in saying that Elder James Jackson was a weaver, his three daughters would increase these numbers to forty families of artisans, including thirteen of weavers. Therefore, it would seem that weaving families constituted one-fifth or one-sixth, not Landsman's one-third, of the total whose social origins can be identified.[30]

Whether or not Landsman is correct in finding that the weaving community dominated the Cambuslang Work in numbers, his argument that this community dictated the converts' understanding of the revival is precarious. In seeking to establish the distinctiveness of lay goals, he exaggerates the distance between the clerical and the lay understanding of conversion and misinterprets some major elements of Reformed spirituality.

Landsman contends that the leaders of the artisan community, especially the weavers, made use of the revivals to promote their own goal of fortifying the self-conscious identity of the members of their craft. He may mean to imply as well that the revival was successful among this group because it promoted personal characteristics useful to it. In any case, to find that the lay leadership had its own agenda in the revivals does not preclude the possibility that the revivalist preachers shared much, if not all, of that agenda. Nor does it mean that the goals of the clerical leadership and those of the lay leadership, however different they may have been, were at odds. There is little evidence of overt conflict between the evangelical laity and the evangelical clergy.[31]

That the laity took a leadership role in the revivals is not surprising, for the canard of the Scots being a "priest-ridden" people, raised as a straw man by Landsman, was long ago unmasked by Scottish historians, who recognized the strength of lay power in the Scottish church. Henry Grey Graham wrote in 1899 "the reverse is far nearer the truth, and the ministers may rather be called a 'people-ridden clergy.' "[32] The moral authority of "the Men," lay leaders of the evangelical movement in the Highlands, is fully described in the literature.[33] Revivalist clergy were generally willing to accept, even to encourage, a leadership role for laity in the revivals, while they opposed tendencies among the laity toward behavioral excesses, misinterpretations of the nature of conversion, and doctrinal error. In his defense of the revivals, John Erskine (1721?–1803), then a student at Edinburgh University and later to be prominent in the Popular Party, spoke to this point. To the objection that, whereas the preaching of the Gospel is the appointed means for the conversion of sinners, the revival in a great measure in some places is carried on by private persons exhorting and instructing others, he answered that, although some imprudent particulars may result, Christians that possess such abilities are obliged to exhort and rebuke each other. "There are clear Promises," he said, "that in the latter Days the Zeal of private Christians in exhorting and instructing others will be blest with remarkable Success."[34] American revivalists shared this attitude. New Jersey minister John Rowland wrote:

> One great means that the Lord used there to prevent backsliding, was the care and diligence of some of the Christian people in conversing with the convinced; for several of the Christians were so engaged in deep concern for the work of God, that they could not rest satisfied until they had reason to hope, that the souls who were convinced from one time to another, were also come through to sound conversion.[35]

Henry Messenger and Elias Haven, pastors in Wrentham, Massachusetts, recommended the practice of Christian conference during fellowship meetings.[36] Jonathan Edwards began his *Thoughts on the Revival* by condemning "ministers' assuming, or taking too much upon them, and appearing as though they supposed that they were the persons to whom it especially belonged to direct and determine."[37] By those ministers, he meant primarily those who opposed the revival because of the unusual manner in which it operated. In Edwards' view, it was the Holy Spirit who had chosen to advance his purposes in that manner. The revivalist clergy, claiming the revival not as their own work, but as God's, were willing to allow God liberty to make use of channels in addition to themselves.

Landsman fails to demonstrate convincingly his central thesis that "under the guidance of those master weavers, the Cambuslang converts forged a concept of conversion that diverged dramatically from that prescribed by the preachers." After showing that the awakened at Cambuslang received spiritual guidance from lay persons as well as from clergy, and having asserted that prominent among those lay persons were master weavers and their wives, Landsman identifies beliefs and attitudes in the narrations of the converts he considers at odds with the teachings of the clergy. He then infers that the divergence was a result of the instruction of the lay exhorters. This leads Landsman to the further conclusion that the lay exhorters had greater influence over the conversions than did the preachers. The lay understanding of the conversion experience, however, differed little from the teaching of their pastors, and any divergence of lay views from those held by the clergy cannot be shown to be owing to the influence of the master weavers.

The primary point of divergence Landsman finds is the converts' seeking "sure, external signs of their election, through voices and visions and signs manifest in their own lives."[38] The one piece of evidence he offers that the master weavers "explicitly encouraged" this process is a report of a disparager of the revivals that on one occasion weaver Robert Bowman and shoemaker Ingram More sought to prolong and intensify the experience of a woman who had fainted during a sermon by urging her to hear the chariot wheels of the coming Christ.[39] Yet, there is direct evidence that the lay leaders, like the clergy, believed that such external signs were unauthentic and imaginary, and certainly no evidence of election. In 1751, the Cambuslang kirk session issued an attestation to the lasting effects of the revival of 1742. The document states:

> Though the most of the subjects of the awakening, whose exercise contained a mixture of strong fancy and imagination, are relapsed to their former sinful courses: yet, there are several instances of persons, whose exercises were mixed with fanciful apprehensions, and which they gave out to be real representations of objects and visions, are of the number of those persevering in a justifiable Christian profession, and unblemished conversation.

Among the five elders who signed, recognizing that visions were a corrupt admixture that proved irrelevant to whether a conversion was genuine, were none other than two of Landsman's principal lay leaders from the artisan community, master weaver Bartholomew Somers and shoemaker Ingram More.[40]

Rather than a spurning of the preachers' directives, the visions of the laity were more likely an indirect product of their preaching. In their sermons, the preachers encouraged the people to employ their imaginations: "Suppose you saw the heavens opened, and the Lord descending among you and saying, 'Lo, here is mercy to pardon all your sins, . . ." McCulloch entreated in a typical passage.[41] Leigh Eric Schmidt notes that Scottish Presbyterian sacramental devotional guides, among which some of the most popular were written by John Willison, emphasized the suffering Christ and encouraged the mental visualizing of the Passion. Schmidt quotes a speech of Willison's as he dispensed the communion cup:

> You are now upon Mount *Calvary*, at the Foot of the Cross, near the Wounds: He is saying, Pray, believing Soul, reach hither thy Hand, feel the Prints of the Nails; yea, thrust into my pierced side, and feel my warm bleeding Heart.

Schmidt observes that "visions of Christ's sufferings were among the most common attested to by the saints at Cambuslang." Those visions are powerful evidence of the influence of the clergy over the private, internal experience of the converts. Schmidt comes to a conclusion opposite to Landsman's. While the ministers clearly distinguished between envisioning Christ with the eyes of faith and seeing him with bodily eyes, in their less guarded moments of exhortation they ignored such distinctions. Hence, "visual, dramatically imagistic diction was shared to a large degree by pastors and people." Schmidt concludes that "it would be wrong to see" in the accounts of visions revised by the clerical editors of the McCulloch Manuscripts "a large gap between the mental world of the laity and that of their pastors." Instead, the changes to the manuscript narratives ought to be understood as resulting from the ministers' concern for how the revivals would be viewed by those skeptical of the movement.[42]

Landsman overlooks alternative likely sources for any divergence the experiences of the converts had from the prescriptions of the preachers. First, the narrations were descriptive, not prescriptive: The converts reported their spiritual exercises in terms of their actual experience, not in terms of what they thought the experience ought to have been. Jonathan Edwards may very well have been right in asserting that one should expect enthusiastic episodes from some when the affections of a multitude of people with various psychological constitutions are profoundly moved. And second, the clergy may have been contending not with innovations of the weaving community, but with longstanding traditions of lay piety. For certain, the powerful influence that single scriptural passages had on the awakened was not unusual among a people to whom Bible reading was so important.

Most of the narrators of the Cambuslang conversion accounts neither experienced visions nor heard voices. Almost all, however, tested their spiritual state by individual scriptural texts that came strongly into their minds, which is not the same as hearing voices. Landsman is incorrect in asserting that this practice was new to the revival. In fact, the habit was an ingrained element of Scottish lay piety. Landsman states that similar providences had been traditionally applied by Scots to the nation as a whole rather than to an individual's religious condition.[43] Yet, Robert Wodrow objected as early as 1709 to his people's testing themselves "mostly with respect to the places of Scripture that have been *borne in* upon them, and will receive no satisfaction or comfort till these or some new Scriptures be borne in upon them, to the raising of their affections."[44] Presbyterian immigrants brought this tradition with them to America. During the Great Awakening, Samuel Blair combatted the practice among some of his people in his Scots-Irish congregation at New Londonderry, Pennsylvania:

> They would be looking and expecting to get some texts of scripture applied to them for their comfort; and when any scripture text which they thought was suitable for that purpose came to their minds, they were in hopes it was brought to them by the Spirit of God, that they might take comfort from it.[45]

Nor is it clear that all the Scottish revivalist clergy objected to this custom. Willison noted as a positive indication of the validity of the Cambuslang Work that:

> Though I conversed with a great number, both men and women, old and young, I could observe nothing visionary or enthusiastic about them; for their discourses were solid, and experiences scriptural; and all the comfort and relief they got from trouble, still came to them by some promise or word of Scripture cast into their minds, and it was pleasant to hear them mention the great variety of these words up and down the Bible.[46]

John Robe, in contrast to the clerical editors of the McCulloch Manuscripts, did not eliminate references to this custom in the narratives of the converts he published. His journal on "G.H.," whom he holds up as an example of deep piety, for instance, contains phrases such as "the following scripture came into her mind with great power," "all the while that word was strongly enforced upon her," and "the following scriptures were impressed upon her."[47] The piety of the evangelical laity and the piety of the evangelical clergymen were not distinct categories. Rather, there was a continuum that ranged from traditional folk customs at one end to learned attitudes taught in the divinity schools at the other. Along this continuum, clerical piety would be found concentrated toward the latter end, and lay piety toward the former, but there was considerable overlapping.[48] Thus, although the applying of strongly apprehended Scriptures to oneself does indicate that Scottish popular religion did not conform in all things to the teaching of the clergy, the practice was neither a recent innovation nor a peculiarity of the emerging artisan community, and it does not indicate a new emphasis on private concerns as opposed to national.

On the whole, the Scottish clergy were successful in controlling those lay tendencies of which they disapproved. While it is true that the clerical leaders of the revival objected to converts' reliance on visions and voices, the Scots, as Clarke Garrett points out, came nowhere near the kind of enthusiastic behavior that was associated with other contemporary pietistic movements.[49]

Landsman says that the Scottish laity thought of conversion as a psychological event, in contrast to the clergy.[50] This is not true. The clergy of the Scottish revivals, like Jonathan Edwards in New England, recognized that conversion was a powerful psychological transformation that influenced the physical as well as spiritual nature, whence they believed unauthentic visions and voices sometimes arose. If worldly cares produce crying, fainting, and trembling, observed John Erskine, why should deep inward concern about religion not do the same?[51] The defenders of the revivals were willing to assert that God could make use even of dreams to push forward the conversion process.[52] It can be concluded, as Garrett noted, that it was because the clerical leaders of the Scottish revival followed Edwards in treating visions as neither of satanic nor of divine origin, but as natural psychological responses to the powerful emotions stimulated by the revival, which neither proved nor disproved a work of God, that these lay tendencies were contained.

Landsman attempts to minimize the influence of the preachers over the spiritual experiences of the converts by asserting that despite all their efforts, McCulloch and his colleagues failed to make their hearers fear hell. "Indeed," he says, "in several places in the manuscript, we find the rather surprising spectacle of converts stating categorically that they did not fear Hell while their clerical editors argued just as strenuously in the margins that they really did."[53] The five instances Landsman cites in his note are weak supports for this conclusion. Mary Mitchell was awakened by McCulloch's sermon on the frail thread of life and the danger of dropping at any time into hell. The thought that she might awaken in hell disturbed her sleep.[54] Landsman cites this same passage again when discussing instances in which narrators *did* express a fear of hell.[55] The case of John McDonald directly undercuts Landsman's thesis. McDonald was awakened by a sermon by McCulloch on Christ's being "revealed from heaven in flaming fire to take vengeance on them that know not God." The reading at family worship by his master, weaver Bartholomew Somers, of the psalm passage "O who shall stand if Thou O Lord, shouldst mark iniquity" struck him with terror and sent him to bed "all a trembling." And a few days later, "After [elder John Bar] had talked to me about my souls case . . . I fell under a deep sense of the wrath of God for my sins."[56] Thus, in McDonald's case, the lay leaders reinforced the fear of hell awakened by the preacher. In neither of these passages of McDonald's or Mitchell's narrations did the clerical editors note any objections in the margins. In two instances in which the editors did make marginal comments, it was a matter of questioning the internal consistency of the narratives. Jean Robe stated that when she heard some persons crying out under fears of hell, she began to fear that her convictions were not of the right sort, "for I had not at that time, nor had I ever formerly, any dread of hell." Yet, as the editors note, earlier she had spoken of great distress, seeing herself condemned, "which looks like the fear of hell."[57] When Margaret Lap asserted that she could not say "I was . . . at any time under terrors of Hell," the editors remark that earlier Whitefield's preaching on the danger of going to sleep without Christ only to "awaken in Hell before next day" had put her into great confusion and made her "much afraid to go to bed."[58] The fifth instance Landsman cites, that of Rebecca Dykes, is subject to interpretation. The passage reads as follows:

> I was then made sensible that I was ane unbeliever, and that the wrath of God was abiding upon me {yet I do not remember that ever I had mind of hell: but} I was made also [preceding word interlined] to remember my sins whereby I had offended a Holy God, and to see that I was abominable and to wonder that he had not cutt me off before that time.[59]

The portion between the braces was marked by two of the editors for deletion, without comment as to their objection. Yet, even without the interlined word *also*, the passage can be read to mean not that the narrator *never* "had a mind of hell," but, rather, that she never thought about hell without also thinking about her sins.

When the Cambuslang converts told McCulloch they were not so much afraid of hell as they were ashamed of their sins, they were indicating that they had gone beyond what Reformed theologians called legal fear to what they termed evangelical sorrow for sin, a further step along the path of conversion, and a distinction Landsman ignores.[60] For most revivalists, an intense realization of one's sinfulness, acknowledgement of the justice of one's damnation, and fear of punishment were only the first steps in awakening, being followed by sorrow for sin, despair of help aside from Christ, shame because of one's impieties, and hatred of sin. From his analysis of McCulloch's published sermons, Robert Rutter concludes that for McCulloch "the discovery of sin . . . was a progressive realization, something more akin to a deep sorrow for sin for having offended God than a deep painful terror like [Gilbert] Tennent's." The general pattern followed by the Cambuslang converts was to move from fear to shame: "Towards the Beginning of my awakening I was mostly affected with the fear of hell; But afterwards a sense of sin and dishonouring God, was far more grieving to me than any fear I had of hell." The converts, rather than rejecting the clergy's teaching, were interpreting their experiences within the conventional theology.[61]

Landsman mistakenly sees the temptation to suicide exhibited by several of the Cambuslang converts as evidence that the lay subjects of the revival were not afraid of hell, as the preachers thought they should be.[62] While this conclusion might follow logically, it is not true historically. As John Owen

King explicates, temptation to suicide was a conventional element in the Reformed morphology of conversion and had more to do with authentication of the spiritual struggle than with anticipation of life after death.[63] The temptation to suicide occurs frequently in seventeenth-century English conversion narratives, where it is related to despair of salvation.[64] Suicide was a prevalent temptation among subjects of the Great Awakening in America.[65] Nathan Cole, of Connecticut, was tempted to kill himself in order to see if he were really converted, yet he had also longed to be annihilated soul and body so that he would not have to suffer in hell's fire.[66] Susanna Anthony was a teenaged Quaker in Newport, Rhode Island, who, awakened by Whitefield's and Tennent's preaching, joined the Congregational church. She was tempted to suicide because she thought she would never be saved, dreaded to continue to live and blaspheme God, and feared to increase God's wrath against her.[67]

The Cambuslang converts felt anguish for their particular transgressions as well as for a graceless disposition, as a male convert related:

> I got not only a sight of my sins in the general, but all my particular sins were brought to my rememberance and view, sometimes particular iniquities stareing me in the face more than others, and not only sins of my life, but sins of my heart.[68]

Landsman finds that the Cambuslang converts' sense of their inherent sinfulness was more because of a sinful disposition than because of overt sinful acts and that as such it was opposed to the clergy's teaching.[69] On the contrary, such an attitude is Reformed orthodoxy and consistent with Scottish Calvinism. Henry Grey Graham observed the common phrases of Scottish Calvinist preachers: "while in the Doric sough of the age the preacher, with a scornful sniff, exclaims, 'O dull duties! O poor professions! O filthy raggs of my righteousness!'"[70] This attitude comports perfectly with George Whitefield's teaching that one could be visibly a great saint, faithful to all outward duties, and still be in God's sight a sinner worthy of damnation.

The shame and sense of unworthiness of the awakened at Cambuslang, of which Landsman makes much to prove that the lay artisan leaders were more influential than the preachers, is the same as that which runs through the conversion narratives of the seventeenth century. Reformed spirituality demanded introspection. Converts as well as those undergoing convictions were required to pursue a course of self-examination, the purpose of which was to discover the remnants of indwelling corruption in order to cooperate with the Holy Spirit to mortify those remnants, which would not be entirely eliminated during life.[71] Sincere converts were expected to think of themselves as humbled sinners. Owen C. Williams finds complaints of the inability to find words to express one's spiritual experiences common in seventeenth-century

English spiritual autobiographies, and Patricia Caldwell underscores the sense of sin and shame in the New England conversion narratives of the seventeenth century and identifies as a peculiarly American phenomenon the narrators' shame at their inability to express themselves freely or properly concerning their spiritual states, a sense which Landsman finds among the Cambuslang converts.[72] Self-loathing is a recurrent theme in seventeenth-century conversion accounts in New England.[73] The same shame and sense of unworthiness is found in the Great Awakening in America, where it is called Christian humility: "Much of their exercise was in self-abasing and self-loathing; and admiring the astonishing condescension and grace of God towards such vile and despicable creatures," wrote Samuel Blair of his converts in New Londonderry, Pennsylvania; and in meetings of youths in Northampton, Massachusetts, led by Jonathan Edwards, "many seemed to be very greatly and most agreeably affected with those views which excited humility, self-condemnation, self-abhorrence, love and joy."[74] The Cambuslang converts' shame corresponds closely to Edwards's description of "pure Christian humility":

> The eminently humble Christian . . . complains most of himself, and cries out of his own coldness and lowness in grace, and is apt to esteem others better than himself, and is ready to hope that there is nobody but what has more love and thankfulness to God than he, and can't bear to think that others should bring forth no more fruit to God's honor than he. . . . The eminently humble Christian is ready to look upon himself as not worthy that others should be united to him, to think himself more brutish than any man, and worthy to be cast out of human society, and especially unworthy of the society of God's children.[75]

The sense of shame was surely a consequence of the ministers' preaching.

"Frivolity and idleness in general were sources of shame" among the Cambuslang converts, notes Landsman. He documents that many supplanted frivolous reading with religious books and vain company with serious conversation.[76] Yet he fails to note that the clergy condemned frivolity and idleness just as firmly as the master weavers may have. The covenant of renewal in which Jonathan Edwards led his congregation in 1742 included the following engagements:

> [The youth] do promise never to allow ourselves in any youthful diversions and pastimes, in meetings or companies of young people, [anything] that we in our consciences, upon sober consideration, judge not well to consist with, or would sinfully tend to hinder the devoutest, and most engaged spirit in religion. . . . [And the entire congregation] devote our whole lives to be laboriously spent in the business of religion: ever making it our greatest business, without backsliding from such a way of living; nor hearkening to the solicitations of our sloth and other corrupt inclinations, or the temptations of the world, that tend to draw us off from it.

In Northampton, as elsewhere in New England, the awakening produced

> a very great alteration among the youth of the town, with respect to reveling, frolicking, profane and unclean conversation, and lewd songs; instances of fornication have been very rare; there has also been a great alteration amongst both old and young with respect to tavern-haunting.[77]

Sobriety and diligence were values promoted by the clergy throughout the evangelical community. It has not been demonstrated that the master weavers were more successful in inculcating these values than were the preachers.

Another source of shame Landsman identifies among the Cambuslang converts is fear of dishonoring the awakening. He fails to explain, however, why the master weavers would have been more concerned about protecting the revival from defamation than the clergy would have been. Much of the ministers' editing of the Cambuslang conversion narratives had nothing to do with removing instances of doubtful orthodoxy, but rather were intended to protect the revival from ridicule. For instance, the editors routinely excised passages that referred to visions or fainting, even where the narrators were stating that they never had visions or fainting spells.[78] When the editors cut out a phrase that had McCulloch "standing on a chair" while exhorting in the manse, they could have had no other motive than that of avoiding embarrassment for the movement.[79] If converts were ashamed of personal weaknesses that were liable to bring disgrace onto the converts as a group, their attitude was just as likely to have been reinforced by the preachers as by lay exhorters.

The guiding hand of the preacher was evident not only in each of the stages leading to conversion, but also in the ways the laity verbalized their joyful, intimate communion with the Savior after conversion. Consider nineteen-year-old seamstress Catherine Stuart, who described her conversion in terms of a courtship, wedding, and consummation in physical union with Christ. She adopted as her own the sensual language of the Song of Songs. Awakened by Whitefield's preaching, she became one of his devotees, hearing him at every opportunity. When on one occasion he spoke on the text "thy Maker is thy Husband" (Isa. 54:5), she reported "I could say that I desired that my maker might be my Husband: But I could not say that he was become my Husband." During another preacher's sermon on the text "Saw ye him whom my soul Loveth," she "thought I had the witness of the Spirit with my Spirit, that Christ was the Object of my Soul's Love." After this, two phrases came into her mind in the form of marriage proposals from Christ: "Have I not called thee as a woman forsaken and a Wife of youth: for a small moment have I hid my face from thee but with great kindness will gather thee"; and "I will betroth thee unto me in loving kindness." She then "longed for night, being in expectation of enjoying much nearness to him & communion with him." Finally, she was able to describe her conversion as an exchange of wedding vows: She devoted herself

> in the most solemn & serious manner to God in Christ, Receiving the Lord Jesus Christ in all his offices, and giving myself away to him to be saved by him in his own way. . . . As I was pressing after further Confirmations of the Love of God, it was said to me, yet a little while and I will receive thee unto myself. . . . When I was coming away that word came into my heart to great sweetness, The Lord hath make with me ane Everlasting covenant well ordered in all things & sure: and now I could say my maker is my husband, The Lord of hosts is his Name and with many such words I came home in holy triumph.

She found confirmation of her new relationship with Christ during a Communion service. While she was seeking an invitation to go forward to take the elements,

> That word came into my Heart, The Voice of the Charmer Charming sweetly Tho' ye have been among the pots, like doves ye shall appear whose wings with silver & with gold whose feathers Covered are.

She received the sacrament as a seal of her covenant with God, and heard the evening sermon on the words "I will never leave thee nor forsake thee" as further confirmation. That night she "longed much for Death, & to be with Christ in heaven." She described her spiritual experience of the Monday following in the terms of a lovers' tryst:

> At night that word, Yet a little while and I will receive thee to myself came in again & greatly rejoiced my heart . . . and in secret duty that word came to me with a heart overcoming delight fear not for thou art a Chosen Vessel to me. On Monday Evening in the fields, I got much nearness to God & that word came home to my heart with great sweetness, O My Elect one whom I have chosen; & overpower'd me with wonder at divine love and free grace.

At another Communion service, awaiting Christ's invitation to come forward to the table, she thought of him as addressing her as a lover:

> These words ravished my heart, and took away my spirits, Arise my fair one & Come away, and then my frame of mind had such ane Effect upon my body that I could scarce walk, I was swallowed up in love & Enflamed affection to the Redeemer. . . . When I was rising up from the table that promise from Mal: 3.17 was sealed upon my heart, Ye shall be mine in the day when I make up my Jewells.

Without embarrassment, she spoke to McCulloch of her union with Christ in frankly sexual imagery. In private devotions she envisioned physical intimacy with Christ: "It had been part of my soul exercise, to have earnest longing desires immediately to be uncloathed and to be with Christ."[80]

One might be tempted to interpret Catherine Stuart's experience as the product of the frustrations of a sexually repressed young woman, particularly since she uses scriptural references to women deprived of husbands: "For the Lord hath called thee as a woman forsaken and grieved in Spirit," and "Thou shalt forget the shame of thy youth, & thou shalt not remember the reproach of thy widowhood any more." Yet, her language is the very language used by the revivalist preachers to explain the soul's closure with Christ. Although the clerical editors liberally marked passages in her relation for deletion, they made no objection to her sensual imagery. Whitefield's sermon on the text "Thy Maker is thy Husband" was one of his stock pieces, and it had a pronounced affect on a number of the Cambuslang converts. The images Whitefield evoked in it were not new but reinforced those of popular Scottish devotional literature, which described the mystical union of the convert with the Savior as a marriage of Christ with his beloved.[81] Presbyterians in the Middle Colonies carried on this tradition. Gilbert Tennent, in a sermon titled *The Espousals, or A Passionate Perswasive to a Marriage with the Lamb of God* (New York, 1735), told his audience "Brethren I come a wooing in the Name and Behalf of Christ my great Master the Kings Son," asserting that the church's great business "is to perswade poor Souls to consent to embrace him as their Lord and Husband, in an everlasting Marriage Covenant."[82] And John Rowland spoke to his people of spiritual marriage to Christ.[83] In Cambuslang, McCulloch encouraged converts to use sensual imagery to describe their joy in Christ. Immediately after Catherine Jackson made her conversion public, setting the Cambuslang revival in motion, McCulloch prompted her to explain to those assembled in the manse her relationship with Christ in the language of sensual love, providing her an unmistakable cue to the appropriate Scripture:

> The Minister . . . said to her . . . you see that there are several Daughters of Jerusalem there that will be, it's like, saying in their hearts to you, What is thy Beloved more than another Beloved: Have you any thing to say to commend Christ to them? She immediately turn'd to them and said in the most moving and feeling manner, My Beloved is white and ruddy, the chief among ten thousand, yea he is altogether lovely. . . . I can now say, my Beloved is mine & I am his.[84]

A number of the female narrators among the McCulloch Manuscripts remarked how deeply they were moved by McCulloch's action sermon before the first communion service at Cambuslang in the summer of 1742 on the text "Yea He is altogether lovely, this is my beloved," as well as by Whitefield's sermon that evening on "Thy Maker is thy Husband."[85] McCulloch preached a sermon in February 1743 on the text "I will betroth thee unto me," moving Rebecca Dykes to rejoice in Christ as "my head & Husband."[86] Married as well as single, male as well as female, and middle aged as well as young among the converted at Cambuslang expressed their relationship with God in the terms of marriage and sensual love.[87] Archibald Bell, thirty-eight years of age and married, found his "heart was ravished with the condescending grace of Christ, in that wonderful expression of his Love . . . 'Thou hast ravished my heart my sister my spouse, with one of thine Eyes, with one chain of thy Neck.' "[88] The text "Thy Maker is thy Husband" was very sweet to Alexander Bilsland, forty-seven years of age and married.[89] At the first Communion at Cambuslang in the summer of 1742, in his sermon on "Thy Maker is thy Husband" Whitefield offered to unite those who were willing in marriage to Christ. Eighteen-year-old Thomas Walker reacted by taking "all the Congregation present, as it were, witnesses of my acceptance of him for my Lord & Husband."[90] And after the same sermon twenty-one-year-old Daniel McLartie threw his arms about an acquaintance, declaring that Whitefield "had married my soul to Christ."[91] These men spoke of becoming brides of Christs without any sense of anomaly because they, with the women converts, understood the metaphor as an analogy. Still, the converts took the metaphor seriously, embraced it as their own, and employed it in highly concrete language: Marrin Calendar no longer feared to die but now considered "Death as a messenger to come and call me home to my Lord & Husband, to be where he is."[92] The ministers helped shape the laity's private experience of joy by elaborating and sanctioning the language with which converts made sense of that joy.

The Scottish evangelical clergy perceived no conflict between themselves and their people over management of the revivals and saw no fundamental theological disagreements between themselves and pious lay people. They recognized enthusiastic tendencies among some of the laity, which they fairly successfully controlled through pastoral counseling and theological instruction. Laity and clergy did view the revivals from different perspectives, and thus their perceptions were not identical. The clergy concerned themselves not only with the welfare of individual souls, but also with the spiritual condition of the entire flock under their pastoral care. Lay people looked at the revivals primarily from the viewpoint of individuals seeking an opportunity of good for their own souls, as Elizabeth Jackson was greatly disturbed that others were "getting good" and she was getting none, and Margaret Richie thought that if she could go to New England or have an opportunity to hear Whitefield she might "get good also."[93] Nonetheless, their own conversions, while private, internal affairs, were stimulated, guided, and confirmed by fellowship and through communal rituals.

Revival and Ritual

The purposes of the Communion sacrament of the Scottish Presbyterian tradition included the strengthening of individuals' sense of identity with a godly community and the sanctifying of that community. The purpose of the ritual, Westerkamp explains,

> is to free participants from guilt, shame, and sin and enable them to identify with a community that is loving, good, and protected by God—in the revivalists' own terms, to experience conversion. Participation in the ritual allows participants to attain or retain membership in the desired community, while the community itself is elevated to a holier status.[94]

As Westerkamp and Schmidt explicate, the structure as well as individual elements of the ritual epitomized the transformation of the community into a more godly people. The process took the assembled community by steps through a cleansing from guilt, through reaffirmation of the church's covenant with God, to union with the Savior and joy in the holy love and fellowship of the communion of saints. Before the date set for the beginning of the celebration of the sacrament, the kirk session might identify scandalous persons, obtain their public confessions, and grant public absolution. Receipt of a lead token from the minister, admitting one to the communion table, symbolized one's good standing and increased one's awareness of his or her responsibilities as a member of the church. Fasting in preparation for the service, and preaching on sin and repentance served further to cleanse the community and to remind the people of their covenant with God. Singing psalms involved everyone in a common activity. Sitting at the communion tables, rather than kneeling, and the passing of the bread and wine from one to another exemplified the function of the Lord's Supper as a communal feast. Receipt of the sacrament consisted of more than the individual soul's union with Christ, but was a celebration of the communion of saints.[95]

Both Schmidt and Westerkamp treat the Scottish Communion as a revivalistic form. Schmidt speaks of an annual cycle of spiritual refreshing during the summer Communion season and of spiritual doldrums during winter, and of a long history of sacramental revivalism.[96] Westerkamp is careful to distinguish between revivals, a general revitalization of piety, and revivalism, "deliberate, organized efforts to stimulate that response." According to her, the key features of revivalist rituals are their relative size, duration, and the intensity of emotional response.[97] In the Scottish tradition, Communion services were revivalistic rituals, but they did not encompass the full extent of the pouring out of the Holy Spirit on the community. Even the great archetype, the 1630 communion of Shotts, was said to be "the sowing of a seed through Clidesdeal." William Halley, reporting on the revival in his parish of Muthill, remarked that during the Communion celebration in July 1742 "our conquering Redeemer made some visible inroads upon the kingdom of Satan." He went on to observe, however,

> whatever the Lord was pleased to shed down of the influences of his Spirit upon that solemn occasion, comparatively speaking, may be accounted but a day of small things, in respect of what he has been pleased to do amongst us since. . . . An unusual power hath attended the word preached every Sabbath-day since. . . .[98]

In Scotland, as in New England, the outpouring of the Spirit was seen to discharge itself over an extended period of time, with intervals of especially heavy showers. The two sacramental celebrations of 1742 at Cambuslang were experienced as high points of religious intensity, special times when the outpouring of the Spirit became an abundant cascade. These services were understood as parts of the larger religious awakening, which had had its beginnings the previous autumn, reached its pinnacle in the summer of 1742, and subsequently subsided, with some lasting effects.

Communion celebrations served a similar function among the Scots and Scots-Irish in America during the Great Awakening as they did in Scotland. As in Scotland, Presbyterians of the Middle Colonies customarily employed a period of several days to celebrate the sacrament, surrounding the Sabbath day with days for preparation and thanksgiving.[99] Samuel Blair stated that at New Londonderry, Pennsylvania, "particularly our sacramental solemnities for communicating in the Lord's Supper have generally been very blessed seasons of enlivening and enlargement to the people of God." He related the religious experiences of a young female convert for whom sacramental occasions were particularly meaningful. Gilbert Tennent recounted that "frequently at sacramental seasons in New Brunswick, there have been signal displays of the Divine power and presence. . . . New Brunswick did then look like a field the Lord had blessed; it was like a little Jerusalem, to which the scattered tribes with eager haste repaired at sacramental solemnities." And William Tennent remarked that "the sacramental season was blessed to the refreshing of the Lord's dear people there [of the congregation of Maidenhead and Hopewell, New Jersey], as well as to others of them who came from other places."[100]

Although the Communion service was not celebrated in the same way in New England as it was in Scotland and among the Scots and Scots-Irish Presbyterians in the Middle Colonies, it did function similarly as a time of increased spiritual intensity, particularly among church members. Among the Scots, the sacrament was a communal festival in which people from an entire region might participate, although only a minority might actually receive the sacrament. In New England, the celebration of the Lord's Supper was a more intimate affair, usually restricted to visible saints and held indoors. Even in those churches that practiced open Communion, the service had more the atmosphere of a large family gathering than that of a communal festival. Yet, these occasions were often times of outpourings of grace. Josiah Crocker comments that after his ordination to the pastorate at Taunton, Massachusetts, "there have been several times of refreshing, and some revival of God's work. Several sacramental occasions have been special seasons of refreshment to the people of God."[101] Jonathan Parsons believed that the celebration at Lyme, Connecticut, on 11 October 1741, at which nearly 300 persons received Communion, would be commemorated as an historic occasion, much as the Communion at Shotts was:

> It was a day never to be forgotten. . . . Especially when the Lord's supper was administered, God poured out his Spirit in a wonderful measure. . . . I am persuaded the marvellous grace of God appearing in that season, is admired by hundreds among us here, in time, and will be much admired by the armies of saints and angels throughout eternity.[102]

Throughout the British/American evangelical community, the celebration of the Lord's Supper, however it varied in form, was a revivalistic ritual that served as a conduit for the outpouring of grace. John Porter, of Bridgewater, Massachusetts, observed:

> God is through his abundant goodness, frequently visiting and refreshing by the gracious influences of his Holy Spirit. . . . It is frequent on lecture days and on Lord's days, while we are supplicating the Divine Majesty, singing the high praises of God, hearing his Word, celebrating the Holy Supper, that we see some of the above-mentioned influences.[103]

As this quotation suggests, the Communion celebration was recognized not as a revival of religion in and of itself, but as part of a larger season of grace.

The lay participants understood the Cambuslang Work as God's dispensing his grace more abundantly than usual during a particular period of time, in particular places—a season of grace. They experienced the revival as a process over an extended period of time. These points are reflected in their conversion narratives and can be demonstrated by a close analysis of the first

26 accounts, designated as AA through AZ in volume one, of the McCulloch Manuscripts. The sample, approximately a quarter of the total, appears to be representative of the whole. It consist of accounts by ten men and sixteen women. The men's ages ranged between 15 and 48 years, with a median of 24/28, averaging 29. The women's ages ranged between 14 and 65 years, with a median of 20/21, averaging 26. Seven of the narrators were residents of Cambuslang and sixteen lived elsewhere; whether the remaining three lived in the parish is not clear.

The commencement and climax of the awakening in the Cambuslang and Glasgow area, as viewed by the awakening's subjects, can be located in time through the dates that the narrators identified as when their own awakenings and conversions took place. According to these criteria, the awakening seems to have had two beginnings, September 1741 in Glasgow, and February 1742 in Cambuslang. Seven of the twenty-six said they were awakened to a sense of their own sins by the preaching of Whitefield in Glasgow in September. Another was awakened during the Communion season in Glasgow the following month. Seven were awakened in Cambuslang in February and March of 1742. Three more dated their awakenings in April and May. The awakening of eight took place during the next two months, the last dating from the first of August.

Conversions commenced in February 1742 with six, continued through the next three months, during which six more believed themselves saved, and then climaxed during the late spring and summer, with three in June, five in July, and five in August. Seven associated their conversions with one or the other of the two Communion seasons at Cambuslang in July and August. One narrator does not clearly date his conversion.

The length of time between awakening to a sense of sin and closing with Christ varied considerably among the subjects in the sample. Eliminating the two cases in which the subjects had been under convictions well before September 1741, and the one in which dates are too unclear to make an intelligent guess, leaves twenty-three cases. Of these, the longest periods of time elapsed between awakening and conversion were between five and seven months (six cases). All but one of these was first awakened toward the end of 1741. None who was awakened at that time took less than four months to complete the process—but this is because recorded conversions did not begin until early in 1742. Those awakened later in 1742 tended slightly to complete their conversions more quickly than those awakened earlier. This is true even though many of the narrations continue to the end of 1742 or even well into 1743, so that longer conversions not begun until mid–1742 would not have been eliminated from the sample. Still, there were a number of relatively rapid conversions early in 1742. Of those awakened in February, March, or April, one had a conversion experience within five days, two took three weeks to come through the process, three took between one and four months, and one about five months. No one awakened after April 1742 took more than about six weeks to go from awakening to closing with Christ. Three converts went through the process within twenty-four hours, all in June 1742.

The conversions of the Cambuslang Work came in two separate waves. The first wave consisted primarily of residents of Cambuslang, many of whom had been awakened by Whitefield in September and then had been carefully coached by McCulloch through the autumn and winter. Six of the residents of Cambuslang in this sample, all except 65-year-old Sarah Strang, completed their conversions by mid-April 1742, four of these by mid-February. Four had heard Whitefield preach in Glasgow the previous September and all were regular hearers of McCulloch's. The second wave was made up of nonresidents who were attracted to Cambuslang by news of the awakening and the hope of an opportunity of good for their souls from the shower of grace there. Four of the nonresidents had their conversion experiences during the spring of 1742, the remaining twelve converted in June, July, and August.

The converts spoke of the proceedings at Cambuslang as a discrete phenomenon. Many echoed the phrases "when the awakening brake out at Cambuslang," and "after the work began at Cambuslang," referring to it as "the awakening," and "the work." William Baillie prayed that if the awakening was a delusion he might be kept from it, and if a work of God "that I might be made a subject of it." Many of the people of Cambuslang believed that the outpouring of the Spirit was a unique opportunity of "getting good," and that if one were not in the right place at the right time, the opportunity would be lost. Returning to the manse on Saturday, 20 February 1742, Mary Lap was greeted by a woman who told her that if she had stayed at the manse instead of going home the previous Thursday night, "you might have been converted 'ere now, as well as others." At the manse she found the people asking each other "Hast thou got any thing? Hast thou got any thing?" Many came to Cambuslang for fear of missing out on the grace that was flowing at that place. Margaret Reid, on hearing of the Cambuslang work "fell under great distress, fearing I was past by, and cryed that everybody will get good but I." Robert Barclay came to Cambuslang "where I heard many were getting good, and I thought I myself had as much need as any." John Wier came on 22 February, when "nothing touched myself, and so went home. But after that day, I had still such a strong inclination to see if I could get anything to my soul, that I could not stay away." He faithfully attended sermons on Sabbaths and lectures on Thursday until he finally felt conviction of sin in June, and reconciliation to God during the Communion service in July. Again and again between April and July 1742, James Kirkland returned to Cambuslang to hear sermons, in the hope of conversion.[104]

The structure of the awakening as described by the ministers in the revival narratives reflects the experience of the laity. The process of revival, from preparation and quiet, internal stirrings, to a public beginning, progress to a culmination, and gradual leveling off or decline, mirrored the experience of individual subjects, as well as of the converts in the aggregate. Thus, just as the preaching of the revivalists helped mold the laity's understanding of the work of the Holy Spirit on their souls, the laity's reports of their spiritual exercises shaped the revivalists' understanding of the workings of grace in a community during an outpouring of the Holy Spirit.

11

Between Awakenings

Evangelicals on both sides of the Atlantic believed that they lived at a crucial juncture in the history of redemption in the decades following New England's first Great Awakening. They worked and prayed together for a new outpouring of grace, believing the Awakening would be rekindled, if not in their own generation, in their children's. Their revival ethos placed the church, the community of saints, at the center of society and focused on God's immanence in the world. Members of the British/American evangelical connection continued to exchange observations of God's activity among men and to work together to keep the revival ethos vital. The tradition persisted through the years of the break up of the imperial relationship to flourish again during America's Second Great Awakening and in a reinvigorated evangelical movement in early nineteenth-century Great Britain.

Revivals and Christian History

From the beginning of the British/American eighteenth-century Evangelical Revival, evangelical ministers speculated about the full significance of the revivals and, by the height of the excitement, were encouraging each other with the vision of a new Reformation that would transform society. Once the Awakening had waned, evangelical ministers attempted to assess its meaning not only for their own localities, but also for all of Christianity. Much of the correspondence that crossed the ocean between the ministers after 1743 transmitted an exchange of thoughts on what the simultaneous outpouring of grace in several parts of the Protestant world might portend and what should be expected next. The evangelicals believed both that revivals were nothing new and that a new religious era had begun. Several attempted to demonstrate the truth of these two beliefs by viewing the Awakening in its historical context.[1]

As the conversion narrative describes the effects of the pouring out of the Spirit on the soul, and the revival narrative describes the effects of the pouring out of the Spirit on the community, so, in *The Christian History* (Boston, 1743–1745), Thomas Prince sought to describe the pouring out of the Spirit on the nation.[2] Although modeling *The Christian History* on the evangelical weekly magazines of London and Glasgow, Prince organized his magazine around a larger theme, the history of New England as a religious community.

The London *Weekly History* and the *Glasgow-Weekly-History* were histories in two respects. Primarily, they were journals of current events of religious significance. Secondarily, they contained accounts of similar occurrences in the more distant past. Boston's *Christian History* played both of these roles. Prince, however, intended the magazine to be a history in the sense of a unified story. In telling his story, the decline and revival of piety in New England, Prince follows a clear, chronological scheme. He begins with accounts of the piety of the founders. Next, he gives evidence of the decline of piety in the generations following the first. Then, he offers "some Instances of the transient Revival of Religion in some particular Places in the midst of these Decays." These revivals include the reformation following the Synod of 1679, the revival at Taunton in 1705, and Stoddard's harvests. Only at this point does Prince arrive at contemporary revivals. These he divides into two periods, before and after Whitefield's arrival. The revivals before Whitefield include the Windham revival of 1721, the awakening following the earthquake of 1727, and the Connecticut Valley revivals of 1735. Not until issue number twenty-six, the end of the first half-year of publishing, do accounts of the revivals of 1740–1742 make an appearance in the magazine. Prince brings *The Christian History* to a close with his own account of the revival in Boston. He uses Boston as a paradigm of New England. The history of piety in Boston parallels the history of piety in New England. Fifteen pages of Prince's account describe the growth of opposition, the coming of disorders, and the end of new awakenings. By explaining the end of the revival in Boston, he explains the end of the Awakening in New England.[3]

According to its subtitle, *The Christian History* was to include accounts of revival in Britain as well as America. The magazine contains lengthy accounts of the Scottish revivals, excerpts from the London and Scottish evangelical magazines, and numerous letters from English and Scottish evangelicals. By printing British materials and accounts of Continental Pietism and of missionary activities in the East Indies in addition, *The Christian History* places the American Great Awakening into the context of the pouring out of God's Spirit for the revival and propagation of true religion throughout the globe. The local revivals hold a meaning greater than a mere concurrence. They point to the transformation of individual towns, the British Empire, the Reformed nations, and perhaps the world.

Many New Englanders considered themselves in a special relation to God, chosen to be a peculiarly religious people. By reviving piety through the Great Awakening, God, they believed, was preserving that relationship; but, as well, the evangelicals considered the story of the decay and revival of the power of piety in New England to be but one instance of the larger story of the fortunes of piety within Reformed religion as a whole. William Cooper places it in this context in his preface to Edwards's *Distinguishing Marks*. In Cooper's scheme, in the apostolic age "the power of the divine Spirit so accompanied the ministry of the Word, as that thousands were converted under one sermon." A gradual withdrawing of the influences of the Spirit brought about spiritual decay, until the Reformation, when "the power of divine grace so accompanied the preaching of the Word, as that it had admirable success in the conversion and edification of souls." But again the influences of the Spirit waned, and the churches of the Reformation became barren of conversions. Founded while the Spirit was abundantly active, New England shared in the subsequent decay of religion of the late seventeenth century. Now God had once more answered the prayers of his people, and "the apostolic times seem to have returned upon us."[4]

In Scotland, John Gillies divided history into the same periods. Ordained pastor of College Church, Glasgow, in the midst of the revivals, Gillies shared the evangelical outlook of his father-in-law, John Maclaurin. In his *Historical Collections Relating to Remarkable Periods of the Success of the Gospel, and Eminent Instruments Employed in Promoting It* (2 vols., Glasgow, 1754), Gillies assesses the significance of the British/American Great Awakening by reviewing the history of Christian revivals since the resurrection of Christ. He suggests that redemption history proceeds by means of periodic revivals, that the Revival of the 1740s was a remarkable instance of such a revival, and that the final revival of religion that would inaugurate the millennium is possibly not far off.[5]

Historical Collections is a compilation of various narratives of revivals of religion and evangelical work, and brief accounts of particularly successful evangelists. John Wesley, visiting Glasgow in 1753, assisted Gillies in his selection of material. Among Gillies' sources were Millar's *Propagation of Christianity* and Fleming's *Fulfilling of the Scripture*, as well as selections from Wesley's own compilation, *The Christian Library*. For the eighteenth century, Gillies relies heavily on Prince's *Christian History*, Robe's *Christian Monthly History*, McCulloch's *Glasgow-Weekly-History*, and Wesley's and Whitefield's journals.

In volume one of his compilation, Gillies deals with the entire period from the first century A.D. to the close of the seventeenth century. Here he attempts to show that revivals have been experienced by the church in every century. The New Testament, says Gillies, contains many "historical narratives of the success of the gospel," such as Romans 1:18, 1 Corinthians 1:4–6, and Acts 2, 3. By Gillies's descriptions, the scripture narratives parallel the narratives of the Great Awakening:

> The chief materials of the [New Testament] narrations are such as these: 1. The numbers that were converted. . . . 2. The religious instructions that were the means of this happy change, and the instruments employed in proposing these instructions. 3. The providences that brought them in the way of those that reaped such benefits by them. 4. The earnest prayers that preceded such times, and the uncommon influences of the Holy Spirit that attended them. 5. The blessed fruits of holiness in the lives of the converts.[6]

By implication, Gillies is justifying the publication of his own compilation of revival accounts. Book I provides "a few Hints of the Success of the Gospel from the Beginnings to the Sixteenth Century," and singles out the Waldensians and Wyclif and the Albigensians and Huss for special mention. Book II concerns the Reformation in the sixteenth century. Book III relates the success of the Gospel during the seventeenth century in England, Wales, Scotland, Ireland, New England, and Germany. It deals with the revivals in western Scotland and Ireland in the 1620s and 1630s, John Eliot's work among the Indians, the Societies for Reformation of Manners, and Pietism in Germany.

Book IV, "in the Eighteenth Century," occupies the entire second volume. The work of Danish missionaries in the East Indies opens the volume. Then Gillies begins a compilation of narratives, in approximate chronological order, of the revivals of the century. From Prince's *Christian History* he takes accounts of the Reforming Synod of 1679, of the Taunton revival of 1705, of Stoddard's "harvests," of the Windham revival of 1721, of the earthquake of 1727, of Freehold, New Jersey, in 1731, of the Connecticut Valley in 1735, and then of the American Great Awakening narratives. He devotes one chapter to the Salzburgers, and another to the Oxford Methodists. Chapter six concerns the Scottish revivals of 1742–1743, and includes the history of the Concert of Prayer. Following chapters give accounts of recent missionary undertakings in America, of the revivals in the Netherlands of 1749–1750, and "of Endeavours to convert Jews and Mohammedans."

Gillies's collection gives the impression that the first seventeen centuries since Christ were but a prelude to the extraordinary activity of the Spirit in the eighteenth, that God is working in history, and that his kingdom will spread gradually by means of revivals until it encompasses the globe.

The New England/Scotland Connection

The close cooperation between revivalists in New England and revivalists in Scotland persisted well after mid-century. When the New England revivalists testified to their orthodoxy in 1745, they cited and reprinted the act on preaching of the 1736 General Assembly of the Church of Scotland. James Robe reprinted the New England testimony in the Edinburgh *Christian Monthly History*.[7] The ministers continued to read each other's writings.[8] More formally, they cooperated in promoting missions to the Indians and in an empirewide agreement to pray at regular, stated times for new revivals and for the millennium.

The friendship of Jonathan Edwards with the Scottish revivalists was the most important link in the transatlantic evangelical connection. By 1743 Edwards was corresponding with Maclaurin, McCulloch, and Robe. Willison had published the Scottish edition of *The Distinguishing Marks*, and in 1745 he republished Edwards's *Sinners in the Hands of an Angry God*. In 1746 Thomas Gillespie, who would later found the Relief Church for the benefit of those aggrieved by patronage, began corresponding with Edwards concerning immediate revelation among contemporaries, and concerning certain difficulties arising from Edwards's *Religious Affections*. In 1745 John Erskine, then minister at Kirkintilloch, near Glasgow, acknowledged the influence of Edwards's *Discourses on Various Important Subjects* (Boston, 1738). In 1747 he commenced a correspondence with Edwards that continued until the latter's death in 1758. During the 1740s and 1750s Erskine corresponded with a number of other American evangelical ministers. He would later edit several of Edwards's works.[9] After the Northampton church voted to dismiss Edwards from his charge in 1750, the Scots offered to find him a pastorate in Scotland. Although Edwards replied to Erskine, "My own country is not so dear to me, but that if there were an evident prospect of being more serviceable to Zion's interest elsewhere, I could forsake it," he declined on the grounds that it would be unfair to expect a congregation to accept a preacher they had never heard.[10] The Scots did aid Edwards by collecting funds with which they shipped goods—some Bibles, but mainly woolens and linens—to America to be sold for his benefit.[11]

Edwards and his Scottish friends maintained a constant exchange of religious news and speculations on the prospects for a revival of religion in all parts of the globe. After 1743 Edwards began more and more to view the revivals in New England in their worldwide context.

In the spring of 1743, Edwards's *Thoughts on the Revival* expressed the pinnacle of his confidence that the revivals presaged the millennium.[12] He was optimistic enough to assert that it was "probable that this work will begin in America." America was to fulfill her role of reforming the old world. "They shall not have the honor of communicating religion in its most glorious state to us, but we to them." And New England would play the leading role in the drama.[13] Cambuslang's William McCulloch agreed. Isaiah 59:9, "So shall they fear the name of the Lord from the west, and his Glory from the rising of the sun: When the Enemy shall come in as a flood, the Spirit of the Lord shall lift up a standard against him," had long been a battle cry of the British/American evangelicals. Now McCulloch applied this text to the Awakening:

> The prophet here, under the conduct of the HOLY SPIRIT who chooses all his Words in infinite Wisdom, puts the West before the East: intending, as I conceive, thereby to signify, that the glorious Revival of Religion, and the wide diffusive spread of vital Christianity, the latter Times of the Gospel, should begin in the more westerly parts, and proceed to these more easterly.

From McCulloch's perspective, the prophecy seemed to match recent events. The revival excitement had passed from America to Scotland. The opposition to the revival now seemed to be coming in as a flood to overwhelm the good work,

> But our comfort is that the *Spirit of the LORD of Hosts will lift up a Standard* against all the combin'd Powers of Earth and Hell and put them all to flight. And CHRIST having begun to conquer so remarkably, will go on from conquering to conquer, 'till the whole Earth be filled with Glory.[14]

During the later 1740s Edwards became increasingly less sure of an imminent renewal of the Awakening.[15] By 1750 he was pessimistic: Wickedness was so great throughout Christendom that some very remarkable dispensation of Providence might be expected, "of mercy to an elect number, and great wrath and vengeance towards others."[16] As his disappointment increased, he saw the religious condition of New England more and more as a reflection of the religious condition of the British/American community in general. He wrote in 1748, in reference to England, "if vice should continue to prevail and increase for one generation more, as it has the generation past, it looks as though the nation could hardly continue in being, but must sink under the weight of its own corruption and wickedness." At the same time he did not contrast a virtuous future for America to England's demise, for he continued the passage by observing that "the state of things in the other parts of the British dominions, besides England, is very deplorable. The Church of Scotland has very much lost her glory . . . and how lamentable is the moral and religious state of these American colonies? Of New England in particular?" Edwards's following 200 words elucidate the alarming decay of holiness, growth of religious confusion, and prevalence of immorality in New England. Edwards thought the whole empire was in need of a new outpouring of grace.[17]

In 1743, believing that the millennium would begin in America, Edwards had suggested a day of fasting and prayer for the millennium to be observed throughout America.[18] Five years later he was promoting a Scottish plan for an empire-wide concert to pray for the millennium.

The Concert of Prayer

An agreement of Christians on both sides of the Atlantic to unite at regular intervals in prayer for the revival of religion had had advocates since early in the eighteenth century. We have noted that in 1712 London Dissenters called on British and American Protestants to spend a common hour weekly in prayer for the true church. Cotton Mather promoted the proposal because it fit into his plan for hastening the millennium. In 1726 Robert Wodrow suggested a similar plan to Benjamin Colman. As a young man, Jonathan Edwards considered the united prayers of the whole church to be of much greater significance than the accumulated prayers of individuals.[19] That God usually preceded an outpouring of grace with an outpouring of the spirit of prayer became an evangelical maxim early in the century. Not merely prayers, but prayers by the church as church presaged a revival of religion. Prayer and Christian union were mutually reinforcing signs of God's preparations to renew his grant of grace, as well as signs of the coming kingdom. The Concert of Prayer developed logically from the evangelical culture of the preceding half-century and seemed to bode well for the renewal of the Awakening.

The concert had its beginnings at Glasgow, where, in October 1744, a number of revivalist ministers, with John Maclaurin at their head, agreed to devote part of Saturday evening and Sunday morning every week and more solemnly the first Tuesday of the last month of each quarter of the year to private or public prayer to God for the renewal of the outpouring of his Spirit and for the spread of his kingdom throughout the world. The concert was to continue for an experimental two years. During the winter the ministers promoted the concert through personal conversation and private correspondence in England and America, as well as in Scotland. The religious societies in Glasgow, Edinburgh, Aberdeen, and Dundee concurred in the plan. When John Wesley was invited to join, he suggested to his Scottish correspondent that Jonathan Edwards and Gilbert Tennent also be brought into the scheme. James Robe devoted most of the April 1745 issue of *The Christian Monthly History* to the concert. As a precedent, he mentioned Edwards's proposal in *Thoughts on the Revival* for a day of fasting and prayer throughout British America. Many in Scotland, England, and America, said Robe, had already been invited into the concert.[20]

Edwards induced the prayer societies in Northampton to adopt the quarterly days for prayer and pressed his friends in the ministry to support the concert. He maintained a regular correspondence with his Scottish friends about the undertaking's progress.[21] The agreement was set to expire in October 1746, but, "in consequence of laudable advice from abroad," a dozen Scottish ministers issued a memorial on 26 August 1746 to renew it for seven years more.[22] During 1747 Edwards continued to promote the concert, while composing a treatise on motives to concur in it, his *Humble Attempt*, published in January 1748.[23] Edwards wrote,

> whenever the time comes that God gives an extraordinary spirit of prayer for this promised advancement of his kingdom on earth (which is God's great aim in all preceding providences, and the main thing that the spirit of prayer in the saints aims at), then the fulfilling [of] this event is nigh.

Edwards contended that the recent ragings of Antichrist against the true church, the attempt to restore popish government in Great Britain, "the chief bulwark of the Protestant cause," and the revived persecution of Protestants in France were provocations to pray for the downfall of Antichrist. The dispensations of Providence in granting the victory at Cape Breton and in confounding the naval fleet from France with bad weather came after days of public fasting and prayer. The late revivals of religion throughout Christendom were probably forerunners of the last great outpouring of grace. And finally, the visible unity of God's people, all praying at the same time together for the same thing would promote the millennium by influencing the minds of men. Men would perceive the beauty of unity and thereby be encouraged to promote that unity.[24] For Edwards, the visible unity of the community of saints, worshipping God together, was an image of heaven and a means of bringing the heavenly kingdom to the earth.

Edwards and the Scots continued to encourage the agreement into the 1750s. Several American ministers reported to their Scottish correspondents the spread of the concert.[25] In Scotland James Robe promoted it in his preface to his published sermons in 1749. Here he quoted a letter from a minister in New England that said that "it would tend to cause this concert to prevail much more here, if we could hear that it was spreading and prevailing on your side of the Atlantick."[26] In 1751 Jonathan Edwards suggested to John Erskine that the ministers in the United Netherlands be invited to join the concert.[27] That same year, extracts of two letters from ministers in the Netherlands to a correspondent in Glasgow, very likely John Gillies, reporting revivals in Guelderland, were published in Boston. Both the editor and one of the authors used the awakening in the Netherlands as an encouragement to keep up the concert.[28] On 3 June 1754, the Scots renewed the concert for another seven years.[29] Americans awaited word of the renewal. On 10 July, James Davenport wrote an inquiry to John Gillies about it:

> In some places I have heard they have dropped attendance on the quarterly day of prayer, on a supposition that the concert was not intended to continue above seven years from the first conclusion of it. I should be glad of your tho'ts upon this head and an account of the practice of God's people in Scotland etc. and

would humbly propose that, if there has been an increased neglect, that concert may be publicly renewed.[30]

Jonathan Edwards also wrote to Gillies to suggest the concert be explicitly renewed. He thought the renewal should be for ten or fourteen years, and that a proposal for so doing be printed; the concert should be constantly encouraged by an exchange of news concerning it; and "if the concert shou'd be heartily come into by great numbers in Britain," some should attempt to spread it through European countries.[31] As late as 1760, John Cleaveland's New Light congregation at Chebacco, in Ipswich, Massachusetts, agreed to spend one day a quarter "in a congregational Fasting and Praying for the Outpouring of God's Spirit upon us and upon all Nations, agreeable to the Concert of Prayer, first entered into in *Scotland* some years since."[32] Numerous congregations in America observed the concert through the 1750s.[33]

The Concert of Prayer for the revival of religion embodied the evangelical convictions that God directs the course of history in intimate detail as well as in the grand scheme and that God ordinarily makes use of the church, his people, to accomplish his aims. Just as salvation comes to individual souls through the means administered by the church, so too the redemption of the world comes through the community of the saved.

A Network for Transatlantic Evangelical News

From late 1747 on, Jonathan Edwards used his private notebook on the apocalypse to keep a record of events anywhere in the world that might suggest the fulfilling of the scripture promises, whether they be outpourings of grace in revivals or in missionary endeavors, or economic or military setbacks for the Catholic powers, indicating a weakening of Antichrist's kingdom.[34] A similar attitude of anxious observation characterized the Scottish evangelicals in the late 1740s and the 1750s. After 1743, with few revivals to report, the American and Scottish evangelicals focused attention on missions as an endeavor on which God seemed to be pouring out his grace.

James Robe filled his *Christian Monthly History* (Edinburgh, 1743–1746) with reports of missionary activities. The February 1744 issue describes the efforts of Danish missionaries in the East Indies and of the Moravians among the blacks of St. Thomas. All sixty-seven pages of the fifth number (March–April 1744) are devoted to the Society in Scotland for Propagating Christian Knowledge (S.S.P.C.K.). Chartered in 1709, primarily for the support of charity schools, the society had entered the American Indian missionary field in 1730 in compliance with the terms of a bequest from the London Presbyterian minister Daniel Williams. After 1737, when it dismissed three unsuccessful missionaries hired for the northeastern frontier of New England and New York by the board of correspondents in Boston, the society supported missionaries in the Middle Colonies and in North Carolina and Georgia. In his magazine, Robe included letters and journals of the society's missionaries Azariah Horton and David Brainerd alongside letters concerning the state of religion among various Scottish towns and reports of a revival of religion among the English soldiers in Flanders.[35]

Americans, too, pointed to the missions as signs that the Spirit was still active. David Brainerd (1718–1747), expelled from Yale College in 1742 when his New Light zeal overran the bounds of propriety, served as a S.S.P.C.K. missionary to the Indians first in western Massachusetts and then in Pennsylvania and New Jersey from 1743 to 1746. In 1746 his journal among the Indians was published, as its preface said, to give some encouragement to those who were awaiting the millennium, "that blessed Time, when the SON OF GOD . . . shall receive the Heathen for his Inheritance." The journal concludes with attestations by William Tennent and William McKnight, who both write of looking forward to the millennium. McKnight describes the work of God among the Indians in terms identical to those used in the narratives of the revivals of the Great Awakening: After a sermon on 8 August 1745,

> while Mr. Brainerd urg'd upon some of them the absolute Necessity of a speedy closure with *Christ*, the Holy Spirit seem'd to be poured out upon them in a plentious Measure, insomuch as the Indians in the Wigwam seem'd . . . utterly unable to conceal the Distress and Perplexity of their Souls, . . . nay, so very strange was the Concern that appeared among these poor Indians in general, that I am ready to conclude, it might have been sufficient to have convinced an Atheist, that the Lord was indeed in the Place.[36]

Brainerd died in Jonathan Edwards's home in 1747, and two years later Edwards edited and published his diary. In his "Reflections and Observations" at the end of the book, Edwards remarks that there was much reason to pray for the conversion of the Indians, and to hope for more extensive and glorious results.[37]

Missions were a necessary and logical companion to revivals in the evangelical scenario for the spread of Christianity across the globe. The Holy Spirit worked through the preaching of the Word for the conversion of sinners in both Christian communities and heathen nations, and the mechanism of the pouring out of the Spirit was the same for the introduction of grace among pagans as it was for the revival of piety among Protestants. J.A. De Jong has shown that the conversion of the nations had begun to be considered in terms of the pouring out of the Spirit at the same time the revival of piety within the Christian world was being spoken of in the same way in the last quarter of the seventeenth century. After the Great Awakening, Jonathan Edwards and his followers saw an intimate connection between revivals, missions, and the millennium.[38]

In the 1750s John Gillies made himself the focal point of the British/American evangelical connection for collecting and publishing religious intelligence. During the fall and winter of 1750–1751 he published a series of twenty-one papers designed to help renew a revival of religion in Glasgow.[39] In several of the papers, Gillies reports news of the success of religion in various parts of the world. He believed the remarkable revivals in parts of the Netherlands had come in answer to the Concert of Prayer in Britain and America.[40] Among his news items we find the following bits of information collected by means of a network of correspondents reaching from the Netherlands to America: in eastern Europe hundreds of Jews converted by the efforts of a traveling English missionary; in New York, the Presbyterian Synod's appeal for ministers to go to Virginia, where there was a considerable concern about religion; some religious concern in the Middle Colonies, as well, especially among the Dutch; concurrence in the Concert of Prayer spreading in the colonies; and glad tidings from American Indian missionaries John Brainerd, Azariah Horton, and Eleazar Wheelock.[41]

During the 1750s Gillies continued to collect information from his network of correspondents, who sent him news specifically for him to publish in a planned appendix to his *Historical Collections*.[42] Published in 1761, Gillies's *Appendix to the Historical Collections* contains accounts chiefly of the success the Gospel had met in the 1750s. This includes the continuation of the revivals in the Netherlands through 1754, missionary work among the Indians and blacks of America, the extension of evangelicalism into Virginia, and the growth of Methodism in England. New England and the Middle Colonies could also report scattered revivals.[43] The many confirmations of God's presence in the world fed the expectation of a renewed general outpouring of grace.

Between the Awakenings

During the last half of the eighteenth century, evangelical religion spread rapidly through portions of the American South, of England and Wales, and of the Scottish Highlands. Local revivals accompanied this spread, but the evangelical movement between 1745 and 1798 produced few revival narratives. British Methodists employed the revival narrative only occasionally.[44] Their narrative literature continued to focus on the activities of particular evangelists and to emphasize the opposition of the "world" to their movement. The rapid spread of evangelical Protestantism in the Scottish Highlands in the latter half of the eighteenth century has been attributed to the need for a new world view to replace the one shattered in the aftermath of "the '45," when the traditional social structure of the clans was destroyed through official policy.[45] John MacInnes, historian of the Evangelical Movement in the Highlands, finds that the Highlands "had nothing comparable to the Kirk of Shotts or Cambuslang till the great revivals of the nineteenth century. On the other hand, there were a number of movements which would be inadequately described as . . . 'awakenings.'" The revivals in the Highlands did not closely follow the model of Cambuslang, where the subjects of revival were already well indoctrinated in the evangelical perspective.[46] Thesc movements did not produce narrative accounts until the turn of the century, when Alexander Stewart described a powerful revival in his parish at Moulin, in Perthshire, about 1798–1799.[47] Evangelicalism spread through portions of the American South in waves, successively increasing the numbers of Presbyterians, Baptists, and Methodists between the 1740s and the 1770s, and of all three denominations in the 1780s.[48] To its leaders, the Great Awakening in the South had more the aspect of the introduction of Christianity than its revivification. This view had some validity for the Afro-American slaves who embraced the movement, but the evangelists held the same view concerning converts of European stock. For example, in phrases that no doubt gave umbrage to local Anglicans, Presbyterian minister Samuel Davies stated in 1751 that "there has been a considerable *Revival* (shall I call it?) or *first Plantation* of Religion in *Baltimore* County."[49] Itinerant evangelists and circuit riders had little opportunity to observe the course of a revival in a single community over a period of time. An important exception during these years was an account of a revival in the parish of Bath, Virginia, by evangelical Anglican priest Devereux Jarratt, published in London in 1778.[50] During the years between the first and second Great Awakenings the principal medium for disseminating news of evangelical activity was private correspondence. The correspondents tended to speak in general terms of the spread of evangelical religion and of the increase of churches and membership, or to focus on specific sacramental celebrations, prayer meetings, love feasts, and the like, on which the Holy Spirit seemed to be poured out. Seldom did the correspondents trace the morphology of a communal revival in the way the narratives printed in the evangelical periodicals of the 1740s did. Not until the 1790s did the evangelicals have any vehicle comparable to those periodicals. Nevertheless, between the awakenings the genre of the revival narrative was kept alive, particularly by those in the New England revival tradition.

During the decades succeeding the 1740s, opinions held in New England on the Great Awakening ranged from complete acceptance, with all its extremes of behavior, as a work of God, to rejection as a total delusion. A large middle group, however, agreed that, although flawed by errors, the revivals had been the result of a extraordinary outpouring of the grace of God. Year by year, New England's evangelicals looked for a renewal of the outpouring of grace they had witnessed in the Great Awakening. The Massachusetts author of *A journal containing some remarks upon the spiritual operations, beginning about the year . . . 1740, or 1741*, eleven pages written in verse, dated as written at Rowley, 18 November 1757, captures their mood. After beginning with a encomium to the Great Awakening, followed by "something of the Essence of the internal Part of the Religion of the People called *New Lights*," that is, principally a description of the stages of conversion, the author writes "a Lamentation for our Backsliding; and also for the Withdrawal of the holy Spirit of God," calling on saints to recall "the Glory of th' former Days" of the Great Awakening. He concludes with an annual survey of the state of religion:

> Now in the Year fifty and three,
> Some gleaning Grapes, methinks I see;
> Again in th' Year fifty and four,
> Comes *Whitefield* brave on this our Shoar.
>
> In Fifty-five, O doleful Sound,
> Terrible shaking of the Ground. . . .[51]

The earth tremors of 1755 did not result in the extensive awakenings that followed the earthquake of 1727, nor did the repeated tours of Whitefield reproduce the Great Awakening. The most that could be reported were occasional, isolated incidences of revival. In 1759 Samuel Davies, for instance, was able to publicize an awakening among the students in the College of New Jersey, and at the same time to report word of a revival in Philadelphia and of "a gracious Work of God at Yale College."[52]

New England, including its cultural extension on Long Island, witnessed a spate of local revivals and the publication of several revival narratives in the mid–1760s. Beginning in March, 1764, Samuel Buell (1716–1798) oversaw a revival in his congregation of East Hampton, Long Island, which he reported in extravagant language at the end of its first week to a fellow cleric in Groton, Connecticut:

> Jehovah himself is come down by way of divine Influences upon, and in the Hearts of People in *East-Hampton*, with amazing Power and Glory, exceeding all I ever before saw, and all I have read or heard of, since the primitive Times of Christianity. . . .

> All the Town seems bowed before this Work of amazing Power and Glory. . . . All the town seems bowed as it were like one Man.

Buell had been studying the millennial prophecies for some years and hoped that in East Hampton "some Gleanings of the new Jerusalem is breaking forth."[53] After publication of this letter, Buell wrote a more detailed account. Buell had resided with Jonathan Edwards during part of the time he was a student at Yale College and his *Faithful Narrative* follows the model of Edwards's *Faithful Narrative*, both in organization and in phraseology. He begins with the standard sketch of the spiritual history of the congregation and then gives a narration of the events of the revival. In orthodox evangelical fashion, a spirit of prayer for the outpouring of the Spirit preceded the revival. The godly had discussed "the absolute Necessity and Importance of the Divine Influences, in order to the Revival of Religion." On a Sunday in March, the spirit of prayer for the divine influences poured forth in a "Celestial Torrent," and the church appointed the following Thursday "to be observed as a Day of Fasting and Prayer, for the out Pouring of the Holy Spirit; on *us* in particular, and the Churches of *Christ* in general." By Monday powerful and public convictions and conversions had begun. Buell then proceeds to a description of the work on the subjects, and, like Edwards, recounts the variety of experiences. In words that echo Edwards's, he enumerates the ways in which the revival was extraordinary. He concludes by associating this revival with the pouring out of the Spirit that was to bring on the millennium. Sharing Edwards's belief that revivals were the engines that drive redemption history, Buell read extraordinary significance into the local revival over which he presided.[54]

Groton, Connecticut, also enjoyed a revival about the same time. In writing its narrative, Jacob Johnson, one of the ministers there, explicates some revival theory. "It has been God's usual way (as appears by sacred and ecclesiastical history)," he wrote, "to revive and carry on his work, at some certain seasons more open and visible than at others." He stated that years of jubilee in the Old Testament were years of "the out-pourings of the holy Spirit," as were the years of Christ's ministry. "The church of Christ, had their special and remarkable seasons of the visitations of divine grace, in, and after the apostle's days," and even during the years of the reign of antichrist (i.e., the papacy). The Protestant Reformation spread light through the nations of Europe. The "last reformation," the Great Awakening, though a glorious work of God, was blemished by imperfections that offended many and retarded the work. But now religion appeared to be reviving again. There were the revival in East Hampton, revivals at the College of New Jersey and Yale College, one at Providence, Rhode Island, and numerous supposed converted in Norwich, Preston, Stonington, and Voluntown, Connecticut, as well as "a very remarkable shaking in this town."[55]

The New Light congregation at Chebacco in Ipswich, Massachusetts, which had been participating in the quarterly days of prayer for the revival of religion since 1760, experienced a revival in 1763–1764. Their minister John Cleaveland's description, *A Short and Plain Narrative of the Late Work of God,* follows the conventional format, except for concluding with a lengthy defense of the church's membership practices.[56]

The American Revolution was not a religious revival. New England's evangelicals embraced the Revolution as the cause of God, believing it a stage in God's design to bring about the millennium, identifying the British government with the tyranny of the antichrist, and praying that the war would serve to purify the nation from its sins. They understood the Revolution as a work of reformation, advancing the nation's corporate redemption. They did not read the Revolutionary upheaval as an outpouring of the Holy Spirit. No narrative account of a new revival appeared in the American press between 1768 and 1799. Southern evangelicals, as well as New Englanders, noted the "restraint of the Holy Spirit" during the war years.[57]

Evangelicals considered politics and warfare as secular not spiritual affairs, but still as affairs directed by divine providence and having religious consequences.[58] They believed that God controls the course of history, that the rise and fall of empires is in his hands, and that, while mortals can cooperate with or oppose the divine will, the destinies of nations are directed by God toward his purposes not men's. Evangelicals hoped that the political revolution was a harbinger of spiritual transformation. They imagined creating a society on which the saints would impose a new moral order. The America of their vision was not an assemblage of individuals set at liberty to pursue their private gain, but a nation in which men and women find happiness in the common good and see beauty in the union of the pious.[59]

In the midst of the Revolutionary conflict, many New Englanders looked back on the Great Awakening as a Golden Age. The community transformed into a holy people, as during a revival, was the ideal to which they aspired. A minor, but revealing instance of this nostalgia for the Awakening involves Nathan Cole, whom we met in an earlier chapter as a lay Separate leader in Connecticut, but whom we meet now as a Separate Baptist. The dark day of 19 May 1780, when the light of the sun was unaccountably obscured for three hours at midday, made Cole reflect that it was nearly "forty years from the time that God poured out his spirit so remarkably in this land & many souls converted to Christ." The prominence of periods of forty days and of forty years in Scripture led Cole to judge that the dark day was a portent of sudden and strange alterations, most likely judgments for the people's wickedness.[60]

Evangelicals were prone to view the sufferings associated with the war more as divine punishment for the sins of Americans than as undeserved evils inflicted by the enemy. From the evangelical perspective, the nation still awaited the gracious outpouring that would make it a New Jerusalem, for the struggle for political liberty had not transformed a nation of patriots into a nation of saints, nor had it purged corruption from the hearts of Americans.

The winning of independence confirmed for pious patriots America's special role in God's plan for redeeming the world: Political and religious liberty would provide a context in which the transformation of society by grace could proceed apace. Independence made the need of large effusions of God's grace in revivals appear all the more urgent, for evangelicals believed that exceptional holiness was required of a free people. Piety alone could prevent a society unrestrained by rulers from disintegrating as individuals pursued self-indulgence and self-aggrandizement. Victory stimulated millennial expectations among Americans holding a broad spectrum of political and religious beliefs. It led evangelicals to look for a renewed outpouring of grace to bring those expectations to fulfillment.[61]

After the war, the evangelical denominations in the South enjoyed a remarkable increase of conversions and growth of membership in the mid–1780s, and there were isolated revivals in the Middle States and New England. Yet, the churches felt the necessity of a general revival of religion and put out calls for concerted prayer for revival. In 1785, for instance, the Presbytery of New Castle, Delaware, exhorted every congregation to set up praying societies, observing that the synod had repeatedly urged "both before and since the commencement of our late troubles, to spend part of the last Thursday in every month in social prayer to God for a revival of his own work in our Church," and that "this has been practiced with great success in several congregations both in Europe and America."[62] An association of Federalist-minded clergy in New Hampshire invoked a concert of prayer for partisan ends in the politically pivotal year of 1787. They called for an agreement to set aside one hour each Sunday evening to ask God to pour out his grace on the country so that all may repent, believe, reform and "that the spirit for true republican government may universally pervade the citizens of the United States . . . that there be no delay in clothing the *Congress* with all necessary powers to act in character as the *Federal Head* of a *sovereign, independent nation*."[63] In 1794 and 1795 Congregational clergy in Connecticut gave consideration to a proposal for a concert of prayer for the outpouring of the Holy Spirit.[64]

The turmoil of the French Revolution and the associated threat of rationalist infidelity perceived in America created a new sense of urgency among New England's evangelical leaders in the 1790s. From 1797 to 1802 the first wave of revivals constituting New England's Second Great Awakening came as if in answer to prayer for deliverance from pressing dangers.[65]

Edwards's interpretation of revivals as a major device by which God advances his purposes in the world remained before the public through his writings: *Thoughts on the Revival* was reprinted in Boston and New York in 1768 and in Boston in 1784, his *History of the Work of Redemption* in Boston in 1782 and in Worcester in 1792, and his *Faithful Narrative* in Elizabeth Town, New Jersey, in 1790.

In 1799 the first revival narratives of New England's Second Great Awakening were published. From 1800 on, New England revival narratives proliferated, dozens being printed in *The Connecticut Evangelical Magazine*. They followed much the same pattern that had been established during the first Great Awakening, tracing the morphology of the local spiritual awakening as a discrete communal phenomenon whose origins, growth, influence, and subsiding could be observed.[66]

As if to emphasize the continuity of the New England revival tradition, in 1800 a Windham, Connecticut, printer published Eliphalet Adams's *A Sermon Preached at Windham, A.D., 1721, on a Day of Thanksgiving . . . Occasioned by a Remarkable Revival of Religion*.[67] The previous year, a printer in Hartford, Connecticut, had taken advantage of the "great demand for HYMNS and SPIRITUAL SONGS" that resulted from the new revivals by publishing a collection of hymns, selected by several Congregational ministers, "suited to persons deeply affected with Evangelical Truth, or anxiously inquiring for Salvation," particularly for use by individuals, families, and private religious meetings. Among the hymns appeared a few that deal directly with the topic of a revival of religion, including "Hoping for a revival" (hymn CCLVII) and "Rejoicing in a revival of religion" (hymn CCII). A line of the latter hymn, "Convinc'd of sin, men now begin To call upon the Lord," echoes Edwards' assertion that the first revival that there ever was occurred in the days of Enos, when men began to call on the name of the Lord, and thus links current revivals with all past revivals. A portion of the income from the sale of each book was to be used to support missionary work on the western frontiers.[68] The publication of this revival hymnal underscores the continued centrality of group singing in the revivals; it also gives a foretaste of what would within a few decades become a vast revival publishing enterprise and an interlocking network of benevolent evangelical projects. The nature of revivalism would undergo great changes in the nineteenth century, but the evangelical understanding of revivals would retain at its core a continuity with the concept as it had come to maturity during the Great Awakening.

12

New England's Revival Tradition and American Revivals

The "New Measures" of the Second Great Awakening transformed revivalism first in America and subsequently in Great Britain. This transformation, which produced "modern revivalism" with its professional revivalists, brought about changes in the concept of a revival during the nineteenth century. The history of those changes is worthy of another volume or more.[1] A consideration of important continuities in the concept of a revival of religion, applicable to nineteenth-century American and British evangelicals in general, that relate to, or derive from, the New England tradition will bring this present study to a close.

Modern Revivalism in America

At the same time that the Second Great Awakening was in its beginning stages in New England, the "Great Revival in the West," 1797–1805, was taking place. John Boles finds that fear of deism and irreligion provided a sense of urgent need of religious revival on the Kentucky frontier, just as it did in New England. He also argues that a "theory of providential deliverance" gave Southern evangelicals firm confidence in the inevitability of revival: God, who controlled all events, including each person's salvation, would unfailingly answer prayer and repentance. As explained by Boles, the theory of providentialism seems to have operated very much as a Southern evangelical counterpart to New England's national covenant, establishing a basis for expectation of revival. The Great Revival had its most conspicuous beginning in Logan County, Kentucky, at a sacramental meeting of Presbyterians in July of 1798. From there it spread through the region. Camp meetings, intended to sustain the revival's momentum, began in 1800. Fairly rapidly, the revival worked itself eastward into the Southern seaboard states. Many Southern evangelical congregations had set aside special times for prayer for the outpouring of grace, in the firm expectation that God would answer. "For the faithful," concludes Boles, "the revival seemed an answer to prayer," and the revival spread because the people expected revival.[2]

After 1800, Baptist and Methodist frontier revivalists adopted the camp meeting, discarded by the Presbyterians, to bring together and convert isolated families scattered across hundreds of square miles of southern back country. In addition to the camp meeting, Methodists employed several overtly revivalistic devises, including the call of the penitent to the altar, itinerants, protracted four day meetings, and women praying in mixed assemblies. Although deriving in a measure from revivalism on the frontier, these methods were common to urban Methodist churches. By the 1820s, seeking similar success, some Congregational and Presbyterian urban churches began imitating the Methodists. In the 1820s Charles Grandison Finney (1792–1875) found conspicuous success in "getting up" revivals in western New York through adaptations of these means. Finney's protracted meetings, anxious seats, prayers for individuals by name, encouragement of women's speaking in assemblies, and professional revivalists were given the name of the "New Measures." By the 1830s, despite resistance by many of New England's church leaders, New Measure revivalism was being practiced in New England.[3]

Until the end of the 1820s, New England's revivals had remained conservative. There were few displays of physical exuberance, and no outdoor assemblies, lay preachers, or itinerants. The revivalist was the local minister, who promoted revival by means of his regular Sabbath preaching, prayer meetings, pastoral visiting, lectures, and days of fasting and prayer. It was not the use of means in itself to which New England's conservatives objected, but to the New Measures in particular and to the belief that artificially generated religious excitements were identical with authentic outpourings of the Holy Spirit.[4]

In the eighteenth century, revival was supposed to be the spontaneous work of the Holy Spirit, which came at the moment God, not a revivalist, chose. No clergyman announced a revival in advance of evidence of a work of the Holy Spirit. No church "held" a revival. Nevertheless, eighteenth-century revivalists laid the foundation for the holding of revivals. Those revivalists defined the phenomena associated with a revival and successfully employed revivalistic techniques. Many eighteenth-century clergymen had no reservations about using appropriate methods to promote a revival. Considered individually, several activities in the revivals were not new. Religious fellowship meetings, which proliferated during the Awakening, were traditional throughout the British world. Field-preaching was well-known in Wales and Scotland. For many congregations in Scotland, the great outdoor communion service was a yearly event. Hence, the participants did not sense that the revivals were contrived. Other forms of behavior in the revivals did possess an aura of novelty: meetings of the spiritually awakened after services and daily religious devotions over an extended period of time, for instance. During the Awakening, these activities, responses to popular demands, still had the quality of spontaneity. During the Second Great Awakening, itinerant evangelist Charles Grandison Finney defended the anxious bench, inquiry and protracted meetings, and similar New Measures against critics who denounced them as affronts to God's free and arbitrary grace. Finney contended that the revival of piety in a community follows scientifically from the proper use of the constituted means as naturally as a crop harvest follows careful planting and cultivation of seeds. He rejected the notion "that revivals came just as showers do, sometimes in one town, and sometimes in another, and that ministers and churches could do nothing to produce them, than they could to make showers of rain come on their town, when they are falling on a neighboring town."[5] The difference between Finney and his eighteenth-century predecessors was not, as he supposed, over the use of means to promote revival. Rather, it was over the expectation that a revival would automatically follow the skillful use of the proper means.

By the 1830s, several identifiable kinds of religious revivalism were being practiced in the United States. William Warren Sweet differentiates three types. The first, which he designates "Presbyterian-Congregational," took place "among a people who have been carefully taught by catechizing," and was based on Calvinist theology. This corresponds to revivalism in eighteenth-century New England. The second, "frontier revivalism," reached out to rough, uneducated people who lacked the benefit of a long-standing local ministry. Sweet's third type of revivalism, "Finney revivalism," differed from "Presbyterian-Congregational" in its modification of Calvinist doctrine and in its New Measures.[6]

The revival narrative, with its focus on the transformation of the community, suited the situation of New England townships and the kinds of revivals they experienced. As Richard Carwardine noted, "the cultural and ethnic homogeneity of a relatively tightly knit community, such as a New England township, with its common religious traditions, its experience and expectation of periodic revivals, and a population small enough for all the families to be acquainted," made it easier than most other places in the United States to engage the whole town.[7] The revival narrative, tracing the progress of grace through a community, hardly suited the situation of frontier revivals, where the camp meetings created communal identity among a widely scattered population. Evangelical clergy in the young urban areas of western New York, on the other hand, did adopt the revival narrative. A profusion of narratives from the 1820s through the 1840s treated the revivals in New York's "Burned-Over District" as outpourings of grace on communities and portrayed the effects as the religious awakening and moral transformation of whole towns.[8]

This continuity in western New York may be expected, since many of the clergy as well as their parishioners had New England backgrounds and maintained strong ties to the rural countryside. Yet, there were significant differences between the social contexts of New England's country townships and western New York's rapidly growing manufacturing and commercial cities. In the latter, the population was considerably more mobile, and there was a much larger class of unattached and propertyless wage earners, a large percentage of whom were unchurched. The social distance between the middle class and the laboring class was greater, as were differences in standards of behavior, particularly in recreational pastimes. As Paul E. Johnson puts it, "master and wage earner inhabited distinct social worlds." Johnson finds that through the extensive evangelical revival in Rochester, New York, in 1830, members of the middle class overcame political and denominational divisions among themselves and forged a united Protestant community, militantly committed to the saving of sinners and the reformation of society, for which temperance headed the list. The revival was a communal event, employing numerous techniques to bring group pressure to bear on people to convert. The evangelically committed applied this pressure first to family members. Faced with a common evangelical front of masters, members of the wage earning class found that a reputation for church going, stability, and sobriety improved their chances for success in finding and keeping jobs and advancing in their employments. Revival was indeed communal, and did change society, but here it was a matter of one part of the community striving to exercise social control over another part. The meaning of community had undergone metamorphosis.[9]

Modern Revivalism in Great Britain

Following the Peace of Paris, British and American evangelicals resumed their transatlantic intercourse with vigor. Over the course of the next several decades the evangelical movement on both sides of the Atlantic profited from an exchange of ideas about moral reform, antislavery, temperance, and missions. The direction of the commerce in new ideas in the realm of revivals, however, was from west to east. After 1830, the New Measures were introduced to Great Britain and began to transform evangelical revivalism there. Until then, revivalism in Great Britain, with the exception perhaps of Wales, had tended to mirror the conservative revivalism of New England's Congregationalists and Presbyterians. By the 1760s the Evangelical Revival was benefiting the Calvinist denominations in England, as people dissatisfied with the Church of England left that denomination for the dissenting churches. Yet, as late as the 1820s revivals were not typical among the churches of the Old Dissent. English Methodists were less revival-oriented than American Methodists. An American Methodist visiting England in the 1830s concluded, "what we in America term revivals are comparatively rare" in Britain.[10] Scotland experienced local revivals in the 1770s and 1780s and at intervals over the next four decades notable awakenings in the Highlands. These were not the results of modern revivalism, however. There were no professional revivalists or itinerants. The local minister was the evangelist of his own parish, although once an awakening began, he might invite neighboring clergy to assist him. The revival was promoted through doctrinal teaching, fellowship meetings, and prayer societies, and the climax was always the annual celebration of the Lord's Supper.[11]

Richard Carwardine has examined the success with which, starting in the 1830s, eastern American revivalists converted British evangelicals to the use of deliberate measures to kindle revivals. British evangelicals were intensely curious about American revivals and engaged visiting American clergymen to lecture about them. During the 1830s, English evangelical Calvinists adopted the aggressive evangelism of New Measure revivalism but in the 1840s drew back because of doctrinal problems with what they viewed as the undervaluation of divine sovereignty. Charles Grandison Finney's *Lectures on Revivals of Religion* (1835), in which he champions the New Measures, found a favorable reception in Wales, where evangelicals were already familiar with itinerancy, lay preaching, and camp meetings. Influenced by Finney's *Lectures*, as well as by religious reformers who returned to Scotland after witnessing revivalism in the United States, Scottish evangelicals began adopting American methods. A wave of revivals washed across Scotland from 1839 until the mid–1840s.[12]

The revival in the parish of St. Peter's, in Dundee, reported by the Reverend Robert Murray McCheyne in response to a set of queries circulated by the Presbytery of Aberdeen, can be taken as representative of the means employed in the Scottish revivals of this period. "The means used were of the ordinary kind," McCheyne writes.

> In addition to the services of the Sabbath, in the summer of 1837, a meeting was opened in the church, on Thursday evenings, for prayer, exposition of the Scripture, reading accounts of missions, revivals of religion, &c. Sabbath schools were formed, private prayer meetings were encouraged, and two week-

> ly classes for young men and young women were instituted with a very large attendance. These means were accompanied with an evident blessing from on high in many instances. But there was no visible or general movement among the people until August 1839, when, immediately after the beginning of the Lord's work at Kilsyth, the word of God came with such power to the hearts and consciences of the people here. . . .[13]

The means McCheyne enumerates were much like those used to promote the Cambuslang Work of 1742. Yet, modern revivalism had introduced a momentous change in the concept of a revival of religion. The revival at St. Peter's, Dundee, broke out with power while McCheyne was in the Holy Land and his place was being filled by William Chalmers Burns (1815–1868), son of the minister at Kilsyth and a recently licensed probationer. It was Burns's activities in promoting revivals that prompted the Presbytery of Aberdeen's circular letter, investigating his work. Several of the ministers who responded to the presbytery's queries remarked that Burns was chief among the instruments in producing revivals in their parishes.[14] Burns "was a new phenomenon," Ian A. Muirhead observes: "He had discovered he had a gift for breaking down audiences; he was the revivalist preacher whose aim was to make revivals happen."[15] British professional revivalists came to agree with Finney that revivals could be brought about, unfailingly, through the proper use of the appropriate means.

By the middle of the nineteenth century, professional revivalists, revival tract societies, protracted meetings, and an array of New Measures were parts of the lives of many of the evangelical churches in Great Britain. Revivals had become institutionalized.

American Revivals

At the same time that American's were exporting modern revivalism, they did not insist on the New Measures as essential to a revival. Rather, they underscored as the essential quality of a revival its communal character. At the request of British acquaintances, American evangelist Calvin Colton wrote a book that explained to the British the *History and Character of American Revivals* (London, 1832). In his account, Colton acknowledges that "the history of religion in England seems to afford examples of many considerable religious excitements," but he is convinced that they were not "of the same character with what I esteem the most genuine American revivals."[16] Positing the following as a typical context of a revival, Colton makes it clear that by an "American revival" he means the outpouring of the Spirit on an entire community:

> Suppose a community . . . under a Christian ministry, for the maintenance of Christian ordinances. We will suppose this community in some degree insulated, like a country, or village parish, or congregation. For there is, doubtless, a difference between a city and country congregation, in the existence and amount of a distinct community of feeling. We will allow, in this case, that the great and fundamental doctrines of Christianity are declared habitually from the pulpit, maintained by the Church, and not opposed especially by the people. But though the people attend regularly on the public means of grace, the interest they manifest towards religion is merely the respect of a decent civility. The Church, perhaps, are orderly, but not liable to the accusation of zeal. And, "like people, like priest." All are decent, and all asleep. The maintenance of civil order is the best that can be said of them. An insulated conversion may now and then occur, and the subject of it be brought into the Church.[17]

In a chapter entitled "Are American Revivals Peculiar to the United States?," Colton argues that the uniqueness of the American experience lay in its communal character:

> Where else shall we find its exact type? A *likeness* we can find. But the peculiar character of these dispensations is to be kept in view. . . . [A revival] is the prevalence of an unseen influence, which seems to charge the whole moral atmosphere of a community at once and thoroughly with a deep religious solemnity. . . . In one week, often in a day or two, a whole community may be seen . . . transformed from a most worldly and reckless condition of the popular mind, to such deep and absorbing thoughts of eternal scenes.[18]

Despite the varieties of religious revivalism practiced across the United States, when nineteenth-century evangelical theorists went to define a revival, it was the New England tradition to which they resorted. Their definition of a revival retained a strong continuity with the concept as it had evolved in New England by the end of the first Great Awakening. The definition and the theory of revivals of religion offered by the *New Schaff-Herzog Encyclopedia of Religious Knowledge* at the turn of the present century, and still held to be valid when reprinted in 1950, could serve as a description of the concept as held in 1850 or 1750. The author of the entry applies the phrase *revival of religion* "to the spiritual condition of a Christian community, more or less limited in extent," in which attention to religion increases, believers grow in grace, backsliders return to duty, and sinners convert. The author begins a section on the "Theory of Revivals" with the proposition that the progress of Christianity "in the individual and in the community is characterized by very obvious fluctuations." Just as an individual may become distracted by worldly concerns from the pursuit of holiness, so may a community. The only remedy for such declension is spiritual revival resulting from "a special and peculiar effusion of the Holy Spirit."[19]

This emphasis on the community was New England's major contribution to the concept of a revival. Viewing themselves as a covenanted people of God, New Englanders perceived a revival to be the community as a moral unit responding to God's grace in a manner parallel to the response of a single soul. In the seventeenth century a revival of religion meant primarily national reformation, involving renewed attention to the duties of religion, more rigorous obedience, and a purging of unrighteousness. Conversions tended to be individual and private. In the eighteenth century, the meaning of the phrase underwent a significant change. A revival came to imply communal conversions, with renewed attention to duty following. Indicative of this change is the transformation of the jeremiad sermon from a warning to the nation to reform if it wanted the blessings of Providence into an assurance that God would produce the required reformation with the outpouring of his Spirit. Another indication of the change is the growing emphasis on the conversionist sermon, which terrorized sinners with the Law and allured them with the free grace of the Gospel. At the outset of the eighteenth century, most noticeably in New England, focus shifted from national reformation to local revival. The ministers switched their efforts from synodical action to the ministry of the individual pastor. Communal revivals became the means for and symbolic of both national reformation and the millennium.

The idea of a revival of religion evolved out of elements of seventeenth-century Reformed theology, under the influence of actual occurrences. The concept was not the invention of New Englanders. Revivalism was as fully a British phenomenon as it was American. English and Welsh Methodists and Scottish Presbyterians developed their own revival traditions. Evangelicals in the Middle Colonies and the South contributed distinctive revivalistic practices and carried revivalism to the frontier. Yet, because of the peculiar importance to New Englanders of covenant theology and their strong sense of corporate identity, New Englanders were the principal developers of the theory of revivals. It was they who fashioned the revival narrative and who proposed revivals as the key to the history and destiny of the church.

The Key to the Church's Past and Future

Charles Grandison Finney championed the institutionalization of the revival on the basis that "almost all the religion in the world has been produced by revivals," and that "it is altogether improbable that religion will ever make progress among *heathen* nations except through the influence of revivals."[20] Finneys' statements of the crucial importance of revivals for the progress of religion were very much like those voiced by Jonathan Edwards nearly a century earlier. By the nineteenth century, a revival-centered version of history was an evangelical commonplace. In his *History of the Work of Redemption* Edwards had written that "from the fall of man to this day wherein we live the Work of Redemption in its effect has mainly been carried on by remarkable pourings out of the Spirit of God."[21] In 1789, fifteen years after these words had been first published in print, Joseph Milner (1749–1797), an Anglican evangelical minister in England, wrote similarly, "the history of the Church is properly nothing else than a history of the effusions of the Spirit of God, and of the effects which they produce in the world."[22] Milner's multi-volume *History of the Church of Christ* (1794–1797), in which he traces the effusions of the Spirit and their effects from Pentecost to the Protestant Reformation, would stand as a major statement of the evangelical view through the first half of the nineteenth century.[23]

If nineteenth-century evangelicals did not always explicitly assert that revivals were the principal engines that drove redemption history, they nearly universally held that revivals had been common occurrences in the relationship between God and his people throughout time and that by such means God would advance his kingdom until the millennium. In *Lectures on Revivals of Religion,* a major statement of New England's conservative evangelicals' position on revivals which was published in 1832 as a corrective to New Measure revivalism, William B. Sprague cites prayers for revival in the Old Testament, instances of God's pouring out his Spirit in both testaments of the Bible, Scriptural prophesies of revivals, and revivals throughout Christian history.[24] In 1840 James Munro, Presbyterian minister at Rutherglen, Scotland, offered "Encouragements from the History of the Church under the Old and under the New Testament Dispensation, to expect, pray and labor for the Revival of Religion." Similarly to Edwards, Munro enumerates all the revivals he can identify in the Old Testament, as well as in the books of the New. He follows Edwards in believing that in the phrase "Then began men to call upon the name of the Lord" (Genesis 4:26), "we have no difficulty in recognizing . . . unequivocal symptoms of what we should in these days term a revival." He finds revivals led by Samuel, Asa, Jehosaphat, Hezekiah, Josiah, Ezra, Nehemiah, John the Baptist, and the Apostles after Christ's Ascension. He then quickly sketches the fortunes of evangelical piety until he reaches the eighteenth-century revivals in England, America, and in particular Scotland. Munro's intention was to demonstrate that throughout its history the church had experienced "remarkable visitations of the Holy Spirit, rapidly increasing the power, and enlarging the sphere of vital godliness," and that God had commonly employed the prayers of the people in the production of such revivals.[25]

By the end of the first Great Awakening, a revival of religion was no longer an amorphous entity discussed indirectly. Now it was understood as a discreet theological phenomenon with specific characteristics. Some theorists found, in the manner of Solomon Stoddard, precedents in Scripture as well as in the history of the Christian church to legitimize contemporary revivals. Led by Jonathan Edwards, some interpreted revivals as the chief mechanism by which God advanced his redemptive purposes and as the key to church history. With the Second Great Awakening, revivals became the subject of tracts, lectures, and treatises.

In 1833, the Reverend Charles Coffin wrote a letter for use in the appendix of the second edition of Sprague's *Lectures on Revivals of Religion.* Coffin (1775–1853), a member of one of the prominent and ancient families of Newburyport, Massachusetts, and a Harvard graduate (1793), had served as president of Greeneville College, in Tennessee, from 1810 to 1827.[26] In the conclusion of his letter, Coffin encapsulated the New England evangelical tradition's vision of the role revivals play in the unfolding of America's destiny. Transferring to the United States the myth of the Puritans' errand into the wilderness, of New England's being founded as a plantation for religion, not for commerce, he asserted the American church's obligation to work for the revival of religion worldwide. He wrote,

> The American church should realize her duty, till one pure and general revival shall spread its blessings over the inhabited globe. Never was there any other country settled, since Canaan itself, so much for the sacred purposes of religion, as our own. Never did any other ancestry, since the days of inspiration, send up so many prayers and lay such ample foundations for the religious prosperity of their descendants, as did our godly forefathers. It is a fact, therefore, in perfect analogy with the course of Providence, that there never has been any other country so distinguished for religious revivals as our own.

Then he issued a call to the church for repentance and reformation. The church

> has only to humble herself for her ingratitude and backslidings, to return from her inexcusable aberrations, and pursue the pure purposes of her pilgrimage, and all the world will soon be made to know that her God is the Lord.

In Coffin's opinion, revivals are essential for the success of the church's every endeavor to fulfill her role in making America a Christian nation and in spreading the kingdom of Christ through the world.

> Our Bible, missionary, education, Sabbath-school, temperance and colonization societies, the supply of our own people with a sufficient number of able and faithful ministers of the New Testament, and with pious and benevolent characters for the thousand other spheres of responsible action, the diffusion of the

> light of life, and the joys of the gospel salvation, through all our numerous habitations; the preservation of our invaluable liberties and free institutions, and all the happy prospects of our most favored country, depend greatly, under God, upon those pure and frequent and spreading revivals of religion, for which all American Christians, of whatever name, should pray and labor and strive and live, with one heart and one soul; and, so far as they possess the mind and spirit of their Master, most certainly will.[27]

The survival of the American republic, itself, depends on the revival of religion in the church, according to Coffin.

In the first Great Awakening, New England's evangelical theorists saw revivals as God's means of fulfilling the national covenant by making national reformation possible. In the Second Great Awakening, revivals continued in theory to fill the same function, but in very altered circumstances. New England's increasing social complexity, diversity, and inequality, and the region's political absorption into a growing and heterogeneous nation after independence diluted the community of feeling that Calvin Colton believed crucial to the atmosphere of a religious revival. The denominational structure of the Protestant church in America meant that no preacher, church, or church body, could pretend to the authority of Jeremiah to speak for the whole nation. Charles Coffin addressed his call to repentance and reform not to the nation, but to the church. On the frontier, revivals helped create communities out of strangers by distinguishing common values, but denominational identification also established boundaries to the holy community and barriers between neighbors. In cities, revivals manufactured corporate purpose, but also illuminated clear divisions between reformers and persons who were objects of reform. Through revivals of religion, evangelical churches sought to subordinate self-aggrandizing individualism to a spirit of community. They expected through revivals to overcome the disintegrative effects of heterogeneity by forging a unified national culture. Many nineteenth-century American evangelicals came to believe that the chief instrument by which the church would achieve its mission of saving sinners, sanctifying society, and establishing the millennial kingdom was the revival.[28]

APPENDIX 1

Occupations of the Cambuslang Converts

A. Females

Name	*Own*	*Father's*	*Husband's*
Alston, Janet			
Anderson, Catherine		portioner	
Anderson, Jean		tenant	
Barry, Janet			carter
Barton, Margaret		tenant	
Borland, Margaret		tenant	
Boyle, Margaret*			shoemaker
Bredon, Janet			soldier
Breehom, El.			
Brownlie, Margaret*		smith	
Buchanan, Agnes		merchant	
Burnsides, Agnes			tenant
Calendar, Marrin			
Cameron, Catherine		gentleman	
Campbell, K., Mrs.			
Carson, Margaret		sailor	
Clark, Margaret			day laborer
Colquhoun, Mary		tenant	
Creelman, Helen			
Davie, Bethea†		weaver	
Dickinson, Jean			
Dykes, Elizabeth	in service		
Dykes, Rebecca			
Finlay, Elizabeth		tenant	
Finlay, Helen		tenant	
Gilchrist, Sarah, Mrs.		schoolmaster	
Givan, Margaret*	seamstress	workman	

*Artisan background.
†Weaving family.

Name	Own	Father's	Husband's
Glasford, A.			
Hamilton, Agnes			
Hay, Jean	in service		
Jackson, Catherine	in service	(elder)	
Jackson, Elizabeth	(spinning)	(elder)	
Jackson, Janet	(spinning)	(elder)	
Lamont, Christine		tenant	
Lap, Mary	in service	collier	
Lennox, Janet*		gardener	
Lyon, Bessie*		cooper	
Matthie, Isabel			
McAlpin, Janet		wool merchant	
Merrilie, Janet	(spinning)		
Mitchell, Mary			
Moffat, Isabel*		shoemaker	
Moffat, Janet*		shoemaker	bleacher
Montgomery, Ann*		shoemaker	
More, Agnes		tenant	
Morton, Jean†			weaver
Park, Janet		packman	
Provan, Isabel		tenant	
Reid, Janet			
Reid, Margaret			
Reid, Rebecca		tenant	
Reston, Janet			
Richie, Margaret	(spinning)		
Robe, Jean	in service (spinning, milking, harvesting)		
Ronald, Jean			
Scot, Mary			
Shaw, Margaret		tenant	
Shaw, Mary*		ship carpenter	
Shearer, Helen			
Sinclair, K.			
Smith, Margaret			
Strang, Sarah			
Struthers, Janet			
Stuart, Catherine*	seamstress		
Tenant, Catherine*		maltman	
Tenant, Janet†		weaver	
Turnbull, Janet*	in service	tailor	

Name	*Own*	*Father's*	*Husband's*
Walker, Jean*		shoemaker	
Walson, Isabel*			tailor
Wark, Jean†	in service	weaver	
Wier, Mrs.			bailie
Wylie, Anne	in service		
Young, Agnes*		smith	

B. Males

Name	*Own*	*Father's*
Aiken, John†	in service, weaver	
Alge, Duncan†	journeyman weaver	
Baillie, William		
Barclay, Thomas*	shoemaker	
Bell, Archibald*	tailor	
Bilsland, Alexander*	shoemaker	
Causlam, William	tenant	
Cunningham, Charles*	shoemaker	
Falls, Andrew		
Forbes, Daniel	sergeant of the *illeg.*	
Foster, Thomas		
Hamilton, Robert†	weaver	
Hepburn, John		
Jack, James		
Jameson, William		
Kirkland, James		
Lang, James†	weaver	weaver
Logan, David	collier (old soldier)	
Malay, John*	gardener	
McDonald, Jo.†	weaver	
McLartie, Daniel†	in service	weaver
Millar, William	agriculture	
Montgomery, William		
Neil, James		
Parker, John*	dyer	
Rogers, Alexander*	apprentice	
Shearer, Robert		
Smith, Archibald*	mason	

Name	Own	Father's
Tassie, George*	tradesman	
Tenant, James		
Thomson, Charles*	shoemaker	
Thomson, Michael*	apprentice	
Walker, Thomas*	apprentice	
Wier, John	tenant	
Wier	bailie	

Derived from William McCulloch, "Examination of Persons under Scriptural Concern at Cambuslang during the Revival in 1741–42 by the Revd. William MacCulloch, Minister at Cambuslang, with Marginal Notes by Dr. Webster and Other Ministers." 2 vols. New College Library, Edinburgh. There are four narrations to which the guides in the back of the volumes do not make it possible to attach names. Since in each neither the name nor the occupation is known, these cases are left off this list. In the guides there is information about the occupational background of five women and one man whose names cannot be attached with certainty to any of the narrations, and those names are included in this list.

APPENDIX 2

Revival Narratives, 1741–1745 In Approximate Order of Composition

I. AMERICAN

Peter Thacher, "Revival of Religion at Middleborough in the County of Plymouth." 21 December 1741. *The Christian History* (Boston) 1: 412–16.

Samuel Allis, "A Brief Account of the Revival of Religion at Somers in the County of Hampshire, in the Province of Massachusetts." 22 May 1742. *The Christian History* 1: 408–12.

William Shurtleff, "Revival of Religion at Portsmouth the chief Town in the Province New-Hampshire in New-England." 1 June 1743. *The Christian History* 1: 382–94; *The Christian Monthly History* (Edinburgh), 2d ser. 4 (July 1745): 116–24; 5 (Aug. 1745): 125–26.

John Cotton, "A general History of the Revival of Religion (in the Summer 1741, and since that Time) at Hallifax, in the County of Plimouth." 26 June 1743. *The Christian History* 1: 259–70.

James Allen, "Brief Account of the Revival of Religion at Brookline." July 1743. *The Christian History* 1: 394–96.

Joseph Parks, "An Account of the late Propagation of Religion at Westerly and Charlestown in Rhode-Island Colony." 1 August 1743. *The Christian History* 1: 201–11.

Henry Messinger and Elias Haven, "An Account of the late Revival of Religion in both the Precincts of Wrentham, in the County of Suffolk in the Massachusets-Province." 12 August 1743. *The Christian History* 1: 236–50; *The Christian Monthly History*, 2d ser. 6 (September 1745): 185–88, 7 (October 1745): 189–200.

Jonathan Dickinson, "An Account of the Revival of Religion at Newark and Elizabeth-Town in the Province of New-Jersey." 23 August 1743. *The Christian History* 1: 252–58.

John Porter, "Revival of Religion in the North Precinct of Bridgewater in the Province of the Massachusetts." 12 October 1743. *The Christian History* 1: 396–408.

Jonathan Edwards, "Continuation of the State of Religion at Northampton in the

County of Hampshire." 12 December 1743. *The Christian History* 1: 367–81; *The Christian Monthly History*, 2d ser. 4 (July 1745): 101–14.

Joseph Parks, farther account of the progress of religion at "Westerly and Charlestown in Rhode-Island Colony." 6 February 1744. *The Christian History* 2: 22–28; *The Christian Monthly History*, 2d ser. 3 (June 1745): 73–79.

John Seccomb, "Revival of Religion at Harvard in the County of Middlesex." 20 February 1744. *The Christian History* 2: 13–21.

John White, "Revival of Religion in the first Precinct of Gloucester in the County of Essex." March 1744. *The Christian History* 2: 41–46.

Peter Thacher, "The Rev. Mr. Thacher's Account of the Revival of Religion at Middleborough East-Precinct." Found among his papers after his death. *The Christian History* 2: 87–92.

George Griswold, "An Account of the Revival of Religion at Lyme East Parish in Connecticut." 3 April 1744. *The Christian History* 2: 105–14.

George Griswold, "An Account of the Beginning of the Revival of Religion at New-London North Parish." 9 April 1744. *The Christian History* 2: 115–18.

Jonathan Parsons, "Account of the Revival of Religion in the West Parish of Lyme in Connecticut." 14 April 1744. *The Christian History* 2: 118–62.

David Hall, "Revival of Religion at Sutton in the County of Worcester, in the Massachusetts-Province." 28 May 1744. *The Christian History* 2: 162–72.

Samuel Blair, "Revival of Religion at New-Londonderry in the Province of Pennsylvania." 6 August 1744. *The Christian History* 2: 242–62. Also published as Samuel Blair, *A Short and Faithful Narrative of the Late Remarkable Revival of Religion in the Congregation of New-Londonderry, and Other Parts of Pennsylvania*. Philadelphia, 1744.

Gilbert Tennent, "A Letter from the Rev. Mr. Gilbert Tennent, late of New-Brunswick in the Province of New-Jersey, now of Philadelphia in the Province of Pennsylvania, relating chiefly to the late glorious Revival of Religion in those Parts of America." 24 August 1744. *The Christian History* 2: 285–98.

William Tennent, "An Account of the Revival of Religion at Freehold and other Places in the Province of New-Jersey." 11 October 1744. *The Christian History* 2: 298–310; *The Christian Monthly History*, 2d ser. 7 (October 1745): 213–18, 8 (November 1745): 220–25.

Nathaniel Leonard, "A brief Account of the late Revival of Religion in Plymouth; the first settled Town in New-England." 23 November 1744. *The Christian History* 2: 313–17.

Josiah Crocker, "An Account of the late Revival of Religion at Taunton, in the County of Bristol." 24 November 1744. *The Christian History* 2: 321–58.

Thomas Prince, "Some Account of the late Revival of Religion in Boston." 26 November 1744. *The Christian History* 2: 374–415; *The Christian Monthly History*, 2d ser. 5 (August 1745): 146–56, 6 (September 1745): 157–83.

John Rowland, *A Narrative of the Rise and Progress of Religion, in the Towns of Hopewell, Amwell and Maiden-Head, in New-Jersey, and New-Providence in Pennsylvania*, second title in Gilbert Tennent, *A Funeral Sermon Occasion'd by the Death of . . . John Rowland* (Boston, 1745), pp. 50–72.

II. SCOTTISH

M. O., *A True Account of the Wonderful Conversions at Cambuslang Contained in a Letter from a Gentleman in the Gorbals of Glasgow, to his Friend at Greenock*. 29 March 1742; also, appended to the Boston and Philadelphia editions of:

[James Robe], *A Short Narrative of the Extraordinary Work at Cambuslang; In a Letter to a Friend*. 8 May 1742. Glasgow, Philadelphia, Boston, 1742.

W. D., *A Short and True Account of the Wonderful Conversion at Kilsyth, in a Letter from a Gentleman there, to a Friend at Glasgow*. 18 June 1742. Glasgow, 1742.

James Robe, *A Faithful Narrative of the Extraordinary Work of the Spirit of God at Kilsyth and other Congregations in the Neighborhood*. Glasgow, 1742–1751. First installment also in *The Christian History* (Boston) 1, nos. 1–12.

William Hally, "An Account of the Revival of Religion at Muthel." 28 September 1742. Originally printed in Robe, *A Faithful Narrative*; reprinted in *The Christian History* (Boston) 1: 311–17.

William Hally, "A farther and more particular Account of the late Revival at Muthell." 29 August 1743. *The Christian History* 2: 183–87.

John Warden, "The Progress, and present State of the Revival in the Parish of Campsy." 16 December 1743. *The Christian Monthly History* (Edinburgh) 2 (December 1743): 34–51; *The Christian History* 2: 195–208.

William Hay, "An Account of the State of the Revival for some Time Past in the Parish of Kirkintilloch." December 1743. *The Christian Monthly History* 4 (February 1744): 41–44.

John Balfour, "Account of the Revival in the Parish of Nig, in East Ross." February 1744. *The Christian Monthly History* 4 (February 1744): 45–48.

John Balfour, "An Account of the Progress of the Revival in the Shires of Ross and Sutherland in the far Northern Parts of Scotland." 22 May 1745. *The Christian Monthly History*, 2d ser. 3 (May 1745): 61–64.

John Sutherland, "Account of the Lord's Work of Grace in the Parish of Golspy in the Shire of Sutherland." 8 August 1745. *The Christian Monthly History*, 2d ser. 5 (August 1745): 130–36.

NOTES

INTRODUCTION

1. See, in particular, Norman Pettit, *The Heart Prepared: Grace and Conversion in Puritan Spiritual Life* (New Haven, Conn.: Yale University Press, 1966), and Thomas A. Schafer, "Solomon Stoddard and the Theology of the Revival," in *A Miscellany of American Christianity: Essays in Honor of H. Sheldon Smith*, ed. Stuart C. Henry (Durham, N.C.: Duke University Press, 1963), pp. 328–61.

2. Gerald F. Moran, "The Puritan Saint" (Ph.D. diss., Rutgers University, 1974).

3. For the morphology of conversion, see Edmund S. Morgan, *Visible Saints: The History of a Puritan Idea* (Ithaca, N.Y.: Cornell University Press, 1963), pp. 66–73, 90–92; for the conversion narrative, see Patricia Caldwell, *The Puritan Conversion Narrative: The Beginnings of American Expression* (Cambridge: Cambridge University Press, 1983), and John Owen King, *The Iron of Melancholy: Structures of Spiritual Conversion in America from the Puritan Conscience to Victorian Neurosis* (Middletown, Conn.: Wesleyan University Press, 1983), pp. 13–82.

4. J. G. A. Pocock, "Languages and Their Implications: The Transformation of the Study of Political Thought," in *Politics, Language and Time: Essays on Political Thought and History* (New York: Atheneum, 1973): 3–41.

5. J. G. A. Pocock, *Virtue, Commerce, and History: Essays on Political Thought and History, Chiefly in the Eighteenth Century* (Cambridge: Cambridge University Press, 1985), p. 58.

6. Perceptual psychologists recognize that a person's perception is not random but influenced by his mental "set": "There may be a pre-established attitude that determines what is to be perceived and how one shall react." The individual's attitude affects "the selection of the objects that will be perceived and to some extent the readiness with which they are perceived. . . . The phenomenon is most clearly shown with respect to objects that we are looking for or meanings that we are seeking to realize from stimulus-situations that are undetermined or vague" (Floyd H. Allport, *Theories of Perception and the Concept of Structure* [New York: John Wiley & Sons, 1955], pp. 65, 85).

7. Morgan, *Visible Saints*, p. 71.

8. King, *The Iron of Melancholy*, pp. 42–43, 47–49, 49. In a different context, Pocock says that language "interacts with experience; it supplies the categories, grammar, and mentality through which experience has to be recognized and articulated. In studying it the historian learns how the inhabitants of a society were capable of cognizing experience, what experience they were capable of cognizing, and what responses to experience they were capable of articulating and consequently performing" (Pocock, *Virtue, Commerce, and History*, pp. 28–29).

9. Pocock, "Languages and Their Implications," pp. 38–39.

10. For the role of fear of sudden death in the revivals, see Michael R. Watts, *The Dissenters, I: From the Reformation to the French Revolution* (Oxford: Clarendon Press, 1978), pp. 412–17; and J. M. Bumsted and John E. Van de Wetering, *What Must I Do to Be Saved? The Great Awakening in Colonial America* (Hinsdale, Ill.: Dryden Press, 1976), p. 9.

11. My understanding of the dynamics of religious constituency has been shaped by the descriptive model of the constituency from which churches recruit members in Robert Currie, Alan Gilbert, and Lee Horsley, *Churches and Churchgoers: Patterns of Church Growth in the British Isles since 1700* (Oxford: Clarendon Press, 1977), pp. 6–7, 42–43.

12. John Demos, "Families in Colonial Bristol, Rhode Island: An Exercise in Historical Demography," *William and Mary Quarterly*, 3d ser. 25 (1968): 40–57; Philip J. Greven, Jr., "Youth, Maturity, and Religious Conversion: A Note on the Ages of Converts in Andover, Massachusetts, 1711–1749," *Essex Institute Historical Collections* 108 (1972): 119–34; Daniel Scott Smith and Michael S. Hindus, "Premarital Pregnancy in America, 1640–1971: An Overview and Interpretation," *Journal of Interdisciplinary History* 5 (1975): 537–70; Watts, *The Dissenters*, pp. 417–21; William F. Willingham, "Religious Conversion in the Second Society of Windham, Connecticut," *Societas* 6 (1976): 109–19. Kathleen Verduin's "'Our Cursed Natures': Sexuality and the Puritan Conscience," *New England Quarterly* 56 (1983): 220–37, sensitively describes the struggles of pious early New Englanders to come to terms with their sexuality.

13. John Bumsted, "Religion, Finance, and Democracy in Massachusetts: The Town of Norton as a Case Study," *Journal of American History* 57 (1971): 817–31; Cedric B. Cowing, "Sex and Preaching in the Great Awakening," *American Quarterly* 20 (1968): 624–44; James A. Henretta, *The Evolution of American Society, 1700–1800: An Interdisciplinary Analysis* (Lexington, Mass.: D.C. Heath & Co. 1973), pp. 132–34, and "The Morphology of New England Society in the Colonial Period," *Journal of Interdisciplinary History* 2 (1971): 379–98; N. Ray Hiner, "Adolescence in Eighteenth-Century America," *History of Childhood Quarterly* 3 (1975): 253–80; Gerald F. Moran, "Conditions of Religious Conversion in the First Society of Norwich, Connecticut, 1718–1744," *Journal of Social History* 5 (1972): 331–43, and "The Puritan Saint," pp. 279–325, 423; Edward Shorter, *The Making of the Modern Family* (New York: Basic Books, 1975), pp. 79–108. For an examination of the effects of land scarcity and increased mobility on intergenerational family relations in early eighteenth-century New England, see Philip J. Greven, Jr., *Four Generations: Population, Land, and Family in Colonial Andover, Massachusetts* (Ithaca, N.Y.: Cornell University Press, 1970), pp. 175–258. Statistical evidence of the decline of parental power is found for the town of Hingham, Massachusetts, where the ability of parents to defer the marriages of sons or to dictate the marriage order of daughters decreased (Daniel Scott Smith, "Parental Power and Marriage Patterns: An Analysis of Historical Trends in Hingham, Massachusetts," *Journal of Marriage and the Family* 35 [1973]: 419–28). On the special ministry directed to young people by pastors who noted their emerging independence, see Patricia J. Tracy, *Jonathan Edwards, Pastor: Religion and*

Society in Eighteenth-Century Northampton (New York: Hill & Wang, 1980), pp. 77–89, 91–92, 106–8, 110–12, 130–31. Roger Thompson documents a youth "counter-culture" in late seventeenth-century Massachusetts and argues that Puritan introspection manifested itself usually only once adult responsibilities were assumed ("Adolescent Culture in Colonial Massachusetts," *Journal of Family History* 9 [1984]: 127–44). On the decline of the legal enforcement of morality, see David H. Flaherty, "Law and the Enforcement of Morals in Early America," in *Law in American History*, ed., Donald Fleming and Bernard Bailyn, vol. 5 of *Perspectives in American History* (Boston: Little, Brown, 1971), pp. 203–53; and R. W. Roetger, "The Transformation of Sexual Morality in 'Puritan' New England: Evidence from New Haven Court Records, 1639–1698," *Canadian Review of American Studies* 15 (1984): 243–57.

14. For the significant changes in the British economy in the first half of the eighteenth century, preceding the momentous transformations of the agricultural and industrial revolutions, see Sir George Clark, *English History: A Survey* (Oxford: Clarendon Press, 1971), p. 350; D. C. Coleman, *The Economy of England, 1450–1750* (London and New York: Oxford University Press, 1977), pp. 125, 171–72; and Paul Mantoux, *The Industrial Revolution in the Eighteenth Century: An Outline of the Beginnings of the Modern Factory System in England*, rev. ed. (New York: Harper & Row, 1965), pp. 64–82, 136–85. For an example of the changes effected in one Midlands village in the first half of the eighteenth century, see William G. Hoskins, *The Midlands Peasant: The Economic and Social History of a Leicester Village* (London: Macmillan & Co., 1957), pp. 211–29.

15. Evelyn Douglas Bebb, *Nonconformity and Social and Economic Life, 1660–1800: Some Problems of the Present as They Appeared in the Past* (London: Epworth Press, 1935), pp. 46–57; John C. C. Probert, *The Sociology of Cornish Methodism to the Present Day*, Cornish Methodist Historical Association, no. 17 (Redruth, 1971), pp. 23–26; Watts, *The Dissenters*, pp. 408–9; Alan D. Gilbert, *Religion and Society in Industrial England: Church, Chapel, and Social Change, 1740–1914* (London: Longman, 1976), pp. 59–61, 64–67.

16. Bebb, *Nonconformity and Social and Economic Life*, p. 42; Robert Currie, "A Micro-theory of Methodist Growth," *Proceedings of the Wesley Historical Society* 36 (1967): 65–73; Alan Everitt, "Nonconformity in Country Parishes," *British Agricultural History Review* 18, suppl. (1970): 178–99; Gilbert, *Religion and Society*, pp. 66–67, 74–85, 94–121.

17. Gilbert, *Religion and Society*, pp. 87–93.

18. Bernard Semmel, *The Methodist Revolution* (New York: Basic Books, 1973), pp. 5, 7–9, 198, and passim. Evangelical Arminianism was a liberal, progressive ideology, says Semmel, in the sense that it assisted the transformation from traditional to modern. In a similar vein, Alan Heimert argues that the postmillennialism of evangelical Calvinism that proceeded from the Great Awakening helped produce liberal Americans who believed in progress, liberty, and equality (*Religion and the American Mind: From the Great Awakening to the Revolution* [Cambridge, Mass.: Harvard University Press, 1966]). It may be that the liberating force had more to do with evangelicalism than with either Arminian or Calvinist theology.

19. Andrew Gibb, *Glasgow: The Making of a City* (London: Croom Helm, 1983), pp. 56–78; Henry Hamilton, *An Economic History of Scotland in the Eighteenth Century* (Oxford: Clarendon Press, 1963), pp. 3–87, 131–59, 249–67, 343–74.

20. Callum G. Brown, *The Social History of Religion in Scotland since 1730* (London and New York: Methuen, 1987), pp. 26, 101–2; see also pp. 26–27, 52, 109–12.

21. For the change from communitarian to individualistic values, see Edward M. Cook, Jr., "Social Behavior and Changing Values in Dedham, Massachusetts, 1700 to 1775," *William and Mary Quarterly*, 3d ser. 27 (1970): 546–80; and Henretta, "The Morphology of New England Society in the Colonial Period," p. 396. For an analysis of the transformation of the nature of community in a New England town, see Kenneth A. Lockridge, *A New England Town, the First Hundred Years: Dedham, Massachusetts, 1636–1736* (New York: W. W. Norton & Co., 1970). Compare, for the persistence of behavior consistent with communal values in eighteenth-century New England towns, Michael Zuckerman, *Peaceable Kingdoms: New England Towns in the Eighteenth Century* (New York: Alfred A. Knopf, 1970).

22. Richard L. Bushman, *From Puritan to Yankee: Character and the Social Order in Connecticut, 1690–1765* (Cambridge, Mass.: Harvard University Press, 1967), pp. 183–95; Henretta, *Evolution of American Society*, pp. 134–38. Bushman's and Henretta's arguments are congruent with Gilbert's and Semmel's that the Evangelical Revival helped recreate personalities psychologically more autonomous and better suited to modern society. Yet it is also possible to argue convincingly, as Jack P. Greene has, that the revivals failed to produce a satisfactory or lasting psychological adjustment to modern conditions. Successful or not, the revivals can be seen as a response to the new conditions (Jack P. Greene, "Search for Identity: An Interpretation of the Meaning of Selected Patterns of Social Response in Eighteenth-Century America," *Journal of Social History* 3 (1970): 189–220).

23. Watts, *The Dissenters*, pp. 408–9.

24. Moran, "The Puritan Saint," pp. 342–77, and "Conditions of Religious Conversion in the First Society of Norwich, Connecticut"; Gerald F. Moran and Maris A. Vinovskis, "The Puritan Family and Religion: A Critical Reappraisal," *William and Mary Quarterly*, 3d ser. 39 (1982): 29–63; James Walsh, "The Great Awakening in the First Congregation of Woodbury, Connecticut," *William and Mary Quarterly*, 3d ser. 28 (1971): 543–62.

25. Walsh, "The Great Awakening in the First Congregation of Woodbury," p. 555; Gary B. Nash, *The Urban Crucible: Social Change, Political Consciousness, and the Origins of the American Revolution* (Cambridge, Mass.: Harvard University Press, 1979), pp. 204–21.

26. Harry S. Stout and Peter S. Onuf, "James Davenport and the Great Awakening in New London," *Journal of American History* 70 (1983): 556–78, 562, 563. For the constituency of the radical revival, see also Peter S. Onuf, "New Lights in New London: A Group Portrait of the Separatists," *William and Mary Quarterly*, 3d ser. 37 (1980): 627–43; and Leigh Eric Schmidt, "'A Second and Glorious Reformation': The New Light Extremism of Andrew Croswell," *William and Mary Quarterly*, 3d ser. 43 (1986): 214–44.

27. Rhys Isaac, *The Transformation of Virginia, 1740–1790* (Chapel Hill, N.C.: University of North Carolina Press, 1982), esp. pp. 161–205.

28. John Sutherland, 8 August 1745, *The Christian Monthly History*, no. 5 (1745), pp. 130–36, reprinted in John Gillies, comp., *Historical Collections Relating to Remarkable Periods of the Success of the Gospel, and Eminent Instruments Employed in Promoting It*, 2 vols. (Glasgow, 1754; revised by Horatius Bonar, 1845) rev. ed. in 1 vol. as *Historical Collections of Accounts of Revival* (Fairfield, Pa.: Banner of Truth Trust, 1981), p. 457.

29. William McCulloch, "Examination of Persons under Scriptural Concern at Cambuslang during the Revival in 1741–42 by the Revd. William MacCulloch, Minister at Cambuslang, with Marginal Notes by Dr. Webster and Other Ministers" ("McCulloch Mss."), 2 vols., New College Library, Edinburgh.

30. T. C. Smout, "Born Again at Cambuslang: New Evidence on Popular Religion and Literacy in Eighteenth Century Scotland," *Past and Present* 97 (1982): 114–27, presents a similar analysis of the McCulloch Mss.

31. Watts, *The Dissenters*, pp. 421–28.

32. See Moran, "The Puritan Saint," and "Conditions of Religious Conversion in the First Society of Norwich, Connecticut"; Moran and Vinovskis, "The Puritan Family"; and Walsh, "The Great Awakening in the First Congregation of Woodbury."

33. Philip J. Greven, Jr., *The Protestant Temperament: Patterns of Child-rearing, Religious Experience, and the Self in Early America* (New York: Alfred A. Knopf, 1977), pp. 21–148.

34. Self-annihilation, Greven's term, here, of course, is intended to be taken psychologically, not physically.

35. William Kemp Lowther Clarke, *Eighteenth Century Piety* (London: Society of Promoting Christian Knowledge, 1944).

36. For pietism as a reform movement, see F. Ernst Stoeffler, *The Rise of Evangelical Pietism* (Leiden: E. J. Brill, 1971), pp. 22–23.

37. Patricia U. Bonomi and Peter R. Eisenstadt, "Church Adherence in the Eighteenth-Century British American Colonies," *William and Mary Quarterly*, 3d ser. 39 (April 1982): 245–86; Greven, *The Protestant Temperament*, pp. 21–148; Charles E. Hambrick-Stowe, *The Practice of Piety: Puritan Devotional Disciplines in Seventeenth-Century New England* (Chapel Hill, N.C.: University of North Carolina Press, 1982); Harry S. Stout, *The New England Soul: Preaching and Religious Culture in Colonial New England* (New York: Oxford University Press, 1986); Marilyn J. Westerkamp, *Triumph of the Laity: Scots-Irish Piety and the Great Awakening* (New York: Oxford University Press, 1988). The starting point for this revision is Robert G. Pope, *The Half-way Covenant: Church Membership in Puritan New England* (Princeton, N.J.: Princeton University Press, 1969); see also Pope, "New England versus the New England Mind: The Myth of Declension," *Journal of Social History* 3 (1969–70): 95–108.

38. See note 24.

39. Westerkamp, *Triumph of the Laity*.

40. Henry Grey Graham, *The Social Life of Scotland in the Eighteenth Century* (1899; reprint ed., London: A & C Black, 1928), pp. 366–71; John MacInnes, *The*

Evangelical Movement in the Highlands of Scotland, 1688–1800 (Aberdeen: The University Press, 1951), pp. 98–103, 197–220.

CHAPTER 1

1. *Oxford English Dictionary*, s.v. "Revival"; W. J. Cooper, *Scottish Revivals* (Dundee, 1918), p. 3.

2. Samuel Torrey, *An Exhortation unto Reformation* (Cambridge, Mass., 1674), p. 34.

3. Solomon Stoddard, "The Benefit of the Gospel," in *The Efficacy of the Fear of Hell, to Restrain Men from Sin* (Boston, 1713), p. 55.

4. Charles E. Hambrick-Stowe, *The Practice of Piety: Puritan Devotional Practice in Seventeenth-Century New England* (Chapel Hill, N.C.: The University of North Carolina Press, 1982), esp. pp. 195–287, discusses the cyclical character of Puritan religious experience.

5. Samuel Torrey, *A Plea for the Life of Dying Religion* (Boston, 1683), p. 20.

6. Gerald R. Cragg, *Puritanism in the Period of the Great Persecution, 1660–1688* (Cambridge: Cambridge University Press, 1957).

7. J. H. S. Burleigh, *A Church History of Scotland* (London: Oxford University Press, 1960), pp. 233–57.

8. Robert Fleming, *Fulfilling of the Scripture* (Rotterdam, 1669; 2d and 3d pts. [1677?]; all three pts., London, 1681; 3d ed., 1681; 4th ed., 1693; 5th ed., 1726). I have used the Boston editon of 1743.

9. Ibid., pp. xii–xiv.

10. Ibid., p. 176.

11. Ibid., pp. 287–351, 344.

12. Ibid., pp. 132–34.

13. Ibid., p. 393.

14. Ibid., p. 394.

15. Leigh Eric Schmidt, *Holy Fairs: Scottish Communions and American Revivals in the Early Modern Period* (Princeton, N.J.: Princeton University Press, 1989), pp. 21–32.

16. John Howe, "The Prosperous State of the Christian Interest . . ." in *Works* (New York, 1835), 1:562–607. The sermons were delivered at a Wednesday lecture series at Cordwainers' Hall, London, but were not published until 1725. In his preface to Jonathan Edwards's *Distinguishing Marks* (Boston, 1741), the Boston minister William Cooper quoted an extended passage from these sermons concerning the pouring out of the Spirit on preachers at the millennium.

17. Howe, "The Prosperous State of the Christian Interest," p. 564.

18. Ibid, pp. 574–79, 575.

19. Ibid., pp. 580–82, 579.

20. Ibid., pp. 602–4. For a later expression of Howe's ideas about the nature, necessity, and promise of an outpouring of the Holy Spirit, see Howe, "A Sermon on the Thanksgiving Day December 2, 1697," in *Works*, 2:925–31.

21. Owen remarked, "You know that, for many years, upon all these occasions,

without failing, I have been warning you continually of an approaching calamitous time, and considering the sins that have been the causes of it" ("The Use and Advantage of Faith in a Time of Public Calamity," [1680] in *The Works of John Owen*, ed. William H. Goold, [New York, 1852], 9:490–98); and "It hath been . . . my constant course . . . to treat in particular about our own sins, our own decays, our own means of recovery" ("Seasonable Words for English Protestants," preached on a fast day, 22 December 1681, published 1690, and reprinted in *Works of John Owen*, 9:2–17). For other jeremiads, see: "An Humble Testimony Unto the Goodness and Severity of God in His Dealing with Sinful Churches and Nations," (1681) in *Works of John Owen*, 8:594–658; "The Sin and Judgment of Spiritual Barrenness," in *Works of John Owen*, 9:179–97; "God's Withdrawing His Presence, the Correction of His Church," (1675) in *Works of John Owen*, 9:296–307; "Perilous Times," (1676) in *Works of John Owen*, 9:320–34; "The Use of Faith in a Time of General Declension in Religion," (1680) in *Works of John Owen*, 9:510–16.

22. Owen, "The Use of Faith in a Time of General Declension in Religion," 9:490–516.

23. Owen, ibid., 9:514.

24. Owen, "An Humble Testimony Unto the Goodness and Severity of God," 8:610, 639.

25. Owen, "Seasonable Words for English Protestants," 9:6.

26. Owen, "An Humble Testimony Unto the Goodness and Severity of God," 8:648.

27. Owen, "Seasonable Words for English Protestants," 9:16.

28. Sacvan Bercovitch, *The American Jeremiad* (Madison, Wis.: The University of Wisconsin Press, 1978), pp. 3–61, esp. 6–10.

29. Increase Mather, "To the Reader," in *An Exhortation unto Reformation*, by Samuel Torrey.

30. Ibid.

31. John Higginson, *The Cause of God and His People in New England* (Cambridge, Mass., 1663), p. 11.

32. Torrey, *A Plea for the Life of Dying Religion*, p. 17.

33. Torrey, *An Exhortation unto Reformation*, p. 8.

34. Ibid., p. 40.

35. S. H. [Samuel Hooker?], "To the Christian Reader," in *The Way of Israel's Welfare*, by John Whiting (Boston, 1686).

36. Torrey, *A Plea for the Life of Dying Religion*, p. 2.

37. John Davenport, *God's Call to His People to Turn unto Him* (Cambridge, Mass., 1669), p. 11.

38. Whiting, *The Way of Israel's Welfare*, p. 22.

39. Samuel Willard, *Reformation the Great Duty* (Boston, 1694), p. 18.

40. Samuel Willard, *The Only Sure Way to Prevent Threatened Calamity*, third title in *The Child's Portion* (Boston, 1684), p. 184.

41. Higginson, *The Cause of God*, p. 23; also, for example, "In the sense of our own nothingness, and believing Gods power and grace: it becometh us suitably to apply our selves unto him" (Samuel Hooker, *Righteousness Rained from Heaven* [Cambridge, Mass., 1677], p. 4).

42. "God's usuall way which he will bless for the converting or turning of elect sinners to himself, is by sending his Ministers with a Message from himself to them, in their Preaching Gods Word unto them" (Davenport, *God's Call*, p. 6).

43. Willard, *The Only Sure Way*, p. 177; or, as John Norton wrote, "When Sion for its sins is become an Out-cast (a subject of contempt) God takes occasion from her Calamity to give her Repentance, that so he may bring her the Blessing of his own People" ("Sion the Outcast," in *Three Choice and Profitable Sermons* [Cambridge, Mass., 1664], p. 3).

44. Davenport, *God's Call*, p. 12.

45. Torrey, *An Exhortation unto Reformation*, p. 14.

46. "Though their Repentance and Reformation could merit nothing, yet if they would sowe in righteousness, and break up their fallow ground, Mourn aright for sins past thorough the forebearance of God, and amend their doings: then they should reap a blessing and inherit prosperity, through the benignity and kindness of God: denoting that they must be beholding to mercy when they had done all they could" (Hooker, *Righteousness Rained from Heaven*, p. 3).

47. William Adams (Boston, 1679).

48. Hooker, *Righteousness Rained from Heaven*, p. 15.

49. Increase Mather, *Pray for the Rising Generation* (Cambridge, Mass., 1678), pp. 14–15.

50. Adams, *The Necessity of the Pouring Out of the Spirit*, p. 3.

51. Ibid., pp. 14–19.

52. Hooker, *Righteousness Rained from Heaven*, pp. 27–28; also, for example: "That all ordinary means . . . have been altogether ineffectual unto a general and saving work of reformation; makes it (at least) a fearful question, whether our degeneracy and apostacy may not prove Judicial, and so perpetual: a question which will not admit of a comfortable resolution, *until God shall pour out his Spirit from on high upon us*. That Sovereign promise of the donation, and effusion of the Spirit, and so of the dispensation of saving grace (as it hath been unto the Churches in all Ages, under their deepest defection,) So it is unto us the main stay of our faith, confidence, and comfort, and that which gives some present reviving, unto our languishing hope; of the resurrection of Religion in these Churches" (Samuel Torrey and Josiah Flint, "To the Reader," in *The Necessity of the Pouring Out of the Spirit*, by William Adams).

53. Adams, *The Necessity of the Pouring Out of the Spirit*, pp. 32–33.

54. Emory Elliott, *Power and the Pulpit in Puritan New England* (Princeton, N.J.: Princeton University Press, 1975).

55. Samuel Torrey, *Man's Extremity, God's Opportunity* (Boston, 1695), pp. 9–10, 60.

56. Owen said that there were special seasons when the Gospel was preached in power and purity, when providential calls joined with Gospel calls, "when God moves, [as he does] at some seasons, more effectually upon your hearts and spirits in the dispensation of the word than at other times" ("The Sin and Judgment of Spiritual Barrenness," 9:195–96).

57. Adams, *The Necessity of the Pouring Out of the Spirit*, p. 35.

58. Ibid; the citation to Fleming is to *Fulfilling of the Scripture*, 2d ed., p. 142.

59. Increase Mather, *Returning Unto God the Great Concernment of a Covenant People* (Boston, 1680), p. 12.

60. Bercovitch, *The American Jeremiad*, pp. 80–86, 84.

CHAPTER 2

1. For the changes in civil enforcement of religious duties, see Timothy H. Breen, *The Character of the Good Ruler: A Study of Puritan Political Ideas in New England, 1630–1730* (New Haven, Conn.: Yale University Press, 1977). For the disappearance of civil prosecutions for premarital sexual intercourse, see David H. Flaherty, "Law and the Enforcement of Morals in Early America," in *Law in American History*, ed. Donald Fleming and Bernard Bailyn, vol. 5 of *Perspectives in American History* (Boston: Little, Brown, 1971), pp. 203–53; and R. W. Roetger, "The Transformation of Sexual Morality in 'Puritan' New England: Evidence from New Haven Court Records, 1639–1698," *Canadian Review of American Studies* 15 (1984): 243–57. For the lessening importance of churches as tribunals, see David Konig, *Law and Society in Puritan Massachusetts: Essex County, 1629–1692* (Chapel Hill, N.C.: University of North Carolina Press, 1979). The conventions of Massachusetts ministers of 1715 and 1725 unsuccessfully sought permission to call a synod for the purpose of considering expedients for reformation and revival. J. William T. Youngs, Jr., points out that the "association movement" in the early eighteenth century increased the formal authority of the ministers in both Massachusetts and Connecticut, but that the extension of formal authority did not result in an extension of effective influence over the laity (J. William T. Youngs, *God's Messengers: Religious Leadership in Colonial New England, 1700–1750* [Baltimore: Johns Hopkins University Press, 1976], pp. 69–78, 92–108).

2. Peter G. Forster, "Secularization in the English Context: Some Conceptual and Empirical Problems," *Sociological Review*, n.s. 20 (1972): 153–68.

3. Tina Isaacs, "The Anglican Hierarchy and the Reformation of Manners," *Journal of Ecclesiastical History* 33 (1982): 391–411; A. Tindal Hall, *Church and Society, 1600–1800* (London: Society for Promoting Christian Knowledge, 1968), pp. 52–53, 71–74.

4. Alan D. Gilbert, *Religion and Society in Industrial England: Church, Chapel, and Social Change, 1740–1914* (London: Longman, 1976), pp. 8–9, 125–43, 205–7.

5. Holden Hutton, *The English Church from the Accession of Charles I. to the Death of Anne (1625–1714)* (London: Macmillan & Co., 1913), chap. 17; John Henry Overton, *Life in the English Church (1661–1714)* (London, 1885), pp. 207–13; Gordon Rupp, *Religion in England, 1688–1791* (Oxford: Clarendon Press, 1986), pp. 290–95; Josiah Woodward, *An Account of the Rise and Progress of the Religious Societies*, 3d ed. (London, 1701).

6. Dudley W. R. Bahlman, *The Moral Reformation of 1688* (New Haven, Conn.: Yale University Press, 1957); Garnet V. Portus, *Caritas Anglicana* (London: A. R. Mowbray & Co., 1912); *Proposals for a National Reformation of Manners* (London, 1694); Rupp, *Religion in England*, pp. 295–98; [Josiah Woodward], *An Account of the*

Progress of the Reformation of Manners, 12th ed. (London, 1704); Woodward, *An Account of the Rise and Progress of the Religious Societies*.

7. Bahlman, *The Moral Reformation of 1688*, pp. 83–97.

8. Daniel Neal, *A Sermon Preach'd to the Societies for Reformation of Manners* (London, 1722), p. 20; Samuel Price, *A Sermon Preach'd to the Societies for Reformation of Manners* (London, 1725), p. 28; John Guyse, *Reformation upon the Gospel Scheme* (London, 1735).

9. Robert Wodrow, *Correspondence*, 3 vols. (Edinburgh, 1842–43), 2: 663–4; John Maclaurin, "The Necessity of Divine Grace to Make the Word Effectual," in *Works* (Glasgow, 1824); Ebenezer Erskine, "The Standard of Heaven Lifted up against the Powers of Hell," in *The Whole Works of Ebenezer Erskine* (Edinburgh, 1793), 2: 93.

10. The proposal is reprinted in *The Diary of Cotton Mather for the Year 1712*, ed. William R. Maniere II (Charlottesville, Va.: University Press of Virginia, 1964), 113–18.

11. See for example, Increase Mather, *Dissertation Concerning the Danger of Apostacy*, second title in *A Call from Heaven to the Present and Succeeding Generations* (Boston, 1679), pp. 68–69, and *The Necessity of Reformation* (Boston, 1679), pp. 14–16; Samuel Torrey, *An Exhortation unto Reformation* (Cambridge, Mass., 1674), pp. 30–35, and *A Plea for the Life of Dying Religion* (Boston, 1683), pp. 33–35; John Whiting, *The Way of Israel's Welfare* (Boston, 1686), pp. 24–31; Samuel Willard, *Reformation the Great Duty* (Boston, 1694), pp. 74–76.

12. Breen, *The Character of the Good Ruler*, pp. 97–100; David D. Hall, *The Faithful Shepherd: A History of the New England Ministry in the Seventeenth Century* (1972; reprint ed., New York: W. W. Norton & Co., 1974), pp. 243–44; Increase Mather, *Renewal of the Covenant* (Boston, 1677).

13. Robert Middlekauff, *The Mathers: Three Generations of Puritan Intellectuals, 1596–1728* (New York: Oxford University Press, 1971), chaps. 7 and 8; Increase Mather, *The Order of the Gospel* (Boston, 1700), p. 11, and *The Blessed Hope* (Boston, 1701), pp. 113–15.

14. Increase Mather, *A Dissertation Wherein the Strange Doctrine Lately Published in a Sermon, the Tendency of Which is, to Encourage Unsanctified Persons (While Such) to Approach the Holy Table of the Lord, Is Examined and Confuted* (Boston, 1708), preface.

15. Ibid., app., p. 92.

16. Increase Mather, *Discourse Concerning Faith and Fervency in Prayer* (Boston, 1710), p. 65.

17. Increase Mather, *Five Sermons* (Boston, 1719), p. 120.

18. Massachusetts Bay Province, *By the Governor and General Court . . . A Proclamation* (Cambridge, 1690); Cotton Mather, *The Present State of New England* (Boston, 1690), *The Serviceable Man* (Boston, 1690), pp. 47, 60; Massachusetts Bay Province, *A Proclamation* (Boston, 1699); Massachusetts Bay Province, *A Declaration against Profaneness and Immoralities* (Boston, 1704); John Norton, *An Essay Tending to Promote Reformation* (Boston, 1708), p. 27; Grindal Rawson, *The Necessity of a Speedy and Thorough Reformation* (Boston, 1709), p. 29; Benjamin Colman, *A Sermon for the Reformation of Manners* (Boston, 1716), p. 25.

19. Cotton Mather, *Methods and Motives for Societies to Suppress Disorders* (Boston, 1703), *A Faithful Monitor* (Boston, 1704), and *Bonifacius* (Boston, 1710), pp. 167–74; Middlekauff, *The Mathers*, p. 273.

20. Cotton Mather, "The Diary of Cotton Mather," *Massachusetts Historical Society Collections*, 7th ser. 7 (1911): 176, *Rules for the Society of Negroes* [2d ed.] [Boston, between 1706 and 1711], *Early Religion Urged* (Boston, 1694), *Private Meetings Animated and Regulated* (Boston, 1706), *Bonifacius,* (Boston, 1710), pp. 82–87, and *Religious Societies* (Boston, 1724); Charles E. Hambrick-Stowe, *The Practice of Piety: Puritan Devotional Practice in Seventeenth-Century New England* (Chapel Hill, N.C.: University of North Carolina Press, 1982), pp. 137–43.

21. Middlekauff, *The Mathers*, pp. 216, 226–27, 239, 276.

22. Ibid., p. 348. Cotton Mather's works on the millennium include: *Things to Be Look'd For* (Cambridge, Mass., 1691), *A Midnight Cry* (Boston, 1692), *Things for a Distress'd People* (Boston, 1696), *Eleutheria* (London, 1698), *An Advice, to the Churches* (Boston, 1702), *Theopolis Americana* (Boston, 1710), *Thoughts for a Day of Rain* (Boston, 1712), *Stone Cut Out of the Mountain* (Boston, 1716), *Malachi* (Boston, 1717), *Columbanus* (Boston, 1722), *Terra Beata* (Boston, 1726), and *The Terror of the Lord* (Boston, 1727).

23. Middlekauff, *The Mathers*, pp. 347–48.

24. Samuel Danforth, *The Duty of Believers to Oppose the Growth of the Kingdom of Sin* (Boston, 1708), pp. 20–21.

25. *The Christian History* (Boston), 1: 108–9.

26. Samuel Danforth, *An Exhortation to All to Use Utmost Endeavours to Obtain a Visit of the God of Hosts, for the Preservation of Religion, and the Church, upon Earth* (Boston, 1714), reprinted in *The Wall and the Garden: Selected Massachusetts Election Sermons, 1670–1775*, ed. A. W. Plumstead (Minneapolis: University of Minnesota Press, 1968), pp. 150–76, 168.

27. Samuel Danforth, "The Building of Sion," in *Bridgewater's Monitor* by James Keith and Samuel Danforth (Boston, 1717), p. 7.

28. Massachusetts Historical Society, *Collections*, 4th ser. 1 (1852): 255–60.

29. For sacramental evangelism, see E. Brooks Holifield, *The Covenant Sealed: The Development of Puritan Sacramental Theology in Old and New England, 1570–1720* (New Haven, Conn.: Yale University Press, 1974).

30. Hall, *The Faithful Shepherd*, pp. 255–59.

31. See James G. Goulding, "The Controversy between Solomon Stoddard and the Mathers: Western Versus Eastern Massachusetts Congregationalism" (Ph.D. diss., Claremont, 1971), and Paul R. Lucas, "'An Appeal to the Learned': The Mind of Solomon Stoddard," *William and Mary Quarterly*, 3d ser. 30 (1973): 257–92.

32. Solomon Stoddard, *An Appeal to the Learned* (Boston, 1709), preface.

33. Solomon Stoddard, *The Efficacy of the Fear of Hell, to Restrain Men from Sin* (Boston, 1713), p. 9.

34. Solomon Stoddard, "The Benefit of the Gospel," in *The Efficacy of the Fear of Hell*, pp. 54, 55.

35. Lucas, "'An Appeal to the Learned,'" pp. 283–92.

36. See Stoddard, "The Benefit of the Gospel," sermon 7, p. 193.

37. Quotations from Increase Mather, *The Mystery of Israel's Salvation* (London, 1669), p. 62, and *A Dissertation Wherein the Strange Doctrine*, p. 93. See also *A Dissertation Wherein the Strange Doctrine*, app., *A Dissertation Concerning the Future Conversion of the Jewish Nation* (London, 1709), and *A Discourse Concerning Faith and Fervency in Prayer*.

38. *The Christian History* (Boston), 1: 108–9; Danforth, "An Exhortation to All," pp. 150–76.

39. Samuel Willard, "To the Reader," in *Man's Extremity, God's Opportunity*, by Samuel Torrey (Boston, 1695).

CHAPTER 3

1. John Howe, "The Prosperous State of the Christian Interest Before the End of Time, by a Plentiful Effusion of the Holy Spirit," in *Works* (New York, 1835), 1: 575.

2. Quoted in John Waddington, *Congregational History, 1700–1800* (London, 1876), p. 23.

3. Horton Davies, *Worship and Theology in England: From Watts and Wesley to Maurice, 1690–1850* (Princeton, N.J.: Princeton University Press, 1961), chap. 4.

4. R. Tudor Jones, *Congregationalism in England, 1662–1962* (London: Independent Press, 1962), p. 130.

5. William Nicholls, *Defense of the Doctrine and Discipline of the Church of England* (London, 1715), p. 335, excerpted in Waddington, *Congregational History*, p. 21.

6. Davies, *Worship and Theology in England*, p. 54.

7. See Olive M. Griffiths, *Religion and Learning: A Study in English Presbyterian Thought* (Cambridge: Cambridge University Press, 1935); Roger Thomas, "Presbyterians in Transition," in *The English Presbyterians: From Elizabethan Puritanism to Modern Unitarianism*, ed. C. Gordon Bolam et al. (Boston: Beacon Press, 1968), pp. 113–74; and Jeremy Goring, "The Break-Up of the Old Dissent," in ibid., pp. 175–218.

8. In numerous compositions, Isaac Watts, for instance, attempted to map a straight path between excessive reliance on reason and a blind faith in devout raptures. See his *Discourses of the Love of God, and Its Influence on All the Passions: With a Discovery of the Right Use and Abuse of Them in Matters of Religion* (London, 1729), and *The Strength and Weakness of Human Reason* (London, 1731); and Arthur Paul Davis, *Isaac Watts: His Life and Works* (New York: The Dryden Press, 1943), pp. 135, 222. For a Scottish attempt to justify emotional religion in terms of contemporary philosophy, see John Maclaurin, "On the Scripture Doctrine of Divine Grace," written in or about 1732 but not published until 1755 in his *Sermons and Essays* (Glasgow, 1755).

9. Jones, *Congregationalism in England*, pp. 141–42; Geoffrey Fillingham Nuttall, *Richard Baxter and Philip Doddridge: A Study in a Tradition* (London: Oxford University Press, 1951).

10. The discourses were published posthumously with a preface by Isaac Watts. Watts reprinted the discourses in 1735, with a letter on evangelical preaching by the

German Pietist August H. Francke. In 1740 Benjamin Colman reprinted in Boston the 1735 version.

11. John Guyse, *Christ the Son of God* (London, 1729), and *The Scripture Notion of Preaching Christ Further Clear'd and Vindicated* (London, 1730); Samuel Chandler, *A Letter to the Reverend Mr. John Guyse* (London, 1730), and *A Second Letter* (London, 1730); see also John Henley, *Samuel Sleeping in the Tabernacle* (London, 1730).

12. Starting in 1727, London merchant William Coward (ca. 1648–1738) financed several lecture series. In 1730 Thomas Bradbury began a lecture at Lime Street, and in 1733 Isaac Watts started one at Bury Street. See John Hubbard et al., *Christ's Loveliness and Glory . . . Twelve Sermons, Preach'd at Mr. Coward's Lecture* (London, 1729); Robert Bragge et al., *A Defense of Some Important Doctrines of the Gospel in Twenty-Six Sermons. Most of Which Were Preached at Lime-Street Lecture*, 2d ed., 2 vols. (Glasgow, 1773); and Isaac Watts et al., *Faith and Practice Represented in Fifty-Four Sermons*, ("Preached at Bury-Street, 1733,") 3d ed. (Edinburgh, 1792).

13. Jones, *Congregationalism in England*, p. 140; R. W. Dale and A. W. W. Dale, *History of English Congregationalism*, 2d ed. (London, Hodder and Stoughton, 1907), pp. 558–59; *Rules Agreed upon to Be Observed, with Relation to the Encouragement of Young Men, Who Are Enclined to Give Themselves up to the Work of the Ministry* (London, 1732).

14. Strickland Gough, *Some Observations on the Present State of the Dissenting Interest* (London, 1730); Philip Doddridge, *Free Thoughts on the Most Probable Means of Reviving the Dissenting Interest* (London, 1730); David Some, *The Methods to Be Taken by Ministers for the Revival of Religion* (London, 1730); Abraham Taylor, *Of Spiritual Declensions* (London, 1732); Isaac Watts, *An Humble Attempt towards the Revival of Practical Religion* (London, 1731). For analyses of this debate, see Dale and Dale, *History of English Congregationalism*, pp. 550–59; and Michael R. Watts, *The Dissenters: From the Reformation to the French Revolution* (Oxford: Clarendon Press, 1978), pp. 382–93.

15. Excerpts from Doddridge's *Free Thoughts* can be found in Waddington, *Congregational History*, pp. 288–91.

16. Isaac Watts, "An Humble Attempt Towards the Revival of Practical Religion," in *The Works of the Reverend and Learned Isaac Watts, D. D.*, 6 vols. (London, 1810–11) 3: 3.

17. Ibid., pp. 14–15.

18. See the *Correspondence* of Robert Wodrow, 3 vols. (Edinburgh, 1842–43), for one contemporary Scot's opinion on the state of religion. The most complete statement of the detrimental religious effects of union with England is [John Willison], *A Fair and Impartial Testimony* (Edinburgh, 1744).

19. Ebenezer Erskine, "The Standard of Heaven Lifted up against the Powers of Hell," in *The Whole Works of Ebenezer Erskine* (Edinburgh, 1793), 2:66–124; [John Maclaurin], *Observations upon Church Affairs* (Edinburgh, 1734); John Willison, *The Church's Danger and the Minister's Duty* (Edinburgh, 1733); and the act on preaching of the 1736 General Assembly, printed in John Macleod, *Scottish Theology in Relation*

to Church History since the Reformation (Edinburgh: Publications Committee of the Free Church of Scotland, 1943), pp. 169–71.

20. Stewart Mechie, "The Theological Climate in Early Eighteenth Century Scotland," in *Reformation and Revolution: Essays presented to The Very Reverend Principal Emeritus Hugh Watt, D.D., D.Litt. on the Sixtieth Anniversary of his Ordination* (Edinburgh: The Saint Andrew Press, 1967), pp. 258–72.

21. Ibid., p. 268.

22. James Walker, *Theology and Theologians of Scotland 1560–1750* (1872; 2d ed. rev. 1888; reprint ed., Edinburgh: Knox Press, 1982), pp. 86–94.

23. Macleod, *Scottish Theology*, p. 141–48.

24. Mechie, "Theological Climate," p. 267.

25. See Ebenezer Erskine, "The Assurance of Faith, Opened and Applied," and "The Necessity and Profitableness of Good Works Asserted," in *The Whole Works*, vol. 1; and Ralph Erskine, "Law-death, Gospel-life; or The Death of Legal Righteousness, the Life of Gospel Holiness," in *A Collection of Sermons*, by Ebenezer Erskine and Ralph Erskine, vol. 2 (London, 1757). The Erskines's sermons were admired, and Ralph Erskine's *Gospel Sonnets* were very popular in New England.

26. Macleod, *Scottish Theology*, pp. 180–81.

27. John McKerrow, *History of the Secession Church* (1839; 3d ed., London, n.d.), pp. 11–18.

28. Henry Sefton, "'Neu-lights and Preachers Legall': some observations on the beginnings of Moderatism in the Church of Scotland," in *Church, Politics and Society: Scotland 1408–1929*, ed. Norman Macdougall (Edinburgh: John Donald, 1983), pp. 186–96.

29. Richard B. Sher, *Church and University in the Scottish Enlightenment: The Moderate Literati of Edinburgh* (Edinburgh: Edinburgh University Press, 1985), pp. 23–36, 32, 35.

30. McKerrow, *History of the Secession*, pp. 8–9, 21–26.

31. Ebenezer Erskine, "The Standard of Heaven," 2:99–101.

32. Ibid., p. 91.

33. J. H. S. Burleigh, *A Church History of Scotland* (London: Oxford University Press, 1960), pt. 4, chap. 2; Ian D. L. Clark, "From Protest to Reaction: The Moderate Regime in the Church of Scotland, 1752–1805," in *Scotland in the Age of Improvement: Essays in Scottish History in the Eighteenth Century*, ed. N. T. Phillipson and Rosalind Mitchison (Edinburgh: Edinburgh University Press, 1970), pp. 200–24; Andrew L. Drummond and James Bulloch, *The Scottish Church, 1688–1843: The Age of the Moderates* (Edinburgh: Saint Andrew Press, 1973), chaps. 1–3; McKerrow, *History of the Secession*; Macleod, *Scottish Theology*, pp. 105–204; Richard Sher and Alexander Murdoch, "Patronage and Party in the Church of Scotland, 1750–1800," in *Church, Politics and Society: Scotland 1408–1929*, ed. Norman Macdougall (Edinburgh: John Donald, 1983), pp. 197–220.

34. McKerrow, *History of the Secession,* pp. 31–64; Sher and Murdoch, "Patronage and Party," p. 208.

35. McKerrow, *History of the Secession,* pp. 72–78.

36. Ibid., pp. 83–91.

37. Ibid., pp. 94–112; Drummond and Bulloch, *The Scottish Church*, pp. 50–51.

38. Willison, *The Church's Danger*, p. 8.

39. Ibid., pp. 43–44.

40. Ibid., p. 52.

41. John Maclaurin, "The Necessity of Divine Grace," and "The Knowledge of Christ Crucified the Sum and Substance of Saving Knowledge," both in *Works* (Glasgow, 1824).

42. Wodrow, Jr., to Colman, 1 April 1734, Colman Papers, Massachusetts Historical Society, Boston.

43. Robert Wodrow, *Analecta*, 3: 155, quoted in John MacInnes, *The Evangelical Movement in the Highlands of Scotland, 1688–1800* (Aberdeen: The University Press, 1951), p. 73.

44. The act is printed in Macleod, *Scottish Theology*, pp. 169–71.

45. J. M. Bumsted and John E. Van de Wetering, *"What Must I Do to Be Saved?" The Great Awakening in Colonial America* (Hinsdale, Ill.: Dryden Press, 1976), pp. 41–46; Robert Middlekauff, *The Mathers: Three Generations of Puritan Intellectuals, 1596–1728* (New York: Oxford University Press, 1971), pp. 305–67.

46. George W. Harper, "Clericalism and Revival: The Great Awakening in Boston as a Pastoral Phenomenon," *New England Quarterly* 57 (1984): 554–66.

47. Increase Mather, Preface, to *God Brings to the Desired Haven*, by Thomas Prince (Boston, 1717).

48. See Gerald J. Goodwin, "The Myth of 'Arminian-Calvinism' in Eighteenth-Century New England," *New England Quarterly* 41 (1968): 213–37.

49. Thomas Foxcroft, *A Practical Discourse Relating to the Gospel-Ministry* (Boston, 1718), pp. 5–8.

50. Clifford K. Shipton, *Sibley's Harvard Graduates* (Boston: Massachusetts Historical Society, 1937), 5:50–51.

51. Cotton Mather, *The Minister* (Boston, 1722), pp. 28–31.

52. Carl Bridenbaugh, *Mitre and Sceptre: Transatlantic Faiths, Ideas, Personalities, and Politics, 1689–1775* (New York: Oxford University Press, 1962), pp. 32–37, 61–67.

53. *The Diary of Cotton Mather for the Year 1712*, ed. William R. Maniere II (Charlottesville: University Press of Virginia, 1964), pp. 86, 101, 113–18.

54. Wodrow, *Correspondence*, 1:46–61.

55. Cotton Mather, *Selected Letters of Cotton Mather*, comp. Kenneth Silverman (Baton Rouge: Louisiana State University Press, 1971), pp. 89–90.

56. See Anne Stokely Pratt, *Isaac Watts and His Gifts of Books to Yale College* (New Haven, Conn.: Yale University Library, 1938), and Davis, *Isaac Watts*, pp. 49–52.

57. Benjamin Colman, *Some Glories of our Lord and Saviour Jesus Christ* (London, 1728); William Harris, *Practical Discourses on . . . Representations of the Messiah, throughout the Old Testament* (London, 1724); and probably Isaac Watts, "The Atonement of Christ," sermons 34–36 in *Sermons on Various Subjects, Divine and Moral*, 3 vols. (London, 1721, 1723, 1729).

58. See for example Cotton Mather to Wodrow, 12 November 1719, and Increase Mather to Wodrow, 11 November 1719 in Wodrow, *Correspondence*, 2: 501–3, 498.

59. The Wodrow–Mather correspondence, which began in 1712, can be found in Wodrow, *Correspondence*.

60. The surviving letters of Wodrow to Colman are printed in Wodrow, *Correspondence*; those of Colman to Wodrow in *Proceedings of the Massachusetts Historical Society*, 77 (1965) (hereafter cited as Mass. Hist. Soc.).

61. Ibid., p. 111.

62. Wodrow, *Correspondence*, 2: 420–21.

63. Ibid., 3: 267–69.

64. Wodrow to Colman, 10 August 1721, ibid., 2:597–98.

65. Wodrow, Jr., to Colman, 1 April 1734, Colmon Papers, Mass. Hist. Soc., Boston.

66. Wodrow, Jr., to Colman, 8 August 1735, ibid.

67. David D. Hall, *The Faithful Shepherd: A History of the New England Ministry in the Eighteenth Century* (1972; reprint ed., New York: W. W. Norton, 1974), pp. 185–90.

68. Perry Miller, *The New England Mind: From Colony to Province* (1953; reprint ed., Boston: Beacon Press, 1961), chap. 20; J. William T. Youngs, Jr., *God's Messengers: Religious Leadership in Colonial New England, 1700–1750* (Baltimore: Johns Hopkins University Press, 1976), chap. 5, pp. 92–108.

69. James W. Schmotter, "Ministerial Careers in Eighteenth-Century New England: The Social Context, 1700–1760," *Journal of Social History* 9 (1975): 249–67.

70. Youngs, *God's Messengers*, pp. 36–63.

71. Hall, *The Faithful Shepherd*, pp. 223, 260–65, 270–75.

72. Paul R. Lucas, "'An Appeal to the Learned': The Mind of Solomon Stoddard," *William and Mary Quarterly*, 3d ser. 30 (1973): 257–72.

73. Solomon Stoddard, *Falseness of the Hopes of Many Professors* (Boston, 1708), p. 16.

74. Solomon Stoddard, *The Inexcusableness of Neglecting the Worship of God* (Boston, 1708), p. 6.

75. Solomon Stoddard, *A Guide to Christ* (Boston, 1714), p. 9.

76. Solomon Stoddard, *The Defects of Preachers Reproved* (New London, 1724), p. 9.

77. Ibid., pp. 25–26.

78. Ibid., p. 13.

79. Ibid., pp. 23–25.

80. Solomon Stoddard, "The Benefit of the Gospel," in *The Efficacy of the Fear of Hell, to Restrain Men from Sin* (Boston, 1713), pp. 34–35.

81. "Boanerges" is the surname Jesus gave James and John, sons of Zebedee, Mark 3:17. Stoddard, "The Benefit of the Gospel," pp. 35–37.

82. Ibid., pp. 39–43.

83. Ibid., pp. 67–71.

84. Ibid., p. 64.

85. Solomon Stoddard, *A Treatise Concerning Conversion* (Boston, 1719), p. 18.

For Stoddard's view of the morphology of conversion, see Thomas A. Schafer, "Solomon Stoddard and the Theology of the Revival," in *A Miscellany of American Christianity*, ed. Stuart C. Henry (Durham, N.C.: Duke University Press, 1963), pp. 328–61, and Eugene E. White, *Puritan Rhetoric: The Issue of Emotion in Religion* (Carbondale and Edwardsville, Ill.: Southern Illinois University Press, 1972), pp. 33–41.

86. Lucas, "'An Appeal to the Learned.'"

87. Jonathan Marsh, *An Essay to Prove the Thorough Reformation of a Sinning People* (New London, 1721).

88. Ibid., p. 40.

89. Philip F. Gura, "Sowing the Harvest: William Williams and the Great Awakening," *Journal of Presbyterian History* 56 (1978): 332.

90. Appended to Solomon Stoddard, *The Presence of Christ with the Ministers of the Gospel* (Boston, 1718).

91. Lucas, "'An Appeal to the Learned,'" pp. 285–87.

92. Marsh, *An Essay to Prove the Thorough Reformation of a Sinning People*, pp. 38–39.

93. Youngs, *God's Messengers*, p. 63.

94. Benjamin Lord, *True Christianity Explained* (New London, 1727), p. 36.

95. Ibid., p. 23.

96. Ibid., p. 78.

97. Eliphalet Adams, *The Gracious Presence of Christ with the Ministers* (New London, 1730), *Ministers Must Take Heed* (New London, 1726), and *The Work of Ministers* (New London, 1725); Isaac Chauncy, *The Faithful Evangelist* (Boston, 1725); Benjamin Lord, *The Faithful and Approved Minister* (New London, 1727); William Williams, *The Great Concern of Christians* (Boston, 1723), *The Great Duty of Ministers* (Boston, 1726), *The Honor of Christ Advanced by the Fidelity of Ministers* (Boston, 1728), and *The Office and Work of Gospel Ministers* (Boston, 1729).

98. Cotton Mather, *Companion for Communicants* (Boston, 1690), p. 131. For Mather's rejection of preparationism, see Middlekauff, *The Mathers*, chaps. 13 and 14.

99. William Williams, *The Great Salvation Revealed and Offered in the Gospel* (Boston, 1717), p. 34.

100. Isaac Chauncy, "To the Candid Reader," in *The Loss of the Soul* (Boston, 1732).

101. William Williams, *The Duty and Interest of a People* (Boston, 1736), p. 9. See also, Williams, *The Great Duty of Ministers*, p. 24.

102. For an analysis of William Williams's role in the development of revivalism in the Connecticut Valley, see Gura, "Sowing the Harvest." Gura accurately points out that historians have allowed Stoddard to overshadow other important figures in the movement: "Stoddard was setting the pace, but there were other strong personalities ready to share the work of spreading the Good News" (p. 331).

103. For preparationism, see Norman Pettit, *The Heart Prepared: Grace and Conversion in Puritan Spiritual Life* (New Haven, Conn.: Yale University Press, 1966).

104. Lord, *True Christianity Explained*, p. 45.

105. Marsh, *The Great Care and Concern of Men under Gospel-Light* (New London, 1721), p. 17.
106. Adams, *The Work of Ministers*, p. 15.
107. Adams, *Ministers Must Take Heed*, p. 25.
108. Lord, *The Faithful and Approved Minister*, pp. 20–24.
109. Williams, *The Great Salvation*, pp. 25, 32–34, 90, 148.
110. Marsh, *The Great Care and Concern*, p. 12.
111. Ibid., p. 6.
112. Ibid., p. 13.
113. Ibid., pp. 18–19.
114. Ibid., pp. 20–21.
115. Ibid., pp. 22–23.
116. Ibid., p. 23.
117. Ibid., p. 24.
118. Schafer, "Solomon Stoddard and the Theology of the Revival," pp. 354–56.
119. Youngs, *God's Messengers*, pp. 78–88.
120. Samuel Whitman, *Practical Godliness* (New London, 1714).
121. Samuel Whitman, *The Happiness of the Godly* (New London, 1727).
122. Samuel Whitman, *A Discourse of God's Omniscience* (New London, 1733), p. 26.
123. Shipton, *Sibley's Harvard Graduates*, 4:317; Noah Porter, *Half-Century Discourse* (Hartford, Conn., 1857), p. 14.
124. William Williams, *A Painful Ministry* (Boston, 1717), pp. 2–3.
125. John Bulkley, *The Usefulness of Reveal'd Religion* (New London, 1730), pp. 35–36.

CHAPTER 4

1. James F. Maclear, "'The Heart of New England Rent': The Mystical Element in Early Puritan History," *Mississippi Valley Historical Review* 42 (1956): 621–52. For the influence of the intellect over the emotional lives of New England's Puritans, see Robert Middlekauff, "Piety and Intellect in Puritanism," *William and Mary Quarterly*, 3d ser. 22 (1965): 457–70.

2. For the relation between the new psychology and the development of rational and evangelical preaching styles in early eighteenth-century New England, see Eugene E. White, *Puritan Rhetoric: The Issue of Emotion in Religion* (Carbondale and Edwardsville: Southern Illinois University Press, 1972), pp. 1–64; and Norman S. Fiering, "Will and Intellect in the New England Mind," *William and Mary Quarterly*, 3d ser. 29 (1972): 515–58. For the relation between Enlightenment epistemology and pietist experientialism, see J. M. Bumsted and John E. Van de Wetering, *What Must I Do to Be Saved? The Great Awakening in Colonial America* (Hinsdale, Ill.: Dryden Press, 1976), pp. 35–39. For Jonathan Edwards's psychological system, see Perry Miller, *Jonathan Edwards* (New York: W. Sloane Associates, 1949), esp. chapter titled "The Will," pp. 235–64; and John E. Smith, "Editor's Introduction," in *Religious Affections*, ed. John E. Smith, vol. 2 of *The Works of Jonathan Edwards* (New Haven,

Conn.: Yale University Press, 1959), pp. 1–43. For the rational and evangelical preaching styles and an analysis of George Whitefield's powers as a preacher, see Horton Davies, *Worship and Theology in England: From Watts and Wesley to Maurice, 1690–1850* (Princeton, N.J.: Princeton University Press, 1961), pp. 64–74, 143–83.

3. August Hermann Francke, *Pietas Hallensis* (London, 1705); Geoffrey F. Nuttall, "Continental Pietism and the Evangelical Movement in Britain," in *Pietism und Reveil*, ed. J. Van den Berg and J. P. van Dooren (Leiden: E. J. Brill, 1978), pp. 207–36; Ernst Benz, "The Pietist and Puritan Sources of Early Protestant World Missions (Cotton Mather and A. H. Francke)," *Church History* 20 (1951): 28–55, and "Ecumenical Relations between Boston Puritanism and German Pietism: Cotton Mather and August Hermann Francke," *Harvard Theological Review* 54 (1961): 159–93. See also Karl Zehrer, "The Relationship between Pietism in Halle and Early Methodism," *Methodist History* 17 (1979): 211–24.

4. For the pastoral style of seventeenth-century Puritan ministers, see David D. Hall, *The Faithful Shepherd: A History of the New England Ministry in the Seventeenth Century* (1972; reprint ed., New York, W. W. Norton & Co., 1974), pp. 48–71.

5. For example, Samuel Danforth, *An Exhortation to All to Use Utmost Endeavours to Obtain a Visit of the God of Hosts, for the Preservation of Religion, and the Church, upon Earth* (Boston, 1714), reprinted in *The Wall and the Garden: Selected Massachusetts Election Sermons, 1670–1775*, ed. A. W. Plumstead (Minneapolis: University of Minnesota Press, 1968) pp. 150–76, p. 175: "When the Lord of hosts intends a remarkable visit to his vine on earth, he is wont to raise up ministers and fill those that are the standing ministry in his churches in a more than common measure with his Holy Spirit. . . . They have at such a time peculiar impressions from above, both as to the subjects they are to handle and as to the manner of treating of them; their faculties and abilities are quickened and enlivened, their ministerial gifts and graces enlarged, their good affections and zeal encreased, and their watchfulness over their flocks to promote their spiritual good more abundant than before; . . . the bent of their souls is to promote religion, to convert and edify souls, and they are made resolute in this work, to pursue and prosecute it, notwithstanding all the oppositions and discouragements that attend them therein."

6. Benjamin Lord, *True Christianity Explained*, (New London, 1727), pp. 21–23, 48.

7. William Williams, *The Great Salvation Revealed and Offered in the Gospel*, (Boston, 1717), pp. 54–56.

8. John Maclaurin, "The Necessity of Divine Grace," in *Works* (Glasgow, 1824), p. 541.

9. John Willison, *A Sermon Preached Before His Majesty's High Commissioners* (Edinburgh, 1734), p. 10.

10. Jonathan Marsh, *An Essay to Prove the Thorough Reformation of a Sinning People* (London, 1721), p. 40.

11. Isaac Watts, "An Humble Attempt towards the Revival of Practical Religion," in *The Works of the Reverend and Learned Isaac Watts, D. D.*, 6 vols. (London, 1810–1811), 3: 18.

12. Willison, *A Sermon Preached Before His Majesty's High Commissioners*, p. 10.

13. Isaac Watts, Preface, to *Two Discourses*, by John Jennings (London, 1723 and 1735; Boston, 1740), p. ix.

14. Jennings, *Two Discourses*, pt. 2.

15. Benjamin Lord, *The Faithful and Approved Minister* (New London, 1727), pp. 36–37.

16. Jennings, *Two Discourses*, pp. 23–25; Isaac Chauncy, *The Faithful Evangelist* (Boston, 1725), p. 27.

17. Jennings, *Two Discourses*, p. 31.

18. Watts, "An Humble Attempt towards the Revival of Practical Religion," 3:25.

19. Chauncy, *The Faithful Evangelist*, p. 28.

20. Watts, "An Humble Attempt towards the Revival of Practical Religion," 3:30.

21. Cotton Mather wrote that "there may be some Difference between the *fair using of Notes*, and the dull Reading of them" (*The Minister* [Boston, 1722], p. 34). On the other hand, John Hancock, of Lexington, Massachusetts, thought the people should not lay an extra burden on ministers by insisting they preach without notes: "I suppose if every candidate for the ministry must tarry till he is endued with this power to deliver all his discourses to the people by the strength of his memory, some of them must even tarry till their dying day" (*A Sermon Preached at the Ordination of Mr. John Hancock . . . by His Father* [Boston, 1726], p. 29).

22. William Williams, *The Great Concern of Christians* (Boston, 1723), p. 20.

23. Lord, *The Faithful and Approved Minister*, pp. 38–39.

24. Watts, "An Humble Attempt towards the Revival of Practical Religion," 3:8.

25. Isaac Watts and John Evans, "To the Reader," in *The Example of St. Paul, Represented to Ministers and Private Christians* [by James Murray] (London, 1726).

26. John Willison, *The Church's Danger and the Minister's Duty* (Edinburgh, 1733), p. 37.

27. Watts, "An Humble Attempt towards the Revival of Practical Religion," 3:38.

28. This paragraph is based on Thomas A Schafer, "Solomon Stoddard and the Theology of the Revival," in *A Miscellany of American Christianity: Essays in Honor of H. Shelton Smith*, ed. Stuart C. Henry (Durham, N.C.: Duke University Press, 1963), pp. 328–61.

29. Archibald Campbell, *Discourse Proving that the Apostles Were No Enthusiasts* (London, 1730), pp. 2–3.

30. Wodrow to Colman, 24 September 1730, in Robert Wodrow, *Correspondence*, 3 vols. (Edinburgh, 1842–43), 1:55–56.

31. [John Maclaurin], *Observations Upon Church Affairs* (Edinburgh, 1734).

32. John Maclaurin, "On the Scripture Doctrine of Divine Grace," in *The Works of the Rev. John Maclaurin* (Edinburgh, 1860), 1: 319–485; originally published in John M'Laurin, *Sermons and Essays* (Glasgow, 1755).

33. Ibid., p. 319.

34. Ibid., pp. 327–28.

35. Ibid., p. 419.

36. Ibid., p. 351.

37. Ibid., p. 420.

38. Ibid., p. 429.

39. Ibid., pp. 132–33.
40. Ibid., pp. 439–50.
41. Ibid., p. 459.
42. Ibid., p. 469.
43. Isaac Watts, "Sermons on Various Subjects, Divine and Moral," in *Works of the Reverend and Learned Isaac Watts,* 1: 11.
44. Ibid., p. 21.
45. For instance, his *Strength and Weakness of Human Reason* (London, 1731). See also Arthur Paul Davis, *Isaac Watts: His Life and Works* (1943), pp. 135, 222.
46. Watts, "Discourses of the Love of God . . ." (London, 1729), in *Works of the Reverend and Learned Isaac Watts*, 2: 631–742.
47. Ibid., p. 633.
48. Ibid., "Discourse II," pp. 643–57.
49. Ibid., p. 669.
50. Ibid., p. 660.
51. Ibid., p. 659.
52. Ibid., p. 671.
53. For religious song and singing in eighteenth-century New England, see Henry Wilder Foote, *Three Centuries of American Hymnody* (Cambridge, Mass.: Harvard University Press, 1940), chap. 3. For England, see Harry Escott, *Isaac Watts, Hymnographer: A Study of the Beginnings, Development, and Philosophy of the English Hymn* (London: Independent Press, 1962). For changing attitudes regarding religious singing and the emotions, see Joyce Irwin, "The Theology of 'Regular Singing,'" *New England Quarterly* 51 (1978): 176–92. For the ways in which regular singing could, in the hopeful thinking of the clergy, revive religion, see Laura L. Becker, "Ministers vs. Laymen: The Singing Controversy in Puritan New England, 1720–1740," *New England Quarterly* 55 (1982): 79–94 (quotation on p. 85).
54. Escott, *Isaac Watts, Hymnographer*, p. 77.
55. Ibid., pp. 107–11.
56. Ibid., p. 107.
57. Isaac Watts, Preface, to *Hymns and Spiritual Songs* (London, 1707).
58. Watts, "Short Essay Toward the Improvement of Psalmody," in *Hymns and Spiritual Songs*.
59. Watts's principal hymns and psalms are contained in his *Hymns and Spiritual Songs* (1707), his *Divine Songs Attempted in Easy Language for the Use of Children* (1715), and his *Psalms of David Imitated in the Language of the New Testament, and Applied to the Christian State and Worship* (1719). Several other hymns appear scattered in others of his works. These pieces are lyric verse designed to be sung to standard hymn and psalm tunes, chiefly in the three standard psalm meters.
60. Doddridge to Watts, May 1731, in Isaac Milner, *The Life, Times, and Correspondence of the Rev. Isaac Watts* (London, 1845), p. 493.
61. For example, "Diary of Cotton Mather," 5 May 1683, Massachusetts Historical Society *Collections*, 7th ser. 7 (1911): 57–59.
62. Jonathan Edwards, "Personal Narrative," in *Jonathan Edwards, Representative Selections*, ed. Clarence H. Faust and Thomas H. Johnson, rev. ed. (New York: Hill and Wang, 1962), p. 61.

63. Benjamin Colman, Preface, to *Two Discourses*, by John Jennings (Boston, 1740).

64. "Diary of Cotton Mather," 7th ser. 8 (1912): 142.

65. For example, "Diary of Samuel Sewall," 18 September 1711, *Massachusetts Historical Society Collections*, 5th ser. 6 (1879), p. 323.

66. Ebenezer Turell, *The Life and Character of the Reverend Benjamin Colman* (Boston, 1749), pp. 175–77.

67. Foote, *Three Centuries of American Hymnody*, chap. 3; Ola Elizabeth Winslow, *Meetinghouse Hill, 1630–1783* (New York: Macmillan Co., 1952), chap. 10.

68. Jonathan Edwards, "Unpublished Letter of May 30, 1735," in *The Great Awakening*, ed. C. C. Goen, vol. 4 of *The Works of Jonathan Edwards* (New Haven, Conn.: Yale University Press, 1972), p. 105.

69. Jonathan Edwards, "A Faithful Narrative," in ibid. p. 151.

70. [Solomon Stoddard], *Cases of Conscience About Singing Psalms* (Boston, 1723; reprinted in Samuel Hopkins Emery, *The Ministry of Taunton* (Boston, 1853), 1:269–81; quotations from pp. 271–72, 282.

71. Cotton Mather, *The Accomplished Singer* (Boston, 1721), pp. 21–22.

72. Escott, *Isaac Watts, Hymnographer*, p. 82.

73. Quoted in ibid., pp. 82–83.

74. William Law, *Serious Call* (London, 1729), vol. 4 of *Works* (New Forest, Hampshire, 1892–93), pp. 146–47.

75. Ibid., p. 154.

76. Watts, "Discourses of the Love of God," 2:674.

77. Benjamin Colman, *Some Glories of Our Lord and Saviour Jesus Christ* (London, 1728), p. 287.

78. Ibid., pp. 290–1.

79. See James Davenport's public confession (*Boston Gazette*, 18 July 1744); Benjamin Colman, *Letter from the Reverend Dr. Colman of Boston, to the Reverend Mr. Williams of Lebanon, Upon Reading the Confession and Retraction of the Reverend Mr. James Davenport* (Boston, 1744), p. 8; and Jonathan Edwards, "Some Thoughts concerning the Present Revival of Religion in New England," in *The Works of Jonathan Edwards* (Yale), 4:491–93.

80. Edwards to Benjamin Colman, 22 May 1744, in *Proceedings of the Massachusetts Historical Society*, 2d ser. 10 (1896): 429.

81. Millar Patrick, *Four Centuries of Scottish Psalmody* (London: Oxford University Press, 1949), pp. 105–63, 209–19; David Johnson, *Music and Society in Lowland Scotland in the Eighteenth Century* (London: Oxford University Press, 1972), pp. 164–84.

CHAPTER 5

1. Adams, "To the Reader," in *Sensible Sinners Invited to Come to Christ*, by Eleazar Williams (New London, 1735), pp. iv–v.

2. Jonathan Edwards, "An Account of the late and wonderful Work of God," appended to William Williams, *The Duty and Interest of a People* (Boston, 1736).

3. Isaac Watts to Benjamin Colman, 28 February 1737, *Proceedings of the Massachusetts Historical Society*, 2d ser. 9 (1895): 352–53.

4. Isaac Watts and John Guyse, Preface, to "A Faithful Narrative of the Surprizing Work of God in the Conversion of Many Hundred Souls in Northampton and the Neighboring Towns and Villages," by Jonathan Edwards, in *The Great Awakening*, ed. C. C. Goen, vol. 4 of *The Works of Jonathan Edwards* (New Haven, Conn.: Yale University Press, 1972), p. 130.

5. Ibid., p. 131.

6. Eliphalet Adams exclaimed: "And, Oh! that this thoughtful, serious, religious disposition, Ever producing good fruits, might spread more and more, not only through this Land, but the world too, till the whole Earth shall be full of the Knowledge of the Lord, as the waters cover the Sea" ("To the Reader," in *Sensible Sinners*, by Eleazar Williams, p. vi). Concern with the millennium permeates Adams's writings. See, for instance, *A Discourse Shewing That so Long as there Is Any Prospect* (New London, 1734), pp. 17–18. Watts and Guyse write similarly: "We are taught also by this happy event how easy it will be for our blessed Lord to make a full accomplishment of all his predictions concerning his kingdom, and to spread his dominion from sea to sea through all the nations of the earth" (Preface, to "A Faithful Narrative" [Yale], 4 132.) See also p. 137, and cf. the preface to the Boston edition of 1738, *The Works of Jonathan Edwards* (Yale), 4:141.

7. *Dictionary of Welsh Biography, Down to 1940*, Honorable Society of Cymmrodorion (London: B. H. Blackwell, 1959).

8. R. Tudor Jones, *Congregationalism in England 1662–1962* (London: Independent Press, 1962), pp. 114–15.

9. Giles Firmin, *A Brief Review of Mr. Davis's Vindication Giving no Satisfaction* (London, 1693), p. 30.

10. *An Account of the Doctrine and Discipline of Mr. Richard Davis* (London, 1700).

11. Firmin, Preface, to *A Brief Review*.

12. *The Sense of the United Nonconforming Ministers, in and about London, concerning Some of the Erroneous Doctrines and Irregular Practices of Mr. Richard Davis* (London, 1693), p. 6.

13. The covenant and its explanation are printed in *An Account of the Doctrine of Davis*; quotations from pp. 6, 13; see also pp. 15, 22.

14. Quoted in Firmin, *A Brief Review*, p. 29.

15. For a fuller description of the politics of the break up of the United Brethren, see Jones, *Congregationalism in England*, pp. 114–19.

16. *An Account of the Doctrine of Davis*.

17. Michael R. Watts, *The Dissenters: From the Restoration to the French Revolution* (Oxford: Clarendon Press, 1978), pp. 386–93.

18. Isaac Watts, "A Guide to Prayer," in *The Works of the Reverend and Learned Isaac Watts, D.D.* 6 vols. (London, 1810–11), 3: 180–81.

19. Watts and Guyse, Preface, to "A Faithful Narrative" (Yale), 4:132.

20. Robert Wodrow, *Correspondence*, 3 vols. (Edinburgh, 1842–43), 1:46–61.

21. Leigh Eric Schmidt, *Holy Fairs: Scottish Communions and American Revivals*

in the Early Modern Period (Princeton, N.J.: Princeton University Press, 1989), pp. 36, 40, 41–50.

22. Cf. ibid., pp. 48–50, where Schmidt emphasizes the continuity of the cycle of renewal the summer sacramental seasons represented.

23. Alexander R. MacEwen, *The Erskines* (New York: Charles Scribner's Sons [1900]), pp. 30–49.

24. Andrew L. Drummond and James Bulloch, *The Scottish Church, 1688–1843: The Age of the Moderates* (Edinburgh: Saint Andrew Press, 1973), pp. 49–50; see also MacEwen, *The Erskines*, pp. 12–13.

25. John MacInnes, *The Evangelical Movement in the Highlands of Scotland 1688–1800* (Aberdeen: The University Press, 1951), pp. 156, 158; see also John Balfour, "Account of the Revival in the Parish of Nig," *The Christian Monthly History* (Edinburgh) 4 (February 1744):45–58.

26. See preface to the Boston edition of 1738, "A Faithful Narrative" (Yale), 4:142.

27. Gerald F. Moran, The Puritan Saint" (Ph.D. diss., Rutgers University, 1974), pp. 42, 104–33, 237; Solomon Stoddard, *The Defects of Preachers Reproved* (New London, 1724), p. 26.

28. Daniel Wadsworth, *Diary of Rev. Daniel Wadsworth* (Hartford, 1894), p. 7.

29. Robert G. Pope, *The Half-Way Covenant: Church Membership In Puritan New England* (Princeton, N.J.: Princeton University Press, 1969); Pope summarizes his argument in chap. 10; quotations from pp. 274–75; on Puritan tribalism, see Edmund S. Morgan, *The Puritan Family: Religion & Domestic Relations in Seventeenth-Century New England* (1944; new ed., rev. and enl., New York: Harper & Row, 1966), pp. 168–86.

30. Pope, *The Half-Way Covenant*, pp. 119–20, 187–89, 241–51.

31. Cotton Mather, Preface, to *Piety Encouraged*, by Samuel Danforth (Boston, 1705).

32. *The Christian History* (Boston) 1: 111.

33. George Leon Walker, *History of the First Church in Hartford, 1633–1883* (Hartford, 1884), pp. 247–49; *Historical Catalogue of the First Church in Hartford 1633–1885* (Hartford, 1885), pp. 38–52.

34. Many of these revivals are mentioned in *The Christian History* (Boston); others by Mary Hewit Mitchell, *The Great Awakening and Other Revivals in the Religious Life of Connecticut*, (New Haven: Tercentenary Commission of the State of Connecticut) 26 (1934): 8–9; and still others by James Walsh, "The Great Awakening in the First Congregational Church of Woodbury, Connecticut," *William and Mary Quarterly*, 3d ser. 28 (1971), 547.

35. Samuel Danforth, "An Exhortation to All to Use Utmost Endeavours to Obtain a Visit of the God of Hosts," in *The Wall and the Garden: Selected Massachusetts Election Sermons, 1670–1775*, ed. A. W. Plumstead (Minneapolis: University of Minnesota Press, 1968), p. 176.

36. Eliphalet Adams, *The Gracious Presence of Christ with the Ministers* (New London, 1730), p. 25.

37. Eliphalet Adams, Preface, to *A Sermon Preached at Windham, July 12th. 1721. On a Day of Thanksgiving for the Late Remarkable Success of the Gospel among Them*

(New London, 1721); Francis Manwaring Caulkins, *History of Norwich, Connecticut* (Hartford, 1874), p. 315; *Manual of the First Congregational Church of Norwich, Conn.* (Norwich, 1868).

38. Adams, *A Sermon Preached at Windham*, pp. 7–9.

39. Jonathan Marsh, *An Essay to Prove the Thorough Reformation of a Sinning People* (New London, 1721), p. 52.

40. Ibid., p. 37.

41. Ibid., p. 38.

42. Cotton Mather, Preface, to *Piety Encouraged*, by Samuel Danforth.

43. Adams, Preface, to *Sermon Preached at Windham*.

44. Edwards, "A Faithful Narrative," (Yale), 4: 146.

45. Ibid., p. 190.

46. Ibid., p. 57.

47. Sereno E. Dwight, *The Life of President Edwards* (New York, 1820), p. 21.

48. Jonathan Parsons, "An Account of the Revival of Religion at Lyme West Parish," *The Christian History* (Boston), 2 (1744): 118–25.

49. C. C. Goen, "Editor's Introduction," in *The Works of Jonathan Edwards* (Yale), 4: 4–7.

50. For Parsons's atypicality, see Gerald J. Goodwin, "The Myth of 'Arminian-Calvinism' in Eighteenth-Century New England," *New England Quarterly* 41 (1968): 219–20.

51. Boardman's letter was printed in Charles Chauncy, *Seasonable Thoughts on the State of Religion in New-England* (Boston, 1743), pp. 202–9; David Ferris, *Memoirs of the Life of David Ferris* (Philadelphia, 1825), pp. 14–25; see also the report by the Quaker John Woolman, quoted in Samuel Orcutt, *History of the Towns of New Milford and Bridgewater, Connecticut, 1703–1882* (Hartford, 1882), pp. 109–10.

52. Boardman's letter in Chauncy, *Seasonable Thoughts*, pp. 204–5.

53. Four of the Seceders were involved in establishing the worship of the Church of England in New Milford in 1743. Ferris later became a minister or Public Friend of the Society of Friends (Orcutt, *History of New Milford and Bridgewater*, pp. 108–17).

54. Eliphalet Adams, *The Work of Ministers* (New London, 1725), p. 22.

55. Adams, *The Gracious Presence of Christ with the Ministers*, p. 25.

56. Edwards, "A Faithful Narrative" (Yale), 4: 189.

57. Benjamin Colman, Preface, to *God's Awful Determination*, by John Cotton (Boston, 1728), p. iii.

58. Thomas Foxcroft, *The Voice of the Lord* (Boston, 1727), p. 37.

59. For example, William Cooper, *The Danger of a People's Loosing the Good Impressions Made by the Late Awful Earthquake* (Boston, 1727).

60. Benjamin Colman, *The Judgments of Providence* (Boston, 1727), pp. 84–85. See also Thomas Prince, *Earthquakes the Work of God* (Boston, 1727), p. 43; and Joseph Sewall, *The Duty of a People* (Boston, 1727), pp. 25–26.

61. Foxcroft, *The Voice of the Lord*, p. 38.

62. Ibid., p. 44.

63. Sewall, *The Duty of a People*, p. 26; Cotton Mather, *The Terror of the Lord*, 3d ed. (Boston, 1727), p. 3.

64. Joseph Sewall, *Repentance the Sure Way* (Boston, 1727), p. 40.

65. Foxcroft, *The Voice of the Lord*, p. 48.

66. Prince, *Earthquakes*, app.

67. William Williams (Jr.), *Divine Warnings* (Boston, 1728), p. iv.

68. Nathaniel Gookin, Dedication, to *The Day of Trouble Near* (Boston, 1728).

69. John Brown, letter, in *A Holy Fear of God*, by John Cotton (Boston, 1727), app. pp. 4–7.

70. Samuel Phillips, *Three Plain Practical Discourses* (Boston, 1728), p. 34.

71. Jonathan Edwards, *The Future Punishment of the Wicked Unavoidable and Intolerable*, in *Works* (New York: Robert Carter and Brothers, 1864), 4: 263.

72. Martin E. Lodge, "The Crisis of the Churches in the Middle Colonies, 1720–1750," *Pennsylvania Magazine of History and Biography* 95 (1971): 195–220.

73. F. Ernst Stoeffler argues that part of the essence of pietism was its position as a reform movement. It only existed in opposition to a prevailing norm (*The Rise of Evangelical Pietism* [Leiden: E. J. Brill, 1971], pp. 22–23). See also Charles Hartshorn Maxson, *The Great Awakening in the Middle Colonies* (1920; reprint ed., Gloucester, Mass.: Peter Smith, 1958); James Tanis, *Dutch Calvinistic Pietism in the Middle Colonies: A Study in the Life and Theology of Theodorus Jacobus Frelinghuysen* (The Hague: Martinus Nijhof, 1967); and Leonard J. Trinterud, *The Forming of an American Tradition: A Re-examination of Colonial Presbyterianism* (Philadelphia: Westminster Press [1949]).

74. Theodorus Jacobus Frelinghuysen, *A Clear Demonstration of A Righteous and Ungodly Man* trans. Hendrik Visscher (New York, 1731), p. ii.

75. Ibid., p. 75.

76. John Gillies, comp., *Historical Collections Relating to Remarkable Periods of the Success of the Gospel, and Eminent Instruments Employed in Promoting It*, 2 vols. (Glasgow, 1754; revised by Horatius Bonar, 1845), rev. ed. in 1 vol. as *Historical Collections of Accounts of Revival* (Fairfield, Pa.: Banner of Truth Trust, 1981), p. 424. For Frelinghuysen's influence on Gilbert Tennent, see Milton J. Coalter, Jr., *Gilbert Tennent, Son of Thunder: A Case Study of Continental Pietism's Impact on the First Great Awakening in the Middle Colonies* (Westport, Conn: Greenwood Press, 1986), pp. 12–25.

77. Gilbert Tennent, *A Solemn Warning to a Secure World* (Boston, 1735), p. 100.

78. Gilbert Tennent, "The Legal Bow Bent," Preface, in *Sermons on Sacramental Occasions by Diverse Ministers*, by Gilbert Tennent et al. (Boston, 1739).

79. Tennent, *A Solemn Warning*, pp. 97–98.

80. Ebenezer Pemberton, *A Sermon Preached at the Ordination of Reverend Mr. Walter Wilmot* (Boston, 1738), p. 9; for the preaching and doctrines of the Middle Colony revivalists, see Glenn T. Miller, "God's Light and Man's Enlightenment: Evangelical Theology of Colonial Presbyterianism," *Journal of Presbyterian History* 51 (1973): 97–115.

81. Gilbert Tennent, "The Unsearchable Riches of Christ," in *Sermons on Sacramental Occasions*, pp. 26–27.

82. Ebenezer Pemberton, *A Sermon Preach'd before the Commission of the Synod of Philadelphia, April 20th, 1735* (New York, 1735).

83. Edwards, "A Faithful Narrative" (Yale), 4: 155–56.

84. Preaching at Perth Amboy on 29 June 1735, Gilbert Tennent announced, "Others are battering the Heavens night and day, with Tears, and Sighs and Groans, as I am informed, they are comming to Christ in Flocks in New England, at this time; and will you forever lye still in your beds, of carnal Security, over the dreadful steeps of Damnation" (*The Necessity of Religious Violence* [New York [1735]], p. 43).

85. Tanis, *Dutch Calvinistic Pietism*.

86. George Whitefield, *George Whitefield's Journals* (London: Banner of Truth Trust, 1960), pp. 351–52.

87. *The Christian History* (Boston), 2 (1744): 292–93.

88. Herman Harmelink III, "Another Look at Frelinghuysen and His 'Awakening,'" *Church History* 37 (1968): 423–38.

89. *The Christian History* (Boston), 2: (1744): 300–1.

90. See, for instance, Tennent, *A Solemn Warning*, pp. 92–93.

91. Gillies, *Historical Collections* (1989 reprint ed.), pp. 425, 426.

92. *The Christian History* (Boston), 2: (1744) 293–94.

93. It is more likely the Middle Colony revivalists would treat the communal significance of the revivals in their sermons intended for publication, their "political" sermons, than in their unpublished preaching, directed primarily to the concerns of individuals.

94. Whitefield, *Journals*, pp. 352, 486.

95. "I hope that more then 300 souls were savingly brought home to Christ in this town in the space of half a year" (Edwards, "A Faithful Narrative" [Yale] 4: 158).

CHAPTER 6

1. C. C. Goen, "Editor's Introduction," in *The Great Awakening*, by Jonathan Edwards, vol. 4 of *The Works of Jonathan Edwards* (New Haven, Conn.: Yale University Press, 1972), pp. 26–27.

2. Isaac Watts and John Guyse, Preface, to "A Faithful Narrative," in ibid., p. 133.

3. Jonathan Edwards, "A Faithful Narrative" in ibid., pp. 144–211.

4. Ibid., p. 157.

5. Ibid., p. 157–59.

6. Ibid., p. 190.

7. *Surprizing* was Watts's and Guyse's term, not Edwards's. Contrast this point with the interpretation of Goen, "Editor's Introduction" (Yale), 4: 19–25.

8. Watts to Colman, 31 May 1738, *Proceedings of the Massachusetts Historical Society* 2d ser. 9 (1895): 361; see also Watts to Colman, 23 September 1738, and Watts to Williams, 7 June 1738, ibid., pp. 353, 335–36; and Williams to Watts, draft, in Anne Stokely Pratt, *Isaac Watts and His Gifts of Books to Yale College* (New Haven, Conn.: Yale University Library, 1938), p. 55.

9. Peter Gay, *A Loss of Mastery: Puritan Historians in Colonial America* (New York: Random House, 1968), chap. 4; for a corrective to Gay's anti-Edwards bias see John F. Wilson, "Jonathan Edwards as Historian," *Church History* 46 (1977): 5–18.

10. Samuel Chandler, *A Paraphrase and Critical Commentary on the Prophecy of Joel* (London, 1735), pp. v, 105.

11. William Adams, *The Necessity of the Pouring Out of the Spirit* (Boston, 1679), p. 35; William Lowth, *Commentary on the Prophets*, vol. 4 of Symon Patrick, *A Commentary on the Old and New Testaments, with Apocrypha* (London, 1809), p. 442.

12. Chandler, *Paraphrase*, p. 101: "As this prophecy is cited by the Apostle *Peter*, and the effusion of the Spirit at the feast of Pentecost is said to be *that which was spoken by the prophet Joel*, Acts ii. 16. it may not be improper to enter into a more particular explication of it than I have given in my notes; because I think it literally fulfilled by that event to which the Apostle refers, and by that only; and because I do not find that it hath been fully stated and settled by any writer, that hath fallen into my hands."

13. Perry Miller, *The New England Mind: The Seventeenth Century* (1939; reprint ed., Boston: Beacon Press, 1954), chap. 16.

14. Wodrow to Cotton Mather, 11 December 1714, in Robert Wodrow, *Correspondence,* 3 vols. (Edinburgh, 1842–43), 1: 630.

15. Ebenezer Erskine, "The Standard of Heaven Lifted up against the Powers of Hell," in *The Whole Works of Ebenezer Erskine* (Edinburgh, 1793), 2: 66–124.

16. Robert Bragge, "The Spirit's Standard Lifted up and Displayed against Error," in *A Defense of Some Important Doctrines of the Gospel*, by Bragge et al., 2d ed., 2 vols. (Glasgow, 1773), 1: 13–38.

17. John Willison, *A Sermon Preached before His Majesty's High Commissioners* (Edinburgh, 1734) pp. 5, 9, 16–17; see also Willison, *The Church's Danger and the Minister's Duty* (Edinburgh, 1733).

18. Richard Baxter, *The Glorious Kingdom of Christ Described and Clearly Vindicated* (London, 1691).

19. Daniel Whitby, "Treatise of the True Millennium," in *A Paraphrase and Commentary on the New Testament*, vol. 2 (1703; 2d ed., London, 1706); Lowth, *Commentary on the Prophets*; Moses Lowman, *Paraphrase and Notes on the Revelation of St. John* (1737; 2d ed., London, 1745).

20. Whitby, "Treatise of the True Millennium," p. 728.

21. For example, Lowth, *Commentary on the Prophets*, p. 442.

22. Joseph Sewall, *Christ Victorious* (Boston, 1733).

23. Thomas Prince, *Six Sermons* (Edinburgh, 1785).

24. Alan Heimert, *Religion and the American Mind: From the Great Awakening to the Revolution* (Cambridge, Mass.: Harvard University Press, 1966), chap. 2. For Edwards's debt to Lowman, see Stephen J. Stein, "Editor's Introduction," in *Apocalyptic Writings*, by Jonathan Edwards, vol. 5 of *The Works of Jonathan Edwards* (New Haven, Conn.: Yale University Press, 1977), pp. 55–59.

25. John B. Buchanon, "Puritan Philosophy of History from Restoration to Revolution," *Essex Institute Historical Collections* 104 (1968): 329–48, argues that the Puritans' theology, which was optimistically millenarian, took second place in emphasis to their philosophy, which was pessimistic in its cyclical view of history. In an intricate argument, James West Davidson, *The Logic of Millennial Thought: Eighteenth-Century New England* (New Haven, Conn.: Yale University Press, 1977),

denies that a simple designation of premillennialism as pessimistic and postmillennialism as optimistic is accurate.

26. For the central place of history in Edwards's theology, see Stephen Morrieson Clark, "Jonathan Edwards: The History of Redemption," (Ph.D. diss., Drew University, 1986).

27. Solomon Stoddard, sermon 7 in "The Benefit of the Gospel," in *The Efficacy of the Fear of Hell, to Restrain Men from Sin* (Boston, 1713), pp. 184–200.

28. Jonathan Edwards, *A History of the Work of Redemption,* ed. by John F. Wilson, vol. 9 of *The Works of Jonathan Edwards* (New Haven, Conn.: Yale University Press, 1989), p. 113 (hereafter cited as *History of Redemption*).

29. Ibid., pp. 281–6, 442–454.

30. Stoddard, sermon 7 in "The Benefit of the Gospel," p. 193.

31. Thomas A. Schafer, "Solomon Stoddard and the Theology of the Revival," in *A Miscellany of American Christianity*, ed. Stuart C. Henry (Durham, N.C.: Duke University Press, 1963): pp. 326–61.

32. Stein, "Editor's Introduction," vol. 5.

33. Edwards, *History of Redemption*, 9: 116.

34. See William J. Scheick, *The Writings of Jonathan Edwards: Theme, Motif, and Style* (College Station: Texas A & M University Press, 1975), pp. 50–58, for comments on the parallels between Edwards's view of the work of redemption of the individual and his view of redemption as it pertains to the whole world.

35. Edwards, *History of Redemption*, 9: 121–22.

36. Ibid., p. 508.

37. Ibid., pp. 141–43.

38. Ibid., p. 195.

39. Ibid., p. 233.

40. Ibid., pp. 266.

41. Ibid., p. 265–66.

42. Ibid., p. 314.

43. Ibid., p. 344–56.

44. Ibid., pp. 376–80.

45. Ibid., p. 459.

46. Ibid., pp. 457–501, 463, 464.

CHAPTER 7

1. Through the years, historians have attempted to explain the triggering of the Revival in 1739–1740 by reference to various crises, demographic, economic, theological, spiritual, and political. They have pointed, for instance to epidemics of diphtheria and scarlet fever that struck down thousands in New England, New York, and New Jersey during the latter half of the 1730s, to the sudden curtailment of paper money in Massachusetts in 1739, to the outbreak of war with Spain that same year, events that aggravated anxiety and increased desire for religious solace. There was little relation, however, between the geographic distribution or timing of the epidemics and the pattern of the revivals. The suppression of the Land Bank may have contrib-

uted to the economic frustrations of the younger generation that found expression in the revivals, but it does not appear to have been a major catalytic force and has little explanatory value outside of Massachusetts. Attempts to relate the appearance of revivals of religion to economic cycles have proved generally fruitless. Michael N. Schute summarizes the arguments that call into question a link between the throat distemper and the Great Awakening, while he shows how a group of evangelical ministers in New Hampshire, campaigning against rationalism, made use of the epidemic to emphasize God's absolute sovereignty and to minimize the power of man's reason ("A Little Great Awakening: An Episode of the Enlightenment," *Journal of the History of Ideas* 37 [1976]: 589–602). Élie Halévy suggests that a "temporary crisis of overproduction" in England in 1739 triggered the Revival there (*The Birth of Methodism in England*, ed. and trans. Bernard Semmel [Chicago: University of Chicago Press, 1971], originally published in *Revue de Paris* [1 and 15 August 1906], pp. 519–39, 841–67). J. D. Walsh effectively challenges Halévy in "Élie Halévy and the Birth of Methodism," *Transactions of the Royal Historical Society*, 5th ser. 25 (1975): 1–20. For the role of the Land Bank crisis in the Great Awakening, cf. John C. Miller, "Religion, Finance, and Democracy in Massachusetts," *New England Quarterly* 6 (1933): 29–58 and John Bumsted, "Religion, Finance, and Democracy in Massachusetts: The Town of Norton as a Case Study," *Journal of American History* 57 (1971): 817–31.

2. C. C. Goen argues that it was the discovery of preachers in New England Congregational and Presbyterian pulpits teaching explicitly Arminian doctrines that prompted orthodox Calvinist ministers to reemphasize the doctrines of human impotence and divine sovereignty, and that the renewed preaching on evangelical themes stimulated the revivals. When Edwards called Arminianism "by name and set his face against it, the Great Awakening in New England was on" (C. C. Goen, "Editor's Introduction," in *The Great Awakening*, by Jonathan Edwards, vol. 4 of *The Works of Jonathan Edwards* [New Haven, Conn.: Yale University Press, 1972], pp. 4–18). The weakest link in Goen's chain of reasoning is his contention that the doctrines of human impotence and divine sovereignty were revived in the 1730s. These doctrines had, rather, been common teaching, repeated often in many, if not most, Congregational and Presbyterian pulpits from Cape Cod to the Berkshires. For the persistence of orthodoxy during the first half of the eighteenth century, see Harry S. Stout, *The New England Soul: Preaching and Religious Culture in Colonial New England* (New York: Oxford University Press, 1986), pp. 148–58.

3. Nehemiah Curnock, ed., *The Journal of the Rev. John Wesley*, standard ed. (New York: Eaton & Mains [1909]), 1:476.

4. J. D. Walsh, "Origins of the Evangelical Revival," in *Essays in Modern Church History, in Memory of Norman Sykes*, ed. G. V. Bennett and J. D. Walsh (New York: Oxford University Press, 1966), pp. 148–53.

5. Richard L. Bushman, "Jonathan Edwards as Great Man: Identity, Conversion, and Leadership in the Great Awakening," *Soundings* 52 (1969): 15–46.

6. W. R. Ward, "Power and Piety: The Origins of Religious Revival in the Early Eighteenth Century," *Bulletin of the John Rylands University Library* 63 (1980): 231–52, 239. See also, Ward, "The Relations of Enlightenment and Religious Revival in

Central Europe and in the English-speaking World," in *Reform and Reformation: England and the Continent, c1500–c1700*, ed. Derek Baker (Oxford: Basil Blackwell, 1979), pp. 281–305, and "Orthodoxy, Enlightenment and Religious Revival," *Studies in Church History* 17 (1982): 275–96.

7. Umphrey Lee, *John Wesley and Modern Religion* (Nashville: Cokesbury Press, 1936); Robert C. Monk, *John Wesley: His Puritan Heritage* (Nashville: Abbingdon Press, 1966); Walsh, "Origins of the Evangelical Revival," pp. 132–62.

8. W. K. Lowther Clarke, *A History of the S. P. C. K.* (London: S.P.C.K., 1959); Holden Hutton, *The English Church from the Accession of Charles I. to the Death of Anne (1625–1714)* (London: Macmillan & Co., 1913), chap. 17; John Henry Overton, *Life in the English Church (1661–1714)* (London, 1885), pp. 207–13; Josiah Woodward, *An Account of the Rise and Progress of the Religious Societies*, 3d ed. (London, 1701).

9. W. K. Lowther Clarke, *Eighteenth Century Piety* (London: S.P.C.K., 1944), chap. 1.

10. Lee, *John Wesley and Modern Religion*, pp. 26–27.

11. F. A. Cavanagh, "Griffith Jones," *Journal of Adult Education* (London), 1, nos. 1 and 2 (1926 and 1927); David Jones, *Life and Times of Griffith Jones* (London: S.P.C.K., 1902); M. G. Jones, *The Charity School Movement: A Study of Eighteenth Century Puritanism in Action* (1938; reprint ed., London and Edinburgh: Frank Cass and Co., 1964), pp. 303–9; John McLeish, *Evangelical Religion and Popular Education: A Modern Interpretation* (London: Methuen & Co., 1969); W. Moses Williams, *The Friends of Griffith Jones: A Study in Educational Philanthropy* (London: Honorable Society of Cymmrodorion, 1939). Jones was promoted to the cure of Llanddowror in 1716 by Sir John Phillips, who would be an encourager of the Oxford Methodists, one of the earliest members of the Methodist Society at Fetter Lane, and a financial backer of George Whitefield. In 1720 Jones married Sir John's sister. In 1722 he published an abridged Welsh translation of *The Whole Duty of Man*.

12. Letters of 30 March 1738 and 16 August 1739 in [Griffith Jones], *Selections from the Welsh Piety*, ed. W. Moses Williams, (Cardiff, 1938), pp. 19–36.

13. Henry Venn, quoted by L. E. Elliot-Binns, *The Early Evangelicals: A Religious and Social Study* (Greenwich, Conn.: Seabury Press, 1953), p. 122.

14. William Law, *A Practical Treatise upon Christian Perfection*, vol. 1 of *Works* (New Forest, Hampshire, 1892–93), pp. 13, 25; Charles Wesley Lowry, Jr., "Spiritual Antecedents of Anglican Evangelicalism," in *Anglican Evangelicalism*, ed. Alexander C. Zabriskie (Philadelphia: Church Historical Society, 1943), p. 58; Norman Sykes, *Church and State in England in the Eighteenth Century* (Cambridge: Cambridge University Press, 1934), pp. 258–61.

15. V. H. H. Green, *The Young Mr. Wesley: A Study of John Wesley and Oxford* (New York: St. Martin's Press, 1961), pp 28–29.

16. Eric W. Baker, *A Herald of the Evangelical Revival: A Critical Inquiry into the Relation of William Law to John Wesley and the Beginning of Methodism* (London: Epworth Press, 1948), pp. 5–18.

17. Ibid., pp. 15–22.

18. Anthony Armstrong, *The Church of England, the Methodists and Society,*

1700–1850 (Totowa, N.J.: Rowman and Littlefield [1973]), p. 125; for individual biographies of early leaders of the Anglican Evangelical movement, see Elliot-Binns, *The Early Evangelicals*.

19. Arnold A. Dallimore, *George Whitefield: The Life and Times of the Great Evangelist of the Eighteenth-Century Revival* (London: Banner of Truth Trust, 1970; reprint ed., 1971), vol. 1; for Whitefield's Calvinism, see Lowry, "Spiritual Antecedents," pp. 48–52.

20. Eifion Evans, *Howell Harris, Evangelist, 1714–1773* (Cardiff: University of Wales Press, 1974); Hugh J. Hughes, *Life of Howell Harris, the Welsh Reformer* (London, 1892); Thomas Rees, *History of Protestant Nonconformity in Wales: From its Rise in 1633 to the Present Time*, 2d ed., rev. (London, 1883).

21. *The Country Parson's Advice to His Parishioners* (London, 1680), p. 81.

22. G. V. Bennett, *White Kennett, 1660–1728, Bishop of Petersborough* (London: S.P.C.K., 1957), pp. 184–90; Edward Carpenter, *The Protestant Bishop, Being the Life of Henry Compton, 1632–1713, Bishop of London* (London: Longmans, Green and Co., 1956), pp. 61–67, 208–32; Norman Sykes, *Edmund Gibson, Bishop of London, 1669–1748* (London: Oxford University Press, 1926), pp. 193–209.

23. Dallimore, *George Whitefield*, 1: 348, 392, 400, 436; Luke Tyerman, *The Life of the Rev. George Whitefield*, 2d ed. (London, 1890), 1: 553; Curnock, *The Journal of the Rev. John Wesley*, 2: 67–68; Howell Harris, *A Brief Account of the Life of Howell Harris* (Trevecka, 1791), p. 113.

24. Curnock, *The Journal of the Rev. John Wesley*, 2: 83–4.

25. Noting the influence on John Wesley of Jonathan Edwards's *Faithful Narrative*, *Distinguishing Marks*, and *Treatise Concerning Religious Affections*, Albert C. Outler asserts, "It is not too much to say that one of the effectual causes of the Wesleyan Revival in England was the Great Awakening in New England" (Outler, ed., *John Wesley* [New York: Oxford University Press, 1964], pp. 15–16).

26. For Wesley's view of the Methodist revival as an outpouring of grace for the reformation of English society at large, see his "Ernest Appeal to Men of Reason and Religion" (1743), in *The Works of John Wesley*, ed. Gerald R. Cragg (Oxford: Clarendon Press, 1975), 11: 88–89.

27. Curnock, *The Journal of the Rev. John Wesley*, 2: 121, 222–23; cf. letter to Samuel Wesley, 10 May 1739, ibid., p. 190n.

28. Ibid., pp. 202–3. Doubters asked "'why were these things only in their private societies? Why were they not done in the face of the sun?'"

29. Dallimore, *George Whitefield*, 1: 385–405.

30. Hearing of Howell Harris's activities in Wales, Whitefield wrote him in December 1738. They met in March 1739. After receiving a letter from Ralph Erskine in May 1739, Whitefield maintained a constant correspondence with Ralph and Ebenezer Erskine until he met them two years later. In England, Whitefield fraternized openly with Dissenters. He sought the friendship of Philip Doddridge and Isaac Watts early in 1739 and preached at the academy of the former. On his return to America in August 1739 he met the Tennents and from New York wrote Benjamin Colman in Boston.

31. Dallimore, *George Whitefield*, 1: 385. He thought the Wesleys' Arminianism and perfectionism hindrances to evangelical cooperation.

32. *The Christian History* (Boston), 2: 358–59.

33. *The Virginia Gazette* no. 72 (16 December 1737), p. 3; no. 75 (6 January 1738), p. 4.

34. Dallimore, *George Whitefield*, 1: 405–9.

35. *The Christian History* (Boston), 2: 358–59.

36. Tyerman, *The Life of the Rev. George Whitefield*, 1: 307–57; Charles Hartshorn Maxson, *The Great Awakening in the Middle Colonies* (1920; reprint ed., Gloucester, Mass.: Peter Smith, 1958), pp. 40–53.

37. [Benjamin Colman et al.], *Three Letters to the Reverend Mr. George Whitefield* (Philadelphia [1739]).

38. Foxcroft to Watts, Foxcroft Papers, Boston University, Special Collections.

39. Thomas Milner, *The Life, Times and Correspondence of the Rev. Isaac Watts, D.D.* (London, 1845), p. 652.

40. *Proceedings of the Massachusetts Historical Society*, 2d ser. 9 (1895): 374–75.

41. George Whitefield, *Three Letters . . . Concerning Archbishop Tillotson* (Philadelphia, 1740).

42. In August 1739, Colman asked Watts for a copy of Jennings's work, which Watts had reprinted in London in 1735 (Milner, *Life of Watts*, p. 640). When Watts obliged, Colman had it reprinted in Boston.

43. Thomas Prince, "The Endless Increase of Christ's Government," in *Six Sermons* (Edinburgh, 1785), p. 36.

44. Thomas Foxcroft, *Some Seasonable Thoughts on Evangelic Preaching* (Boston, 1740), pp. 30–33, 43.

45. *The Weekly History* (London, 1741), 8: 4.

46. *The Christian History* (Boston) 2: 358–59.

47. Isaac Chanler, *New Converts Exhorted to Cleave to the Lord* (Boston, 1740), pp. [vi], 3–4.

48. Josiah Smith, *The Character, Preaching, etc. of the Rev. Mr. George Whitefield* (Boston, 1740), pp. 19–20.

49. *The Christian History* (Boston), 2: 374.

50. Benjamin Colman and William Cooper, "To the Reader," in *The Character, Preaching, etc. of Whitefield*, by Josiah Smith, pp. ii, v.

51. Jonathan Edwards, *A History of the Work of Redemption*, ed. John F. Wilson, vol. 9 of *The Works of Jonathan Edwards* (New Haven, Conn.: Yale University Press, 1989), p. 195.

52. *The Weekly History* (1741), 7: 4.

53. *The Scots Magazine* (Edinburgh), 1 (1739): 64–68, 199–202, 205–10.

54. *The Christian's Amusement containing Letters Concerning the Progress of the Gospel both at Home and Abroad . . .* (London, September 1740–March 1741).

55. *The Weekly History: Or, An Account of the Most Remarkable Particulars Relating to the Present Progress of the Gospel. By the Encouragement of the Rev. Mr. Whitefield* (London, April 1741–November 1742); changed to *An Account of the Most Remarkable Particulars Relating to the Present Progress of the Gospel* (London, 1742–43), also a weekly. In autumn 1743, Lewis began printing *The Christian History or General Account of the Progress of the Gospel in England, Wales, Scotland and*

America, as far as the Rev. Mr. Whitefield, His Fellow Labourers and Assistants are concerned (London, 1743–44) every seven weeks.

56. Susan Durden, "A Study of the First Evangelical Magazines, 1740–1748," *Journal of Ecclesiastical History* 27 (1976): 255–75.

57. *The Weekly History* (1741), 22: 2–3; 26: 2–4.

58. James Ogilvie, of Aberdeen, in October 1742 said he had urged Whitefield "to undertake this Journey, in consequence of a Correspondence with him, now and then, for more than two, or I think three years" (*Glasgow-Weekly-History* [Glasgow, 1742], 28: 6).

59. Dugald Butler, *John Wesley and George Whitefield in Scotland* (Edinburgh and London, 1898), pp. 11–25.

60. *A Letter from a Gentleman in Scotland to His Friend in New England* (Boston, 1743), p. 4.

61. Tyerman, *Life of Whitefield*, 1:495–518.

62. *Glasgow-Weekly-History*, 13:4–7.

63. Tyerman, *Life of Whitefield*, 2:514.

64. *Glasgow-Weekly-History*, 13:4–7.

65. Tyerman, *Life of Whitefield*, 2:21.

66. See letters from Scottish ministers to Whitefield and others printed in *The Weekly History*, nos. 34–36 and 41–46 (1742).

67. Willison to Whitefield, n.d., "It would be good news to hear that Mr. *Tennent* were coming to succeed you in Scotland, as he did in *New England*" (*The Weekly History* [1742], 42: 2–3). Whitefield wrote McCulloch 15 September 1742 that he had received a letter from Tennent that said he could not come to Scotland. (*Glasgow-Weekly-History*, no. 47).

68. John Willison, *The Balm of Gilead for Healing a Diseased Land* (London, 1742); quotations from pp. 2, 11–12, 60–61. Several of the conversion narratives recorded in "Examination of Persons under Scriptural Concern at Cambuslang during the Revival in 1741–42 by the Revd. William MacCulloch, Minister at Cambuslang, with Marginal Notes by Dr. Webster and Other Ministers," 2 vols., New College Library, Edinburgh (hereafter cited as "McCulloch Mss."), note the influence of these sermons.

69. James Meek, "Parish of Cambuslang," in *The Statistical Account of Scotland*, comp. Sir John Sinclair, vol. 5 (Edinburgh: William Creech, 1793), pp. 241–74.

70. Ned C. Landsman, "Evangelists and Their Hearers: Popular Interpretation of Revivalist Preaching in Eighteenth-Century Scotland," *Journal of British Studies* 28 (1989): 120–49.

71. Robert Fleming, *The Fulfilling of the Scripture* (Rotterdam, 1669; 2d and 3d pts. [1677?]; all three pts., London, 1681; 3d ed., 1681; 4th ed., 1693; 5th ed., 1726); I used the Boston edition of 1743, p. 393–94.

72. Arthur Fawcett, *The Cambuslang Revival: The Scottish Evangelical Revival of the Eighteenth Century* (London: Banner of Truth Trust, 1971), pp. 29–31, 44–52; James A. Wilson, *A History of Cambuslang, A Clydesdale Parish* (Glasgow: Jackson, Wylie and Co., 1929), pp. 85–87, 91.

73. Hew Scott, ed., *Fasti Ecclesiae Scoticanae*, vol. 2, pt. 1 (Edinburgh, 1915), p. 273; Henry Grey Graham, *The Social Life of Scotland in the Eighteenth Century* (1899; reprint ed., London: A & C Black, 1928), p. 307.

74. "McCulloch Mss.," 1: 21.

75. Ibid., p. 96.

76. Jonathan Edwards, "Sinners in the Hands of an Angry God," in *Jonathan Edwards: Basic Writings*, ed. Ola Elizabeth Winslow (New York: New American Library, 1966), pp. 150–67, 160, 164–65.

77. *The Glasgow-Weekly-History Relating to the Late Progress of the Gospel at Home and Abroad; Being a Collection of Letters, partly reprinted from the London-Weekly-History, and partly printed first here at Glasgow* (Glasgow, 1742). This ran for one year, with fifty-two numbers.

78. "McCulloch Mss.," 1:21, 102–3; 2:331–43.

79. *A Short Account of the Remarkable Conversions at Cambuslang. In a Letter from a Gentleman in the West-Country to his Friend at Edinburgh* (Glasgow, 1742), p. 4.

80. *The Weekly-History* (1742), 51: 2.

81. [James Robe], *A Short Narrative of the Extraordinary Work at Cambuslang* (Glasgow, 1742).

82. *A Short Account of the Remarkable Conversions*, pp. 4–5.

83. [Robe], *A Short Narrative*; *A Short Account of the Remarkable Conversions*, pp. 5–9; Meek, "Parish of Cambuslang," p. 274. See Landsman, "Evangelists and Their Hearers," pp. 128–32, where he makes a case for the crucial importance of lay leadership of the revival at Cambuslang.

84. *Glasgow-Weekly-History*, no. 17.

85. Ibid., no. 31; and *The Weekly History*, no. 77 (1742).

86. D[uncan] MacFarlan, *The Revivals of the Eighteenth Century, Particularly at Cambuslang* (London and Edinburgh [1845]; reprint ed. Wheaton, Ill.: Richard Owen Roberts, 1980), p. 63.

87. Ibid., pp. 76–77.

88. Among the parishes the awakening touched were: neighboring ones in the Presbytery of Hamilton, including East Kilbride, Blantyre, Bothwell, and Cathcart, as well as several nearby in Glasgow Presbytery; to the west of Glasgow in the Presbytery of Irvine, Kilmarnock, Stewarton, Dreghorn, and Irvine; in Glasgow, Ramshorn, College, Barony, and Tron; to the north of Glasgow, Calder, Baldernock, Kirkintilloch, Campsie, Kilsyth, Cumbernauld, Torphichen, Falkirk, Denny, Larbert, Dunipace, St. Ninians, and Gargunnock; in the Presbytery of Auchterarder, Muthil, Monivaird, Crieff, and Madderty. By 1743 and 1744 the revival movement appeared in the northern counties of Ross and Sutherland, at Rosskeen, Nigg, Rosemarkie, Logie, Alness, Killearn, Cromarty, Kirkmichael, Kilmuir-Easter, and Golspie (Ibid., pp. 218–56).

89. John Sutherland to James Robe, 8 August 1745, in *The Christian Monthly History*, no. 5 (Edinburgh, 1745), p. 130, reprinted in John Gillies, comp., *Historical Collections Relating to Remarkable Periods of the Success of the Gospel, and Eminent*

Instruments Employed in Promoting It, 2 vols. (Glasgow, 1754: revised by Horatius Bonar, 1845), revised ed. in 1 vol. as *Historical Collections of Accounts of Revival* (Fairfield, Pa.: Banner of Truth Trust, 1981), p. 456.

90. Diane Susan Durden, "Transatlantic Communications and Literature in the Religious Revivals, 1735–1745" (Ph.D. diss., University of Hull, 1978), pp. 180–85.

91. [Colman et al.], *Three Letters to the Reverend Mr. George Whitefield*, p. 5.

92. J. William T. Youngs, Jr. *God's Messengers: Religious Leadership in Colonial New England, 1700–1750* (Baltimore: Johns Hopkins University Press, 1976), pp. 116–19.

CHAPTER 8

1. For a similar assessment of the transatlantic connections among the revivalists during the Great Awakening, see Diane Susan Durden, "Transatlantic Communications and Literature in the Religious Revivals, 1735–1745" (Ph.D. diss., University of Hull, 1978). See also Harold P. Simonson, "Jonathan Edwards and His Scottish Connections," *Journal of American Studies* 21 (1987): 353–76.

2. James Fisher, *A Review of the Preface to a Narrative of the Extraordinary Work at Kilsyth* (Glasgow, 1742), pp. 46–47.

3. Alexander Webster, *Divine Influence the True Spring of the Extraordinary Work at Cambuslang* (Edinburgh, 1742), p. 45.

4. *Short Account of the Remarkable Conversions at Cambuslang* (Glasgow, 1742), p. 14; *A Letter from a Gentleman in Scotland to His Friend in New England* (Boston, 1743), p. 4.

5. John Willison, *A Letter . . . to Mr. James Fisher*, 2d ed. (Edinburgh, 1743), p. 32.

6. James Robe, et al., *Narratives of the Extraordinary Work of the Spirit of God* (Glasgow, 1790), pp. 189–90; "Examinations of persons Under Spiritual Concern at Cambuslang, during the Revival, in 1741–42; By the Revd. William Macculloch Minister of Cambuslang, with Marginal Notes by Dr. Webster and other Ministers," 2 vols., New College Library, Edinburgh, 1:398–99, 495–85 (hereafter cited as "McCulloch Mss.").

7. Adam Gib, *A Warning against Countenancing the Ministrations of Mr. George Whitefield* (Edinburgh, 1742); quotation from the title page.

8. *The Scots Magazine* (Edinburgh), 4 (1742): 310–12.

9. *Declaration of the True Presbyterians* (n.p., 1742), p. 6.

10. James Robe, *A Faithful Narrative of the Extraordinary Work of the Spirit of God, at Kilsyth, and Other Congregations in the Neighborhood. With a Preface Wherein There Is an Address to the Brethren of the Associate Presbytery, Anent Their Late Act for a Public Fast* (Glasgow, 1789), second title, in Robe, *et al.*, *Narratives of the Extraordinary Work of the Spirit of God*, p. 55.

11. *Distinguishing Marks* was published in Boston at the close of 1741, in London early in 1742 by Isaac Watts, and during the summer in Glasgow and Edinburgh with an epistle to the Scots reader by John Willison; James Robe *Mr. Robe's First-(Fourth)*

Letter to . . . James Fisher Concerning His Review of the Preface to a Narrative of the Extraordinary Work at Kilsyth, 4 pts. (Glasgow and Edinburgh, 1742, 1743).

12. Fisher, *A Review of the Preface to a Narrative*, pp. 11, 37.

13. James Robe, *Mr. Robe's Second Letter to . . . James Fisher* (Edinburgh, 1743), p. 8; Willison, *A Letter . . . to Mr. James Fisher*, pp. 9–10.

14. Leigh Eric Schmidt, "'A Second and Glorious Reformation': The New Light Extremism of Andrew Croswell," *William and Mary Quarterly*, 3d ser. 43 (1986): 214–44, 239, 242–43.

15. Alexander R. MacEwen, *The Erskines* (New York: Charles Scribner's Sons [1900]), pp. 122–23.

16. *Letter from a Gentleman in Scotland*, p. 3.

17. *A Short Account of the Remarkable Conversions at Cambuslang*.

18. A. M., *The State of Religion in New-England, since the Reverend Mr. George Whitefield's Arrival There* (Glasgow, 1742), 2–3.

19. *A Letter from a Gentleman in Scotland*, pp. 11–14; [Charles Chauncy], *A Letter from a Gentleman of Boston, to Mr. George Wishart, One of the Ministers of Edinburgh, Concerning the State of Religion in New England* (Edinburgh, 1742).

20. Alexander Malcolm to Charles Mackie, 22 July 1741, Historical Manuscript Commission, *Report on the Laing Manuscripts* (London: HMSO, 1925), 2: 327–28.

21. *Colonial Society of Massachusetts Publications* 15 (1925), 316–17.

22. *A Letter from a Gentleman in Scotland*, p. 4.

23. Willison to McCulloch, 22 September 1743, in *The Practical Works of the Rev. John Willison* (Glasgow, Edinburgh, and London, 1844), pp. xviii–xix.

24. *Glasgow-Weekly-History* nos. 35–37 (1742).

25. [James Robe], *A Short Narrative of the Extraordinary Work at Cambuslang in Scotland* (Boston and Philadelphia, 1742); *A True Account of the Wonderful Conversions at Cambuslang* is appended.

26. *The Christian History* (Boston), 1: 79–80.

27. *Boston News-Letter*, 12 May 1743, p. 2: "Just Imported from *Glasgow* And to be Sold on Board the Meriam Brigantine lying at Minot's T. *Boston*. A Variety of Books, never before printed, viz. Mr. Bisset's Letter to a Gentleman in Edinburgh, with Observations on the Conduct of Mr. Whitefield, etc. The true CHRIST, no new CHRIST; in a Sermon, by Mr. Ralph Erskine. The State of Religion in New-England, since Mr. Whitefield's Arrival there: With a Letter from a Gentleman in Boston, to his Friend in Glasgow: With sundry other new Books; All relating to the Conduct of the said Mr. Whitefield."

28. *The Testimony of the Pastors of the Churches in the Province of the Massachusetts-Bay in New-England, at their Annual Convention in Boston, May 15. 1743. Against several Errors in Doctrine, and Disorders in Practice . . .* (Boston, 1743).

29. *The Testimony and Advice of an Assembly of Pastors of the Churches in New-England, At a Meeting in Boston July 7. 1743. Occasion'd By the late happy Revival of Religion in many Parts of the Land . . .* (Boston, 1743).

30. *The Christian Monthly History* (Edinburgh) no. 4 (February 1744), p. 5; see

also no. 3 (January 1744), p. 4, and James Robe, *Mr. Robe's Fourth Letter to Mr. Fisher* (Edinburgh, 1744), pp. 112–13.

31. *Proceedings of the Massachusetts Historical Society*, 2d ser. 9 (1895): 404 (hereafter cited as *Proc. Mass. Hist. Soc.*).

32. Maclaurin to Thomas Prince, 6 April 1744, *The Christian History* (Boston), 2:217.

33. *The Christian History* (London, 1744), 5: 80.

34. M. H. Jones Mss., item 13,674, Calvinist Methodist Archives, National Library of Wales, Aberystwyth, Wales.

35. Michael R. Watts, *The Dissenters: From the Reformation to the French Revolution* (Oxford: Clarendon Press, 1978), 1: 270.

36. Ibid., 1:434–35, 450–64.

37. *Proc. Mass. Hist. Soc.*, 2d ser. 9 (1895): 387.

38. Ibid., p. 394.

39. Ibid., p. 379.

40. Thomas Milner, *The Life, Times and Correspondence of the Rev. Isaac Watts, D.D.* (London, 1845), pp. 637–40.

41. [Benjamin Colman et al.], *Three Letters to the Reverend Mr. George Whitefield* (Philadelphia [1739]), p. 6.

42. Milner, *Life of Watts*, p. 652.

43. Colman to Watts et al., 3 October 1740, Colman Papers, Massachusetts Historical Society, Boston.

44. *Proc. Mass. Hist. Soc.*, 2d ser. 9 (1895): 379, 383–84, 387, 392, 394, 395.

45. Arnold A. Dallimore, *George Whitefield: The Life and Times of the Great Evangelist of the Eighteenth-Century Revival* (London: Banner of Truth Trust, 1970; reprint ed., 1971), 1: 343.

46. Earnest A. Payne, "Doddridge and the Missionary Enterprise," in *Philip Doddridge 1702–1751: His Contribution to English Religion*, ed. Geoffrey F. Nuttall (London: Independent Press, 1951), pp. 93–95.

47. Miscellaneous Eighteenth-Century Papers, item 24.179.9, Dr. Williams's Library, London; see also the letters from John Barker, David Jennings, and Nathaniel Neal, and Doddridge's to Neal in *The Correspondence and Diary of Philip Doddridge* (London, 1830), 4: 256–94.

48. Miscellaneous Eighteenth-Century Papers, item 24.179.9, Dr. Williams's Library, London.

49. *Proc. Mass. Hist. Soc.*, 2d ser. 9 (1895): 395.

50. Ibid., p. 396–97.

51. Ibid., p. 400.

52. *Boston News-Letter*, 26 May 1743, p. 2.

53. *Proc. Mass. Hist. Soc.*, 2d ser. 9 (1895): 404; Durden, "Transatlantic Communications," pp. 132–33.

54. *The Testimony and Advice . . . With a Recommendation of It by the Revd. Dr. Watts* (London, 1744).

55. *Proc. Mass. Hist. Soc.*, 2d ser. 9 (1895): 400–1.

56. Ibid., pp. 403–4.

CHAPTER 9

1. Unknown to James Robe, 15 April 1745, *The Christian Monthly History* (Edinburgh), 2d ser. 1 (1745): 9.

2. Increase Mather, *Dissertation Concerning the Danger of Apostacy*, second title in *A Call from Heaven to the Present and Succeeding Generations* (Boston, 1679), p. 89.

3. *The Christian History* (Boston), 1: 374–75.

4. Ibid., 2: 313–17.

5. Perry Miller, "Jonathan Edwards and the Great Awakening," in *Errand into the Wilderness* (Cambridge, Mass: Harvard University Press, 1956; reprint ed., New York: Harper & Row, 1964), pp. 153–66, esp. 159–61.

6. *The Christian History* (Boston), 1: 239, 259, 384; 2: 42, 89, 375, 378, 399.

7. Marilyn J. Westerkamp, *Triumph of the Laity: Scots-Irish Piety and the Great Awakening* (New York: Oxford University Press, 1988), pp. 121–22.

8. James Fisher, *A Review of the Preface to a Narrative of the Extraordinary Work at Kilsyth* (Glasgow, 1742).

9. John Willison, Preface, to *Distinguishing Marks of a Work of the Spirit of God*, by Jonathan Edwards (Edinburgh, 1742). The preface is dated 23 June 1742.

10. John Willison, *A Letter from Mr. John Willison, Minister at Dundee, to Mr. James Fisher, Minister at Glasgow. Containing Serious Expostulations with Him Concerning His Unfair-Declaring in His Review of Mr. Robe's Preface, &c.*, 2d ed. (Edinburgh, 1743), pp. 18–19.

11. "Examination of Persons under Scriptural Concern at Cambuslang during the Revival in 1741–42 by the Revd. William MacCulloch, Minister at Cambuslang, with Marginal Notes by Dr. Webster and Other Ministers." 2 vols., New College Library, Edinburgh, 1: 162 (hereafter cited as "McCulloch Mss.").

12. Jonathan Edwards, *The Great Awakening*, ed. C. C. Goen, vol. 4 of *The Works of Jonathan Edwards* (New Haven, Conn.: Yale University Press, 1972) pp. 99–110.

13. Jonathan Edwards, "A Faithful Narrative of the Surprizing Work of God in the Conversion of Many Hundred Souls in Northampton and the Neighboring Towns and Villages," in ibid., pp. 209–10.

14. William Cooper, Preface, to "The Distinguishing Marks of a Work of the Spirit of God," in ibid., pp. 224–25.

15. James West Davidson, *The Logic of Millennial Thought: Eighteenth-Century New England* (New Haven, Conn.: Yale University Press, 1977), pp. 136–37.

16. *The Christian History* (Boston), 1: 410–11.

17. Ibid., 2:397.

18. Ibid., 1:405–6.

19. *A Short Account of the Remarkable Conversions at Cambuslang* (Glasgow, 1742), p. 5.

20. "McCulloch Mss." Edited and abridged versions of twenty-three of these personal narratives appear in Duncan MacFarlan, *The Revivals of the Eighteenth Century, Particularly at Cambuslang* (London and Edinburgh [1845]; reprint ed., Wheaton, Ill.: Richard Owen Roberts, 1980), pp. 113–212.

21. James Robe, *A Faithful Narrative of the Extraordinary Work of the Spirit of God, at Kilsyth* (Glasgow, 1789), second title in *Narratives of the Extraordinary Work of the Spirit of God, at Cambuslang, Kilsyth, &c. Begun 1742*, by James Robe et al. (Glasgow, 1790), pp. 61–62.

22. *A Letter from a Gentleman in Scotland to His Friend in New England* (Boston, 1743), pp. 7–8.

23. Edwards, "A Faithful Narrative" (Yale), 4: 144–46; Robe, *Narratives*, pp. 65–66.

24. Edwards, "A Faithful Narrative" (Yale), 4: 146–59; Robe, *Narratives*, pp. 68–83.

25. It is divided into "articles": Article I. "Concerning the methods I have observed in carrying on this Work"; Article II. "Concerning the Fruits of this Dispensation, which are general as to the Body of the People"; Article III. "Concerning those who have been awakened, and appear now to be converted in a silent and unobservable manner, for some months past;" Article IV. "Concerning them who cried out when they were awakened, or made application to me, from time to time, under their spiritual distress; but were not under bodily affections"; Article V. "Concerning these, upon whose bodies, spiritual operations had real and sensible influence in a more unusual way"; Article VI. "Concerning the variety, and number of the persons, who have been under the influence of this blessed work, in this, and some neighboring parishes." The Last Article, "Concerning the perseverance of these who appeared to be hopefully changed, during this extraordinary season of grace," was published in 1751.

26. See John Gillies, comp., *Historical Collections Relating to Remarkable Periods of the Success of the Gospel*, 2 vols. (Glasgow, 1754), 1: 306–27; revised by Horatius Bonar, 1845, rev. ed. in 1 vol. as *Historical Collections of Accounts of Revival* (Fairfield, Pa.: Banner of Truth Trust, 1981), pp. 197–208.

27. Robe, *Narratives*, p. 62.

28. It was accounts of their lives, not revival accounts, for instance, that John Wesley sought from his older preachers for *The Arminian Magazine* in 1778. T. B. Shepherd, *Methodism and the Literature of the Eighteenth Century* (New York: Haskell House, 1966), chap. 7.

29. *The Christian's Amusement*, nos. 14–16 (London, 1741).

30. For example, John Wesley, *Modern Christianity: Exemplified at Wednesbury and Other Adjacent Places in Staffordshire* (Newcastle, 1745), also known as *Sufferings of the Primitive Methodists at Wednesbury*.

31. Charles Chauncy, *Prayer for Help* (Boston, 1737), p. 24.

32. Charles Chauncy, *The Outpouring of the Holy Ghost* (Boston, 1742).

33. Charles Chauncy, *Enthusiasm Described* (Boston, 1742).

34. Charles Chauncy, *Seasonable Thoughts on the State of Religion in New-England* (Boston, 1743), p. 3.

35. Ibid., p. 13. The reference here is to Ezek. 36: 21, 29. For the outpouring of the Spirit, see pp. 11–16.

36. Charles Chauncy, *The New Creature* (Boston, 1741).

37. *The Christian History* (Boston), 2: 117.

38. *The Weekly History* (London, 1742), 64:3.

39. *The Christian History* (Boston), 2: 117.
40. Ibid., 1: 43.
41. *Glasgow-Weekly-History*, 13: 6.
42. *The Christian History* (Boston), 2: 147–48.
43. Ibid., 1: 372.
44. George Whitefield, *George Whitefield's Journals* (London: Banner of Truth Trust, 1960), pp. 354, 364, 410, 415.
45. *The Glasgow-Weekly-History*, 13: 6.
46. Jonathan Edwards, "Some Thoughts Concerning the Revival," in *The Works of Jonathan Edwards* (Yale), 4:346.
47. Ibid., pp. 399–400.
48. Chauncy, *Seasonable Thoughts*, pp. 239–42.
49. *The Testimony of the Pastors of the Churches in Massachusetts* (Boston, 1743), excerpted in Richard L. Bushman, ed., *The Great Awakening: Documents on the Revival of Religion, 1740–1745* (New York: Atheneum, 1970), p. 128.
50. *The Christian History* (Boston), 1: 413.
51. J. William T. Youngs, Jr., *God's Messengers: Religious Leadership in Colonial New England, 1700–1750* (Baltimore: Johns Hopkins University Press, 1976), pp. 112–16.
52. See ibid., pp. 120–27, for the effects of the Great Awakening on clerical authority in New England.
53. John Howe, "The Prosperous State of the Christian Interest Before the End of Time, By a Plentiful Effusion of the Holy Spirit," in *Works* (New York, 1835), pp. 579–82.

CHAPTER 10

1. "Examinations of persons Under Spiritual Concern at Cambuslang, during the Revival, in 1741–42; By the Revd. William Macculloch Minister of Cambuslang, with Marginal Notes by Dr. Webster and other Ministers," 2 vols., New College Library, Edinburgh, 2: 351 (hereafter cited as "McCulloch Mss.").
2. Nathan Cole, "The Spiritual Travels of Nathan Cole," ed. Michael J. Crawford, *William and Mary Quarterly* 3d ser. 33 (1976): 109–10.
3. *The Christian History* (Boston), 2:45.
4. Jonathan Edwards, "Some Thoughts concerning the Present Revival of Religion in New England," in *The Great Awakening*, ed. C. C. Goen, vol. 4 of *The Works of Jonathan Edwards* (New Haven, Conn.: Yale University Press, 1972), pp. 405–7. For a critique of religious singing in the streets, see Benjamin Colman, *Letter from the Reverend Dr. Colman of Boston, to the Reverend Mr. Williams of Lebanon, Upon Reading the Confession and Retractions of the Reverend Mr. James Davenport* (Boston, 1744), p. 8.
5. Cole, "Spiritual Travels," pp. 105–7.
6. Ibid., referring to John 4:32.
7. Samuel Finley, *Christ Triumphing, and Satan Raging,* (Philadelphia, 1741), p. 18; Robert Sherman Rutter, "The New Birth: Evangelicalism in the Transatlantic

Community During the Great Awakening, 1739–1745," (Ph.D. diss., Rutgers University, 1982), pp. 101–5, 124–25.

8. Catherine M. North, *History of Berlin, Connecticut* (New Haven, Conn.: Tuttle, Morehouse & Taylor, 1916), pp. 150–51.

9. Cole, "Spiritual Travels," pp. 102–3.

10. Rutter, "The New Birth," pp. 107–8, citing Tennent's *A Solemn Warning to a Secure World* (Boston, 1735), pp. 23, 37. See also Tennent, "The Unsearchable Riches of Christ. Sermon II," in *Sermons on Sacramental Occasions by Divers Ministers*, by Tennent et al. (Boston, 1739), p. 200. Cf. Ned C. Landsman, "Evangelists and Their Hearers: Popular Interpretation of Revivalist Preaching in Eighteenth-Century Scotland," *Journal of British Studies* 28 (1989): 130, where he implies conflict or tension between clerical and lay counselors.

11. Cole, "Spiritual Travels," p. 97.

12. James Walsh, "The Great Awakening in the First Congregational Church of Woodbury, Connecticut," *William and Mary Quarterly*, 3d ser. 28 (1971): 543–62.

13. John Bumsted, "Religion, Finance, and Democracy in Massachusetts: The Town of Norton as a Case Study," *Journal of American History* 57 (1971): 817–31.

14. Cole, "Spiritual Travels," pp. 97–98, 121–22.

15. Ibid., pp. 104, 108–9, 119–21, 125.

16. Landsman, "Evangelists and Their Hearers," pp. 120–49, and *Scotland and Its First American Colony, 1683–1765* (Princeton, N.J.: Princeton University Press, 1985), pp. 232–42.

17. Landsman, "Evangelists and Their Hearers."

18. Callum G. Brown, *The Social History of Religion in Scotland since 1730* (London and New York: Methuen, 1987), pp. 26–27, 52, 101–2, 109–12.

19. Alastair J. Durie, *The Scottish Linen Industry in the Eighteenth Century* (Edinburgh: John Donald, 1979), pp. 22–28.

20. Rosalind Mitchison, *Lordship to Patronage: Scotland 1603–1745* (London: Edward Arnold, 1983), pp. 102–3.

21. Durie, *The Scottish Linen Industry*, pp. 38, 78; Mitchison, *Lordship to Patronage*, p. 169; R. A. Houston, "Women in the Economy and Society of Scotland, 1500–1800," in *Scottish Society 1500–1800*, ed. by R. A. Houston and I. D. Whyte (Cambridge: Cambridge University Press, 1989), pp. 118–47, 124.

22. Brown, *The Social History of Religion in Scotland*, p. 97.

23. "McCulloch Mss.," 1: 310–11.

24. Landsman, "Evangelists and Their Hearers," pp. 144–46.

25. David Buchan, *The Ballad and the Folk* (London and Boston: Routledge & Kegan Paul, 1972), pp. 18, 77–89, 190–201.

26. See, for instance, T. J. Byres, "Scottish Peasants and Their Song," *Journal of Peasant Studies* 3 (1976): 236–51; and Thomas Crawford, *Society and the Lyric: A Study of the Song Culture of Eighteenth-Century Scotland* (Edinburgh: Scottish Academic Press, 1979).

27. David D. Hall, *Worlds of Wonder, Days of Judgment: Popular Religious Belief in Early New England* (New York: Alfred A. Knopf, 1989), p. 10.

28. Landsman, "Evangelists and Their Hearers," p. 124; James Meek, "Parish of

Cambuslang," in *The Statistical Account of Scotland*, comp. Sir John Sinclair (Edinburgh: William Creech, 1793), 5: 261–62.

29. Landsman, "Evangelists and Their Hearers," p. 131–32.

30. Landsman also says that William Miller [Millar] was of a weaving family. The narrative, however, mentions nothing of his family's occupation, but does refer to the young man's participation in harvesting and shearing (ibid., p. 141; "McCulloch Mss.," 1: 416–19).

31. J. F. Maclear notes this point in questioning the validity of the title of Marilyn Westerkamp's study *Triumph of the Laity, Scots-Irish Piety and the Great Awakening, 1625–1760* (New York: Oxford University Press, 1988), reviewed in *American Historical Review* 94 (1989): 1165–66.

32. Landsman, "Evangelists and Their Hearers," p. 123; Henry Grey Graham, *The Social Life of Scotland in the Eighteenth Century* (1899; reprint ed., London: A & C Black, 1928), pp. 366–71, 366.

33. John MacInnes, *The Evangelical Movement in the Highlands of Scotland, 1688 to 1800* (Aberdeen: The University Press, 1951), pp. 98–99 and ff., also see chap. 6, pp. 197–220; and Steve Bruce, "Social change and collective behaviour; the revival in eighteenth-century Ross-shire," *The British Journal of Sociology* 34 (1983): 554–72.

34. John Erskine, *The Signs of the Times Consider'd* (Edinburgh, 1742), pp. 26–28.

35. John Gillies, comp., *Historical Collections Relating to Remarkable Periods of the Success of the Gospel, and Eminent Instruments Employed in Promoting It,* 2 vols. (Glasgow, 1754; revised by Horatius Bonar, 1845), rev. ed. in 1 vol. as *Historical Collections of Accounts of Revival* (Fairfield, Pa.: Banner of Truth Trust, 1981), p. 339.

36. Ibid., pp. 370–71.

37. Edwards, "Thoughts on the Revival" (Yale), 4:291.

38. Landsman, "Evangelists and Their Hearers," p. 132.

39. Ibid., p. 136, citing *A Short Account of the Remarkable Conversions at Cambuslang* (Glasgow, 1742), pp. 15 ff.

40. "Attestation of the Kirk-Session of Cambuslang, April 30th, 1751," in Gillies, *Historical Collections* (1981 reprint ed.), p. 462.

41. William McCulloch, *Sermons on Several Subjects* (Glasgow: David Niven, 1793), p. 261, quoted in Rutter, "The New Birth," p. 121.

42. Leigh Eric Schmidt, "Scottish Communions and American Revivals: Evangelical Ritual, Sacramental Piety, and Popular Festivity from the Reformation through the Mid-Nineteenth Century" (Ph.D. diss., Princeton University, 1987), pp. 213–24. In the book version of this dissertation, Schmidt revised this section to emphasize that Willison and other ministers, although encouraging intensely visual faith, did not sanction visionary experience (Schmidt, *Holy Fairs: Scottish Communions and American Revivals in the Early Modern Period* [Princeton, N.J.: Princeton University Press, 1989], pp. 145–50).

43. Landsman, "Evangelists and Their Hearers," p. 135.

44. Quoted in Andrew Lang, *A History of Scotland from the Roman Occupation*, 4 vols. (New York: Dodd, Mead, and Co., 1907), 4: 309–10.

45. Gillies, *Historical Collections* (1981 reprint ed.), p. 346.

46. Ibid., p. 436.

47. Ibid., p. 450.

48. For the frequent closeness of the mentality of the learned to that of the folk, see Hall, *Worlds of Wonder,* pp. 71–116.

49. Landsman, "Evangelists and Their Hearers," pp. 132–37; Clarke Garrett, *Spirit Possession and Popular Religion: From the Camisards to the Shakers* (Baltimore: Johns Hopkins University Press, 1987), pp. 93–96.

50. Landsman, "Evangelists and Their Hearers," p. 142.

51. John Erskine, *The Signs of the Times*, pp. 28–30. See also John Maclaurin, "On the Scripture Doctrine of Divine Grace," in *Sermons and Essays* (Glasgow, 1755); and Alexander Webster, *Divine Influence the True Spring of the Extraordinary Work at Cambuslang* (Edinburgh, 1742).

52. Erskine, *The Signs of the Times,* p. 30.

53. Landsman, "Evangelists and Their Hearers," p. 137.

54. "McCulloch Mss.," 1: 95–97.

55. Landsman, "Evangelists and Their Hearers," p. 138.

56. "McCulloch Mss.," 1: 76–77.

57. Ibid., pp. 177–80.

58. Ibid., pp. 9–10.

59. Ibid., pp. 282–83.

60. On the distinction between legal fear and evangelical sorrow, see John Owen King, *The Iron of Melancholy: Structures of Spiritual Conversion in America from the Puritan Conscience to Victorian Neurosis* (Middletown, Conn: Wesleyan University Press, 1983), pp. 70–73, and notes thereto.

61. "McCulloch Mss.," 1: 517; Rutter, "The New Birth," pp. 96, 227, 362.

62. Landsman, "Evangelists and Their Hearers," pp. 137–38.

63. King, *The Iron of Melancholy*, pp. 49–54, and passim.

64. Owen C. Williams, *The Puritan Experience: Studies in Spiritual Autobiography*, (New York: Schocken Books, 1972), p. 42.

65. Jonathan Edwards, "A Faithful Narrative," pp. 205–7, and "Thoughts on the Revival," p. 393, in *The Works of Jonathan Edwards* (Yale), vol 4.

66. Cole, "Spiritual Travels," pp. 94–95, 101–3.

67. Samuel Hopkins, *The Life and Character of Miss Susanna Anthony, Who Died in Newport, (R.I.) June 23, MDCCXCI, in the Sixty Fifth Year of Her Age, Consisting Chiefly in Extracts from Her Writings, with Some Brief Observations on Them* (Worcester, Mass., 1796), pp. 19–28.

68. "McCulloch Mss.," 1:516.

69. Landsman, "Evangelists and Their Hearers," p. 140.

70. Graham, *The Social Life of Scotland*, p. 310.

71. Williams, *The Puritan Experience*, pp. 13–14.

72. Ibid., p. 209; Patricia Caldwell, *The Puritan Conversion Narrative: The Beginnings of American Expression* (Cambridge: Cambridge University Press, 1983), chap. 4, pp. 135–62; Landsman, "Evangelists and Their Hearers," pp. 138–44.

73. See, for instance, the accounts in *Thomas Shepard's Confessions*, ed. George

Solement and Bruce C. Woolley, Publications of the Colonial Society of Massachusetts, *Collections* 53 (Boston, 1981), and the accounts appended to "The Diary of Michael Wigglesworth," ed. Edmund S. Morgan, in ibid., *Transactions* 35 (Boston, 1951): 311–444.

74. Gillies, *Historical Collections* (1981 reprint ed.), pp. 345, 360.

75. Edwards, "Thoughts on the Revival" (Yale), 4: 418, 422.

76. Landsman, "Evangelists and Their Hearers," p. 141.

77. Jonathan Edwards, "To the Rev. Thomas Prince of Boston," in *The Works of Jonathan Edwards* (Yale), 4: 544, 553–54.

78. "McCulloch Mss.," 1: 283.

79. Ibid., p. 178.

80. Ibid., pp. 228–48.

81. Schmidt, *Holy Fairs,* pp. 163–64.

82. Gilbert Tennent, *The Espousals, or a Passionate Perswasive to a Marriage with the Lamb of God* (New York, 1735).

83. Gillies, *Historical Collections* (1981 reprint ed.), pp. 338, 339.

84. "McCulloch Mss.," 2: 271–72.

85. Ibid., 1: 529, 544, 580.

86. Ibid., p. 285.

87. Ibid., p. 261, 285–86, 323–24, 391, 472, 529, 544, 580; 2: 272, 313, 325, 354, 449, 500, 541, 553, 579, 592, 678.

88. Ibid., 1:404.

89. Ibid., p. 132.

90. Ibid., pp. 549–50.

91. Ibid., 2:162–63. When Whitefield published this sermon in Philadelphia in 1746, he remarked on a message sent him by a woman in Scotland from her death bed: "'Tell him,' says she, 'for his Comfort, that at such a Time he married me to the Lord Jesus,'" "Christ the Believer's Husband" (sermon 1 in *Sermons on the Following Subjects*, pp. 44–45.

92. "McCulloch Mss.," 2: 579.

93. Ibid., 1:21, 102–3; 2: 331–43.

94. Westerkamp, *Triumph of the Laity,* p. 10.

95. Ibid., pp. 29–34; Schmidt, *Holy Fairs,* pp. 76–114.

96. Schmidt, *Holy Fairs,* p. 49–50. Perhaps because he became aware of the mistake of equating communion seasons with revivals, for his book Schmidt changed a phrase in his dissertation referring to Presbyterian communion seasons, "a long almost continuous chain of revivals," to "a long and well-developed tradition of revivalism" ("Scottish Communions," pp. 86–87; *Holy Fairs,* p. 49).

97. Westerkamp, *Triumph of the Laity,* pp. 9–10, 28.

98. Gillies, *Historical Collections* (1981 reprint ed.), p. 447.

99. Richard Webster, *A History of the Presbyterian Church in America, from Its Origin until the Year 1760* (Philadelphia, 1857), pp. 122–24.

100. Gillies, *Historical Collections* (1981 reprint ed.) pp. 346–48, 425, 426.

101. Ibid., p. 378.

102. Ibid., p. 389–90.

103. Ibid., p. 366.

104. "McCulloch Mss.," 1: 3–4, 10, 42, 176, 229, 288, 308–14, 317, 346, 358, 363–66.

CHAPTER 11

1. For the intensified millennial hopes in New England resulting from the Great Awakening, see James West Davidson, *The Logic of Millennial Thought: Eighteenth-Century New England* (New Haven, Conn.: Yale University Press, 1977); Alan Heimert, *Religion and the American Mind: From the Great Awakening to the Revolution* (Cambridge, Mass.: Harvard University Press, 1966); and Stephen J. Stein, "Editor's Introduction," in *Apocalyptic Writings*, by Jonathan Edwards, vol. 5 of *The Works of Jonathan Edwards* (New Haven, Conn.: Yale University Press, 1977), pp. 1–93. Expressions of Scottish millennial hopes were ecstatic: John Erskine, *The Signs of the Times Considered* (Edinburgh, 1742); John Willison, *A Scripture Prophecy of the Increase of Christ's Kingdom and the Destruction of Antichrist*, appended to *The Balm of Gilead* (London, 1742).

2. *The Christian History, Containing Accounts of the Revival and Propagation of Religion in Great Britain and America* was published every Saturday from March 1743 through February 1745. Although the magazine was nominally under the editorship of Thomas Prince, Jr. (1722–1748; A.B. Harvard, 1740), the guiding role of the elder Thomas Prince is generally recognized.

3. John E. Van de Wetering, "The Christian History of the Great Awakening," *Journal of Presbyterian History* 44 (1966): 128–29, argues that Prince ceased publication of *The Christian History* after its second year because opposition to it as an instrument of a party had grown too powerful to resist; however, by Van de Wetering's own evidence, opposition to the magazine had been strong from the beginning. Thomas Prince, Jr.'s, own explanation was "the scarcity of suitable materials" (*The Christian Monthly History* [Edinburgh], 2d ser. 7 [November 1745]: 230–32.

4. William Cooper, "To the Reader," in *The Distinguishing Marks of a Work of the Spirit of God*, by Jonathan Edwards (Boston, 1741), reprinted in *The Works of Jonathan Edwards* (New Haven, Conn.: Yale University Press, 1972), 4:216–17.

5. John Gillies, comp., *Historical Collections Relating to Remarkable Periods of the Success of the Gospel, and Eminent Instruments Employed in Promoting It*, 2 vols. (Glasgow, 1754; revised by Horatius Bonar, 1845), rev. ed. in 1 vol. as *Historical Collections of Accounts of Revival* (Fairfield, Pa.: Banner of Truth Trust, 1981).

6. Gillies, *Historical Collections* (Glasgow, 1754), 1: iv; (rev. ed., 1981), p. v.

7. *The Testimony of a Number of Ministers* (Boston, 1745); *The Christian Monthly History* (Edinburgh), 2d ser. 9 (December 1745): 273–88.

8. The number of Scottish editions of Edwards's works reflects his influence in Scotland. See Thomas H. Johnson, *The Printed Writings of Jonathan Edwards, 1703–1758: A Bibliography* (Princeton, N.J.: Princeton University Press, 1940).

9. G. D. Henderson, "Jonathan Edwards in Scotland," in *The Burning Bush: Studies in Scottish Church History* (Edinburgh: Saint Andrew Press, 1957), pp. 151–62; Henry Moncreiff Wellwood, *Account of the Life and Writings of John Erskine* (Edinburgh, 1818), pp. 144, 160–62, 196, 516–22.

10. Sereno E. Dwight, *The Life of President Edwards* (New York, 1830), p. 412.

11. The letters of John Maclaurin for 1751, in his *Works* (Edinburgh, 1860), 1. xlvii–lix, reveal the deep interest of the Scots in Edwards's circumstances.

12. "'Tis not unlikely that the work of God's Spirit, that is so extraordinary and wonderful, is the dawning, or at least a prelude, of that glorious work of God . . . which . . . shall renew the world of mankind" (Jonathan Edwards, "Some Thoughts concerning the Present Revival of Religion in New England," in *The Great Awakening*, ed. C. C. Goen, vol. 4 of *The Works of Jonathan Edwards* [New Haven, Conn.: Yale University Press, 1972], p. 353).

13. Ibid., pp. 353–38.

14. Dwight, *The Life of President Edwards*, p. 198. Not everyone found Edwards convincing. Charles Chauncy found his evidence "absolutely precarious" (*Seasonable Thoughts on the State of Religion in New England* [Boston, 1743], p. 372). Isaac Watts thought that his "reasonings about America want force" (*Proceedings of the Massachusetts Historical Society*, 2d ser. 9 [1895], 402).

15. In a letter to McCulloch of 23 September 1747, he stated that he was "full of apprehension, that God has no design of mercy to those that were left unconverted, of the generation that were on the stage, in the time of the late extraordinary religious commotion, and striving of God's Spirit" (Dwight, *The Life of President Edwards*, p. 244).

16. Ibid., p. 211.

17. Jonathan Edwards, "An Humble Attempt to Promote Explicit Agreement and Visible Union of God's People in Extraordinary Prayer for the Revival of Religion and the Advancement of Christ's Kingdom on Earth," in *The Works of Jonathan Edwards* 5: 357. In *Religion and the American Mind*, Alan Heimert mistakenly cites Edwards's remark giving Britain "one generation more" before it would "sink under the weight" of its vice as part of *Thoughts on the Revival* in 1743, and then contrasts this view to Edwards's belief that the millennium was about to dawn in America (Heimert, *Religion and the American Mind*, p. 58). In its actual context in 1748, the quotation cannot support such a contrast. Perry Miller notes that Edwards believed that colonial culture was dependent on England. Growth of heresy and decline of piety in England were soon reflected in America ("Jonathan Edwards' Sociology of the Great Awakening," *New England Quarterly* 21 [1948]: 54–56).

18. Edwards, "Thoughts on the Revival" (Yale), 4: 520.

19. *The Diary of Cotton Mather for the Year 1712*, ed. William R. Maniere II (Charlottesville: University Press of Virginia, 1964), pp. 86, 101, 113–18; Robert Wodrow, *Correspondence*, 3 vols. (Edinburgh, 1842–43), 3:267–68; Dwight, *The Life of President Edwards*, p. 102.

20. *The Christian Monthly History* 2d ser. 1 (April 1745).

21. Stein, "Editor's Introduction" 5: 38–39.

22. James Robe, Preface, to *Sermons in Three Parts* (1749; London, 1763); excerpts from this are in Gillies, *Historical Collections* (Glasgow, 1754), 2: 399–40, and (rev. ed., 1981), pp. 462–64.

23. Stein, "Editor's Introduction," 5: 82–89, for the publication history.

24. Edwards, "Humble Attempt," 5: 307–436, 353, 358.

25. "I am much encouraged by the Regard my People pay to the Concert of Prayer,"

wrote a minister from New Jersey in August 1749, "and especially that of late a Number of them are excited to add to their Prayers some more vigorous Attempts for a Reformation than have hitherto been used." They were forming a society for reforming themselves with respect to family government and observation of the Lord's Day, and for informing the civil magistrates of all sins committed which were punishable by law. An unidentified minister near Boston wrote to John Gillies, at Glasgow, on 3 July 1750, that "the Affair of the Concert was prevailing and increasing in those Parts of the Land" (John Gillies, *An Exhortation to the Inhabitants of the South Parish of Glasgow, and the Hearers in the College-Kirk* [Glasgow, 1750–51], pp. 159–60).

26. Robe, *Sermons in Three Parts*, p. xxi.

27. Quoted in Arthur Fawcett, *The Cambuslang Revival: The Scottish Evangelical Revival of the Eighteenth Century* (London: Banner of Truth Trust, 1971), p. 224.

28. *Good News from the Netherlands. Extracts of Letters from Two Ministers of Holland Confirming and Giving Accounts of the Revival of Religion in Guelderland* (Boston, 1751).

29. Gillies, *Historical Collections* (Glasgow, 1754), 2:402n; (rev. ed., 1981), p. 464.

30. John Gillies, comp., *Appendix to the Historical Collections* (Glasgow, 1761), p. 73.

31. Ibid., pp. 149–51.

32. John Cleaveland, *A Short and Plain Narrative of the Late Work of God's Spirit at Chebacco in Ipswich, in the Years 1763 and 1764* (Boston, 1767), p. 4.

33. The evidence presented here supports Alan Heimert's finding that throughout the years of the French and Indian War, "the Concert of Prayer had enjoyed a considerable vogue among American Calvinists generally." Heimert cites a sermon by Robert Smith in 1759 in which he told the Presbyterians of Pennsylvania that the Concert of Prayer would be more effective than British arms for the victory of the true church over the forces of false religion (*Religion and the American Mind,* p. 336). For a different emphasis, which minimizes the success of the Concert, see Stein, "Editor's Introduction," 5: 47–48, esp. 48, n. 4.

34. Stein, "Editor's Introduction," 5: 46–47.

35. For the history of the Society, see *State of the Society in Scotland for Propagating Christian Knowledge* (Edinburgh, 1741).

36. David Brainerd, *Mirabilia Dei inter Indicos: or, An Account of the Revival and Progress of Religion, amongst a Number of the Indians in the Provinces of New Jersey and Pennsylvania* (Philadelphia [1746]), pp. 251–52.

37. Jonathan Edwards, *An Account of the Life of the Late Reverend Mr. David Brainerd* (Boston, 1749), p. 308. After his dismissal from Northampton in 1750, Edwards served as missionary to the Indians at Stockbridge, Massachusetts, until called to the presidency of the college at Princeton in 1757.

38. For the place of missions in the millennial thought of American and British evangelicals in the eighteenth century, see Ruth H. Bloch, *Visionary Republic: Millennial Themes in American Thought, 1756–1800* (Cambridge: Cambridge University Press, 1985), pp. 216–24, and J. A. de Jong, *As the Waters Cover the Sea: Millennial Expectations in the Rise of Anglo-American Missions, 1640–1810* (Kampen, Netherlands: J. H. Kok, 1970), passim, esp. pp. 81, 119–21, 225–26.

39. Gillies, *An Exhortation to the Inhabitants of the South Parish of Glasgow*.

40. For the Dutch revivals, see Hugh Kennedy (pastor of the Scottish congregation at Rotterdam), *A Short Account of the Rise and Continuing Progress of a Remarkable Work of Grace in the United Netherlands* (London, 1752).

41. Gillies, *An Exhortation to the Inhabitants of the South Parish of Glasgow*, pp. 97, 109, 110, 114–18, 156–60, 172–74.

42. In April 1756, Jonathan Edwards replied to Gillies's suggestion that he "write some particular account of affairs, relating to the success of the gospel in America." Edwards left it up to the discretion of Gillies whether to publish a letter of 1757 in the projected appendix. In February 1758, John Erskine sent Gillies two letters from Edwards "which you are at liberty to publish in your appendix" (Gillies, *Appendix to the Historical Collections*, pp. 102–3, 137, 145).

43. Ibid., pp. 101, 141–42.

44. For example, Thomas Coke, *An Account of the Great Revival of the Work of God, in the City of Dublin, which commenced on the 4th of July, 1790* (Richmond, Va.: Reprinted by Augustine Davis, 1791).

45. Steve Bruce, "Social Change and Collective Behaviour: The Revival in Eighteenth-Century Ross-shire," *British Journal of Sociology* 34 (1983): 554–72.

46. John MacInnes, *The Evangelical Movement in the Highlands of Scotland, 1688 to 1800* (Aberdeen: The University Press, 1951), pp. 154–55.

47. Alexander Stewart, *An Account of the Late Revival of Religion*, 3d ed. (Edinburgh, 1802).

48. For evangelicalism in the South in the eighteenth century, see Wesley M. Gewehr, *The Great Awakening in Virginia, 1740–1790* (1930; reprint ed., Gloucester, Mass.: Peter Smith, 1965); and Donald G. Mathews, *Religion in the Old South* (Chicago: University of Chicago Press, 1977).

49. Samuel Davies, *The State of Religion among the Protestant Dissenters in Virginia; In a letter to the Reverend Mr. Joseph Bellamy* (Boston, 1751), p. 37.

50. [Devereax Jarratt], *A Brief Narrative of the Revival of Religion in Virginia. In a Letter to a Friend* (London, 1778), reprinted in Francis Asbury, *An Extract from the Journal of Francis Asbury, Bishop of the Methodist Episcopal Church in America, from August 7, 1771, to December, 29, 1778*, vol. 1 (Philadelphia, 1792), pp. 245–66.

51. [Samuel Chaplin], *A journal containing some remarks upon the spiritual operations, beginning about the year . . . 1740, or 1741* (n.p., 1757), p. 11.

52. Samuel Davies, *Little Children Invited to Jesus Christ. . . . With an Account of the late remarkable Religious Impressions among the Students in the College of New-Jersey*, 5th ed. (Boston, 1765), pp. 19–24. In his *Appendix to the Historical Collections* in 1761, Gillies printed letters from his American correspondents about the revival at the College of New Jersey, in *Historical Collections* (rev. ed., 1981) pp. 521–55.

53. Samuel Buell, *A Copy of a Letter from the Rev. Mr. Buell, of East-Hampton, on Long-Island, to the Rev. Mr. Barber, of Groton, Connecticut* [New London? 1764], pp. 1, 4.

54. Samuel Buell, *A Faithful Narrative of the Remarkable Revival of Religion in the Congregation of East-Hampton on Long-Island, in the Year of Our Lord 1764* (New York, 1766), p.6.

55. Jacob Johnson, *Zion's Memorial of the Present Work of God* ([New London?], 1765), pp. 5–9.

56. Cleaveland, *Short and Plain Narrative*.

57. Charles Royster, *A Revolutionary People at War: The Continental Army and American Character, 1775–1783* (Chapel Hill, N.C.: University of North Carolina Press, 1979), pp. 152–69; Mathews, *Religion in the Old South*, p. 46–47. Royster notes that "American soldiers saw themselves as righteous men fighting for a righteous cause, but not as an army of saints," and that "the evangelical aspiration that encouraged the army's wartime perseverance . . . did not make the army an evangelical institution or camp life a revival" (pp. 164, 165). See also Cushing Strout, *The New Heavens and New Earth: Political Religion in America* (New York: Harper & Row, 1974), pp. 50–76.

In the numerous studies of the relation between the Great Awakening and the American Revolution, two of the principal topics of debate are the effect the Awakening had on the nature and style of colonial politics and the relative importance of evangelicals and religious liberals in the movement of resistance to British power. The interest of the paragraphs in the present work that touch on the Revolution is not the influence of religion on politics, but the reverse, the influence of politics on religion, or, to be more specific, how the Revolution affected evangelicals' understanding of religious revivals. See Bernard Bailyn, "Religion and Revolution: Three Biographical Studies," *Perspectives in American History* 4 (1970): 85–169; Bloch, *Visionary Republic*; Patricia U. Bonomi, *Under the Cope of Heaven: Religion, Society, and Politics in Colonial America* (New York: Oxford University Press, 1986); Richard L. Bushman, *From Puritan to Yankee: Character and the Social Order in Connecticut, 1690–1765* (Cambridge, Mass.: Harvard University Press, 1967); Nathan O. Hatch, *The Sacred Cause of Liberty: Republican Thought and the Millennium in Revolutionary New England* (New Haven, Conn.: Yale University Press, 1977); Heimert, *Religion and the American Mind*; and Harry S. Stout, "Religion, Communications, and the Ideological Origins of the American Revolution," *William and Mary Quarterly,* 3d ser. (1977): 519–41.

58. Although evangelicals strongly supported the Patriot movement, they did not supply any substantial proportion of the leadership of resistance to the British government. Evangelical support contributed notably to Patriot success, but, as John M. Murrin puts it succinctly, "the Awakening did not create the Revolution" (John M. Murrin, "No Awakening, No Revolution? More Counterfactual Speculations," *Reviews in American History* 11 [1983]: 161–71).

59. Cedric B. Cowing, *The Great Awakening and the American Revolution: Colonial Thought in the 18th Century* (Chicago: Rand McNally, 1971), pp. 200–6; Heimert, *Religion and the American Mind*, pp. 296–97.

60. "The Day of darkness in America 1780," in Nathan Cole Mss., Connecticut Historical Society, Hartford, Conn.

61. Bloch, *Visionary Republic*; Hatch, *The Sacred Cause of Liberty.*

62. *An Address from the Presbytery of New-Castle [on] Duties Necessary for a Revival of Decayed Piety* (Wilmington, Del., 1785), pp. 45–56.

63. *A Concert of Prayer. Propounded to the Citizens of the United States of America. By an Association of Christian Ministers* (Exeter, 1787) pp. 9–10.

64. Charles Roy Keller, *The Second Great Awakening in Connecticut* (New Haven, Conn.: Yale University Press, 1942; reprint ed., Archon Books, 1968), p. 50.

65. Ibid., p. 42, and passim.

66. For the formulaic structure of New England revival narratives in the Second Great Awakening, see Richard D. Shiels, "The Connecticut Clergy in the Second Great Awakening" (Ph.D. diss., Boston University, 1976), chap. 6. A convenient compilation of these accounts is Bennett Tyler, comp., *New England Revivals, as they existed at the Close of the Eighteenth, and the Beginning of the Nineteenth Centuries. Compiled Principally from Narratives First Published in the Conn. Evangelical Magazine* (Boston: Massachusetts Sabbath School Society, 1846).

67. The publisher mistakenly attributed the sermon to Samuel Whiting.

68. John Babcock, "Proposals. The present religious state of many towns in New-England," broadside dated 4 July 1799; Nathan Strong, Abel Flint, and Joseph Steward, comps., *The Hartford Selection of Hymns* (Hartford: John Babcock, 1799), pp. iii, 165–66, 208.

CHAPTER 12

1. For works on closely related topics, see, for instance, William G. McLoughlin, *Modern Revivalism: Charles Grandison Finney to Billy Graham* (New York: Ronald Press, 1959); and Timothy L. Smith, *Revivalism and Social Reform: American Protestantism on the Eve of the Civil War* (Abington Press, 1957; reprint ed., New York: Harper & Row, 1965).

2. John B. Boles, *The Great Revival, 1787–1805: The Origins of the Southern Evangelical Mind* (Lexington, Ky.: University Press of Kentucky, 1972), pp. 12–70, 58–59.

3. Richard Carwardine, *Transatlantic Revivalism: Popular Evangelicalism in Britain and America, 1790–1865* (Westport, Conn.: Greenwood Press, 1978), pp. 12–15; Charles Roy Keller, *The Second Great Awakening in Connecticut* (New Haven, Conn.: Yale University Press, 1942; reprint ed., Archon Books, 1968), p. 49.

4. Keller, *The Second Great Awakening in Connecticut*, pp. 53–54; Carwardine, *Transatlantic Revivalism*, p. 4; William B. Sprague, *Lectures on Revivals of Religion* (Albany, 1832).

5. Charles Grandison Finney, *Lectures on Revivals of Religion*, ed. William G. McLoughlin (Cambridge, Mass.: Harvard University Press, 1960), p. 20.

6. William Warren Sweet, *Revivalism in America: Its Origin, Growth and Decline* (New York: Charles Scribner's Sons, 1944), pp. 134–39.

7. Carwardine, *Transatlantic Revivalism*, p. 25.

8. For example, the accounts collected in *A Narrative of the Revival of Religion in the County of Oneida . . . in the Year 1826* (Utica, 1826).

9. Paul E. Johnson, *A Shopkeeper's Millennium: Society and Revivals in Rochester, New York, 1815–1837* (New York: Hill and Wang, 1978), passim; quotation from p. 55.

10. Michael R. Watts, *The Dissenters: From the Reformation to the French Revolution* (Oxford: Clarendon Press, 1978), pp. 450–61; Carwardine, *Transatlantic Revival-*

ism, pp. 60–63, 103, and 103–4, quoting Wilbur Fisk, *Travels in Europe*, 4th ed. (New York, 1838), p. 602.

11. John MacInnes, *The Evangelical Movement in the Highlands of Scotland, 1688–1800* (Aberdeen: The University Press, 1951) pp. 161–66.

12. Carwardine, *Transatlantic Revivalism*, pp. 66–97.

13. Horatio Bonar, "Conclusion," appended to Gillies, comp., *Historical Collections Relating to Remarkable Periods of the Success of the Gospel, and Eminent Instruments Employed in Promoting It,* 2 vols. (Glasgow, 1754; revised by Horatius Bonar, 1845), rev. ed. in 1 vol. as *Historical Collections of Accounts of Revival* (Fairfield, Pa.: Banner of Truth Trust, 1981), p. 559.

14. Ibid., pp. 558–60.

15. Ian A. Muirhead, "The Revival as a Dimension of Scottish Church History," *Records of the Scottish Church History Society* 20 (1978): pp. 182–83, 188–89, 195–96.

16. Calvin Colton, *History and Character of American Revivals of Religion* (London, 1832; reprint ed., New York: AMS Press, 1973), pp. 167–68.

17. Ibid., pp. 67–69.

18. Ibid., pp. 161–62.

19. *The New Schaff-Herzog Encyclopedia of Religious Knowledge*, 13 vols. (New York and London: Funk and Wagnalls, 1908–14; reprint ed., Grand Rapids, Mich.: Baker, 1949–50), 10: 9–10. Although this work is based on the third edition of the *Real-Encyklopedia für Protestantische Theologie und Kirche*, founded by Johann Jakob Herzog, ed. Albert Hauck, 24 vols. (Leipzig: J. C. Hinrichs, 1896–1913), the entry on revivals is entirely American. The German work, under the entry "Entweckung," the German term for revival or awakening, primarily treats the term in reference to its meaning as a stage in conversion. See also the first edition of the *Real-Encyklopedia*, 22 vols. (Hamburg: R. Besser, 1854–68), s.v. "Entweckung".

20. Finney, *Lectures on Revivals of Religion*, pp. 9, 12.

21. Jonathan Edwards, *A History of the Work of Redemption,* ed. John F. Wilson, vol. 9 of *The Works of Jonathan Edwards* (New Haven, Conn.: Yale University Press, 1989), p. 143.

22. Joseph Milner, *Essays on Several Subjects, Chiefly Tending to Illustrate the Scripture-Doctrine of the Holy Spirit* (1789), p. 167, quoted by J. D. Walsh, "Joseph Milner's Evangelical Church History," *Journal of Ecclesiastical History* 10 (1959): 178.

23. In *The Sacred Cause of Liberty: Republican Thought and the Millennium in Revolutionary New England* (New Haven, Conn.: Yale University Press, 1977), Nathan O. Hatch traces the development among New England's Congregational clergy of a version of "civil millennialism," in which the spread of Christ's church is directly coupled with political liberty rather than with revivals. In *Visionary Republic: Millennial Themes in American Thought, 1756–1800* (Cambridge: Cambridge University Press, 1985), Ruth H. Bloch demonstrates that millennial thought in America possessed a much greater variety than Hatch implies, and that in American the evangelical emphasis on the spread of piety through the dispensations of divine grace, rather than by human political actions, remained a vibrant interpretation of millennialism.

24. Sprague, *Lectures on Revivals of Religion,* pp. 29–34.

25. James Munro, "Encouragements from the History of the Church under the Old and under the New Testament Dispensation, to expect, pray and labor for the Revival of Religion," in *Revival: Scriptural and Historical*, by John G. Lorimer and James Munro (Strathpine North, Australia: Covenanter Press, 1977), pp. 83–142, 83, 84.

26. Biographical information on Coffin is from Wayne Dobson, archivist of Tusculum College, Greeneville, Tenn., telephone conversation with the author, 14 March 1990, and Allen E. Ragan, *A History of Tusculum College, 1794–1944* (Bristol, Tenn.: Tusculum Sesquicentennial Committee, 1945), pp. 10–11.

27. Coffin to Sprague, 22 July 1833, in William B. Sprague, *Lectures on Revivals of Religion*, 2d ed. (New York, 1833), pp. 399–400.

28. Donald G. Mathews, *Religion in the Old South* (Chicago: University of Chicago Press, 1977), pp. 54–55; Sidney E. Mead, "The Nation with the Soul of a Church," *Church History* 36 (1967): 262–83; Perry Miller, "From the Covenant to the Revival," in *The Shaping of American Religion*, ed. James Ward Smith and A. Leland Johnson (Princeton, N.J.: Princeton University Press, 1961), pp. 322–68, esp. 351–61; Smith, *Revivalism and Social Reform*; John Edwin Smylie, "National Ethos and the Church," *Theology Today* 20 (1963): 313–21.

BIBLIOGRAPHY

PRIMARY SOURCES

Unpublished Manuscripts

Aberystwyth, Wales. Calvinist Methodist Archives, National Library of Wales. M. H. Jones Mss. Item 13,674.

London. Dr. Williams's Library. Miscellaneous Eighteenth-Century papers. Item 24.179.9.

Boston. Boston University. Special Collections. Foxcroft papers.

Hartford, Conn. Connecticut Historical Society. Nathan Cole Mss.

Edinburgh. New College Library. "Examination of Persons under Scriptural Concern at Cambuslang during the Revival in 1741–42 by the Revd. William MacCulloch, Minister at Cambuslang, with Marginal Notes by Dr. Webster and Other Ministers" [William McCulloch Mss.] 2 vols.

Boston. Massachusetts Historical Society. Colman papers.

Published Manuscripts

Cole, Nathan. "The Spiritual Trials of Nathan Cole." Edited by Michael J. Crawford. *William and Mary Quarterly*, 3d ser. 33 (1976): 89–126.

Colonial Society of Massachusetts Publications 15 (1925).

Doddridge, Philip. *The Correspondence and Diary of Philip Doddridge*. Vol. 4. London, 1830.

Edwards, Jonathan. Letter to Benjamin Colman, 22 May 1744. In *Proceedings of the Massachusetts Historical Society*, 2d ser. 10 (1896): 429.

Historical Manuscript Commission. *Report on the Laing Manuscripts*. Alexander Malcolm, to Charles Mackie, 22 July 1741. 2:327–8. London: HMSO, 1925.

Massachusetts Historical Society. *Collections*, 4th ser. 1 (1852); 7th ser. 7 and 8 (1911 and 1912)

———. *Proceedings*, 2d ser. 9 (1895); 10 (1896); 77 (1965).

Mather, Cotton. "The Diary of Cotton Mather." *Massachusetts Historical Society Collections*, 7th ser. 7 and 8 (1911 and 1912).

———. *The Diary of Cotton Mather for the Year 1712*. Edited by William R. Maniere II. Charlottesville, Va.: University Press of Virginia, 1964.

———. *Selected Letters of Cotton Mather*. Compiled by Kenneth Silverman. Baton Rouge: Louisiana State University Press, 1971.

Sewall, Samuel. "Diary of Samuel Sewall." *Massachusetts Historical Society Collections*, 5th ser. 6 (1879).

Shepard, Thomas. *Thomas Shepard's Confessions*. Edited by George Selement and

Bruce C. Woolley. Publications of the Colonial Society of Massachusetts, *Collections* 53 (1981).

Wadsworth, Daniel. *Diary of Rev. Daniel Wadsworth*. Hartford, 1894.

Wigglesworth, Michael. "The Diary of Michael Wigglesworth." Edited by Edmund S. Morgan. Colonial Society of Massachusetts, *Transactions* 35 (1951): 311–444.

Wodrow, Robert. *Correspondence*. 3 vols. Edinburgh, 1842–43.

Periodicals

An Account of the Most Remarkable Particulars Relating to the Present Progress of the Gospel. London, 1742–43.

Boston News-Letter. 12, 26 May 1743.

The Christian History: Containing Accounts of the Revival and Progress of the Propagation of Religion in Great Britain and America. 2 vols. Boston, 1743–45.

The Christian History or General Account of the Progress of the Gospel in England, Wales, Scotland and America, as far as the Rev. Mr. Whitefield, His Fellow Labourers and Assistants are concerned. London, 1743–44.

The Christian Monthly History: or An Account of the Revival and Progress of Religion, Abroad, and at Home. Edinburgh, 1743–46.

The Christian's Amusement containing Letters Concerning the Progress of the Gospel both at Home and Abroad etc. London, September 1740–March 1741.

The Glasgow-Weekly-History Relating to the Late Progress of the Gospel at Home and Abroad: Being a Collection of Letters, Partly Reprinted from the London-Weekly-History, and Partly Printed First Here at Glasgow. Glasgow, 1742.

The Scots Magazine. Edinburgh, 1739, 1742.

The Virginia Gazette. Williamsburg, 1737, 1738.

The Weekly History: or, An Account of the Most Remarkable Particulars Relating to the Present Progress of the Gospel. By the Encouragement of the Rev. Mr. Whitefield. London, 1741–42.

Printed Works

A. M. *The State of Religion in New-England, since the Reverend Mr. George Whitefield's Arrival There*. Glasgow, 1742.

An Account of the Doctrine and Discipline of Mr. Richard Davis. London, 1700.

Adams, Eliphalet. *A Discourse Shewing That so Long as there Is Any Prospect*. New London, 1734.

———. *The Gracious Presence of Christ with the Ministers*. New London, 1730.

———. *Ministers Must Take Heed*. New London, 1726.

———. *A Sermon Preached at Windham, July 12th. 1721. On a Day of Thanksgiving for the Late Remarkable Success of the Gospel among Them*. New London, 1721. Windham, Conn., 1800.

———. "To the Reader." In *Sensible Sinners Invited to Come to Christ*, by Eleazar Williams. New London, 1735.

———. *The Work of Ministers Rightly to Divide the Word of Truth*. New London, 1725.

Adams, William. *The Necessity of the Pouring Out of the Spirit from on High upon a Sinning Apostatizing People*. Boston, 1679.

An Address from the Presbytery of New-Castle [on] Duties Necessary for a Revival of Decayed Piety. Wilmington, Del., 1785.

Babcock, John. "Proposals. The present religious state of many towns in New-England." Broadside dated 4 July 1799.

Baxter, Richard. *The Glorious Kingdom of Christ Described and Clearly Vindicated.* London, 1691.

Bragge, Robert, *et al. A Defense of Some Important Doctrines of the Gospel in Twenty-Six Sermons. Most of Which Were Preached at Lime-Street Lecture*. 2d ed., 2 vols. Glasgow, 1773.

Brainerd, David. *Mirabilia Dei inter Indicos: or, An Account of the Revival and Progress of Religion, amongst a Number of the Indians in the Provinces of New Jersey and Pennsylvania*. Philadelphia [1746].

Brown, John. Letter. In *A Holy Fear of God*, by John Cotton, appendix, pp. 4–7. Boston, 1727.

Buell, Samuel. *A Copy of a Letter from the Rev. Mr. Buell, of East-Hampton, on Long-Island, to the Rev. Mr. Barber, of Groton, Connecticut*. [New London? 1764].

———. *A Faithful Narrative of the Remarkable Revival of Religion in the Congregation of East-Hampton on Long-Island, in the Year of Our Lord 1764*. New York, 1766.

Bulkley, John. *The Usefulness of Reveal'd Religion*. New London, 1730.

Campbell, Archibald. *Discourse Proving that the Apostles Were No Enthusiasts*. London, 1730.

Chandler, Samuel. *A Letter to the Reverend Mr. John Guyse. Occasioned by His Two Sermons Preached at St. Hellens*. London, 1730.

———. *A Paraphrase and Critical Commentary on the Prophecy of Joel*. London, 1735.

———. *A Second Letter to the Reverend Mr. John Guyse*. London, 1730.

Chanler, Isaac. *New Converts Exhorted to Cleave to the Lord*. Boston, 1740.

[Chaplin, Samuel]. *A journal containing some remarks upon the spiritual operations, beginning about the year . . . 1740, or 1741*. N.p., 1757.

Chauncy, Charles. *Enthusiasm Described*. Boston, 1742.

[———.] *A Letter from a Gentleman of Boston, to Mr. George Wishart, One of the Ministers of Edinburgh, Concerning the State of Religion in New England*. Edinburgh, 1742.

———. *The New Creature*. Boston, 1741.

———. *The Outpouring of the Holy Ghost*. Boston, 1742.

———. *Prayer for Help*. Boston, 1737.

———. *Seasonable Thoughts on the State of Religion in New-England*. Boston, 1743.

Chauncy, Isaac. *The Faithful Evangelist*. Boston, 1725.

———. *The Loss of the Soul*. Boston, 1732.

Cleaveland, John. *A Short and Plain Narrative of the Late Work of God's Spirit at Chebacco in Ipswich, in the Years 1763 and 1764*. Boston, 1767.

Coke, Thomas. *An Account of the Great Revival of the Work of God, in the City of Dublin, which commenced on the 4th of July, 1790*. Reprinted by Augustine Davis. Richmond, Va., 1791.

Colman, Benjamin. *The Judgments of Providence*. Boston, 1727.

———. *Letter from the Reverend Dr. Colman of Boston, to the Reverend Mr. Williams of Lebanon, Upon Reading the Confession and Retraction of the Reverend Mr. James Davenport*. Boston, 1744.

———. Preface, to *God's Awful Determination*, by John Cotton. Boston, 1728.

———. Preface, to *Two Discourses*, by John Jennings. Boston, 1740.

———. *A Sermon for the Reformation of Manners*. Boston, 1716.

———. *Some Glories of our Lord and Saviour Jesus Christ*. London, 1728.

[Colman, Benjamin, Tennent, Gilbert, and Tennent, William], *Three Letters to the Reverend Mr. George Whitefield*. Philadelphia [1739].

Colton, Calvin. *History and Character of American Revivals of Religion*. London, 1832. Reprint ed. New York: AMS Press, 1973.

Cooper, William. *The Danger of a People's Loosing the Good Impressions Made by the Late Awful Earthquake*. Boston, 1727.

———. Preface to "The Distinguishing Marks of a Work of the Spirit of God," by Jonathan Edwards. In *The Great Awakening*. Edited by C. C. Goen. Vol. 4 of *The Works of Jonathan Edwards*. New Haven, Conn.: Yale University Press, 1972.

Cotton, John. *God's Awful Determination*. Boston, 1728.

———. *A Holy Fear of God*. Boston, 1727.

The Country Parson's Advice to His Parishioners. London, 1680.

Danforth, Samuel. "The Building of Sion." In *Bridgewater's Monitor*. By James Keith and Samuel Danforth. Boston, 1717.

———. *The Duty of Believers to Oppose the Growth of the Kingdom of Sin*. Boston, 1708.

———. *An Exhortation to All to Use Utmost Endeavours to Obtain a Visit of the God of Hosts, for the Preservation of Religion, and the Church, upon Earth*. Boston, 1714. Reprinted in *The Wall and the Garden: Selected Massachusetts Election Sermons, 1670–1775*. Edited by A. W. Plumstead, pp. 150–76. Minneapolis: University of Minnesota Press, 1968.

———. *Piety Encouraged*. Boston, 1705.

Davenport, James. [Confession]. *Boston Gazette*, 18 July 1744.

Davenport, John. *God's Call to His People to Turn unto Him*. Cambridge, Mass., 1669.

Davies, Samuel. *Little Children Invited to Jesus Christ. . . . With an Account of the late remarkable Religious Impressions among the Students in the College of New-Jersey*. 5th ed. Boston, 1765.

———. *The State of Religion among the Protestant Dissenters in Virginia; In a letter to the Reverend Mr. Joseph Bellamy*. Boston, 1751.

Declaration of the True Presbyterians. N.p., 1742.

Doddridge, Philip. *Free Thoughts on the Most Probable Means of Reviving the Dissenting Interest*. London, 1730.

Edwards, Jonathan. "An Account of the late and wonderful Work of God." Appended to William Williams. *The Duty and Interest of a People*. Boston, 1736.

———. *An Account of the Life of the Late Reverend Mr. David Brainerd*. Boston, 1749.

———. *Apocalyptic Writings*. Edited by Stephen J. Stein. Vol. 5 of *The Works of Jonathan Edwards*. New Haven, Conn.: Yale University Press, 1977.

———. "The Distinguishing Marks of a Work of the Spirit of God." In *The Great Awakening*. Edited by C. C. Goen. Vol. 4 of *The Works of Jonathan Edwards*. New Haven, Conn.: Yale University Press, 1972.

———. "A Faithful Narrative of the Surprizing Work of God in the Conversion of Many Hundred Souls in Northampton and the Neighboring Towns and Villages." In *The Great Awakening*. Edited C. C. Goen. Vol. 4 of *The Works of Jonathan Edwards*. New Haven, Conn.: Yale University Press, 1972.

———. *The Future Punishment of the Wicked Unavoidable and Intolerable*. In *Works* Vol. 4. New York: Robert Carter and Brothers, 1864.

———. *A History of the Work of Redemption*. Edited and transcribed by John F. Wilson. Vol. 9 of *The Works of Jonathan Edwards*. New Haven, Conn.: Yale University Press, 1989.

———. "An Humble Attempt to Promote Explicit Agreement and Visible Union of God's People in Extraordinary Prayer for the Revival of Religion and the Advancement of Christ's Kingdom on Earth." In *Apocalyptic Writings*. Edited by Stephen J. Stein. Vol. 5 of *The Works of Jonathan Edwards*. New Haven, Conn.: Yale University Press, 1977.

———. "Personal Narrative." In *Jonathan Edwards, Representative Selections*. Edited by Clarence H. Faust and Thomas H. Johnson. Rev. ed. New York: Hill and Wang, 1962.

———. *Religious Affections*. Edited John E. Smith. Vol. 2 of *The Works of Jonathan Edwards*. New Haven, Conn.: Yale University Press, 1959.

———. "Sinners in the Hands of an Angry God." In *Jonathan Edwards: Basic Writings*. Edited by Ola Elizabeth Winslow. New York: New American Library, 1966.

———. "Some Thoughts concerning the Present Revival of Religion in New England." In *The Great Awakening*. Edited by C. C. Goen. Vol. 4 of *The Works of Jonathan Edwards*. New Haven, Conn.: Yale University Press, 1972.

Erskine, Ebenezer. *The Whole Works of Ebenezer Erskine*. Edinburgh, 1793.

Erskine, Ebenezer, and Erskine, Ralph. *A Collection of Sermons*. London, 1757.

Erskine, John. *The Signs of the Times Considered*. Edinburgh, 1742.

Erskine, Ralph. *Faith no Fancy: or A Treatise of Mental Images*. Edinburgh, 1745.

Ferris, David. *Memoirs of the Life of David Ferris*. Philadelphia, 1825.

Finley, Samuel. *Christ Triumphing, and Satan Raging*. Philadelphia, 1741.

Finney, Charles Grandison. *Lectures on Revivals of Religion*. Edited by William G. McLoughlin. Cambridge, Mass.: Harvard University Press, 1960.

Firmin, Giles. *A Brief Review of Mr. Davis's Vindication Giving no Satisfaction.* London, 1693.

Fisher, James. *A Review of the Preface to a Narrative of the Extraordinary Work at Kilsyth*. Glasgow, 1742.

Fleming, Robert. *Fulfilling of the Scripture*. Rotterdam, 1669; 2d and 3d pts. [1677?]; all three pts., London, 1681; 3d ed., 1681; 4th ed., 1693; 5th ed., 1726. I have used the Boston edition of 1743.

Flint, Josiah, and Torrey, Samuel. "To the Reader." In *The Necessity of the Pouring Out of the Spirit*, by William Adams. Boston, 1679.

Foxcroft, Thomas. *A Practical Discourse Relating to the Gospel-Ministry*. Boston, 1718.

———. *Some Seasonable Thoughts on Evangelic Preaching*. Boston, 1740.

———. *The Voice of the Lord*. Boston, 1727.

Francke, August Hermann. *Pietas Hallensis*. London, 1705.

Frelinghuysen, Theodorus Jacobus. *A Clear Demonstration of A Righteous and Ungodly Man*. New York, 1731.

Gib, Adam. *A Warning against Countenancing the Ministrations of Mr. George Whitefield*. Edinburgh, 1742.

Gillies, John, comp. *Appendix to the Historical Collections*. Glasgow, 1761.

———. *An Exhortation to the Inhabitants of the South Parish of Glasgow, and the Hearers in the College-Kirk*. Glasgow, 1750–51.

———, comp. *Historical Collections Relating to Remarkable Periods of the Success of the Gospel, and Eminent Instruments Employed in Promoting It*. 2 vols. Glasgow, 1754. Revised by Horatius Bonar, 1845. Revised ed. reprinted (2 vols. in 1) as *Historical Collections of Accounts of Revival*. Fairfield, Pa.: Banner of Truth Trust, 1981.

Good News from the Netherlands. Extracts of Letters from Two Ministers of Holland Confirming and Giving Accounts of the Revival of Religion in Guelderland. Boston, 1751.

Gookin, Nathaniel. *The Day of Trouble Near*. Boston, 1728.

Gough, Strickland. *Some Observations on the Present State of the Dissenting Interest.* London, 1730.

Guyse, John. *Christ the Son of God*. London, 1729.

———. *Reformation upon the Gospel Scheme*. London, 1735.

———. *The Scripture Notion of Preaching Christ Further Clear'd and Vindicated.* London, 1730.

Hancock, John. *A Sermon Preached at the Ordination of Mr. John Hancock . . . by His Father*. Boston, 1726.

Harris, Howell. *A Brief Account of the Life of Howell Harris*. Trevecka, 1791.

Harris, William. *Practical Discourses on . . . Representations of the Messiah, throughout the Old Testament*. London, 1724.

Henley, John. *Samuel Sleeping in the Tabernacle*. London, 1730.

Higginson, John. *The Cause of God and His People in New England*. Cambridge, Mass., 1663.

Hooker, Samuel. *Righteousness Rained from Heaven*. Cambridge, Mass., 1677.

Hopkins, Samuel. *The Life and Character of Miss Susanna Anthony, Who Died in Newport, (R.I.) June 23, MDCCXCI, in the Sixty Fifth Year of Her Age, Consisting Chiefly in Extracts from Her Writings, with Some Brief Observations on Them*. Worcester, Mass., 1796.

Howe, John. "The Prosperous State of the Christian Interest Before the End of Time, By a Plentiful Effusion of the Holy Spirit." In *Works*, 1:562–607. New York, 1835.

———. "A Sermon on the Thanksgiving Day December 2, 1697," in *Works*, 2:925–31. New York, 1835.

Hubbard, John, Gibbs, Philip, Godwin, Edward, Guyse, John, Hall, Thomas, and Wood, James. *Christ's Loveliness and Glory . . . Twelve Sermons, Preach'd at Mr. Coward's Lecture*. London, 1729.

[Jarratt, Devereax]. *A Brief Narrative of the Revival of Religion in Virginia. In a Letter to a Friend*. London, 1778. Reprinted in Francis Asbury, *An Extract from the Journal of Francis Asbury, Bishop of the Methodist Episcopal Church in America, from August 7, 1771, to December, 29, 1778*, 1:245–66. Philadelphia, 1792.

Jennings, John. *Two Discourses: The First of Preaching Christ; The Second of Experimental Preaching*. London, 1723 and 1735; Boston, 1740.

Johnson, Jacob. *Zion's Memorial of the Present Work of God*. [New London?], 1765.

[Jones, Griffith.] *Selections from the Welsh Piety*. Edited by W. Moses Williams. Cardiff, 1938.

Kennedy, Hugh. *A Short Account of the Rise and Continuing Progress of a Remarkable Work of Grace in the United Netherlands*. London, 1752.

Law, William. *A Practical Treatise upon Christian Perfection*. Vol. 1 of *Works*. New Forest, Hampshire, 1892–93.

———. *A Serious Call to a Devout and Holy Life*. Vol. 4 of *Works*. New Forest, Hampshire, 1892–93.

A Letter from a Gentleman in Scotland to His Friend in New England. Boston, 1743.

Lord, Benjamin. *The Faithful and Approved Minister*. New London, 1727.

———. *True Christianity Explained*. New London, 1727.

Lowman, Moses. *Paraphrase and Notes on the Revelation of St. John*. 1737. 2d ed. London, 1745.

Lowth, William. *Commentary on the Prophets*. Vol. 4 of Symon Patrick, *A Commentary on the Old and New Testaments, with Apocrypha*. London, 1809.

M. O. *A True Account of the Wonderful Conversions at Cambuslang Contained in a Letter from a Gentleman in the Gorbals of Glasgow, to his Friend at Greenock*. Appended to [James Robe]. *A Short Narrative of the Extraordinary Work at Cambuslang in Scotland*. Boston and Philadelphia, 1742.

[Maclaurin, John.] *Observations upon Church Affairs, Addressed to Principal Smith*. Edinburgh, 1734.

———. "On the Scripture Doctrine of Divine Grace." In *Sermons and Essays*. Glasgow, 1755.

———. *Works*. Glasgow, 1824.

———. *The Works of the Rev. John Maclaurin*. Edinburgh, 1860.

Marsh, Jonathan. *An Essay to Prove the Thorough Reformation of a Sinning People*. New London, 1721.

———. *The Great Care and Concern of Men under Gospel-Light*. New London, 1721.

Massachusetts Bay Province. *By the Governor and General Court . . . A Proclamation*. Cambridge, 1690.

———. *A Declaration against Prophaneness and Immoralities*. Boston, 1704.

———. *A Proclamation*. Boston, 1699.

Mather, Cotton. *The Accomplished Singer*. Boston, 1721.

———. *An Advice, to the Churches*. Boston, 1702.

———. *Bonifacius. An Essay upon the Good That Is to Be Divised and Designed by Those Who Desire to Answer the Great End of Life, and to Do Good While They Live*. Boston, 1710.

———. *Columbanus*. Boston, 1722.

———. *Companion for Communicants*. Boston, 1690.

———. *Early Religion Urged*. Boston, 1694.

———. *Eleutheria*. London, 1698.

———. *A Faithful Monitor*. Boston, 1704.

———. *Malachi*. Boston, 1717.

———. *Methods and Motives for Societies to Suppress Disorders*. Boston, 1703

———. *A Midnight Cry*. Boston, 1692.

———. *The Minister*. Boston, 1722.

———. *The Present State of New England*. Boston, 1690.

———. *Private Meetings Animated and Regulated*. Boston, 1706.

———. *Religious Societies*. Boston, 1724.

———. *Rules for the Society of Negroes*. [2d ed.] [Boston, between 1706 and 1711].

———. *The Serviceable Man*. Boston, 1690.

———. *Stone Cut Out of the Mountain*. Boston, 1716.

———. *Terra Beata*. Boston, 1726.

———. *The Terror of the Lord*. Boston, 1727.

———. *Theopolis Americana*. Boston, 1710.

———. *Things for a Distress'd People*. Boston, 1696.

———. *Things to Be Look'd For*. Cambridge, Mass., 1691.

———. *Thoughts for a Day of Rain*. Boston, 1712.

Mather, Increase. *The Blessed Hope*. Boston, 1701.

———. *A Discourse Concerning Faith and Fervency in Prayer, and the Glorious Kingdom of the Lord Jesus Christ, on Earth, Now Approaching*. Boston, 1710.

———. *Dissertation Concerning the Danger of Apostacy*. Second title in *A Call from Heaven to the Present and Succeeding Generations*. Boston, 1679.

———. *A Dissertation Concerning the Future Conversion of the Jewish Nation*. London, 1709.

———. *A Dissertation Wherein the Strange Doctrine Lately Published in a Sermon, the Tendency of Which is, to Encourage Unsanctified Persons (While Such) to Approach the Holy Table of the Lord, Is Examined and Confuted*. Boston, 1708.

———. *Five Sermons*. Boston, 1719.

———. *The Mystery of Israel's Salvation*. London, 1669.
———. *The Necessity of Reformation*. Boston, 1679.
———. *The Order of the Gospel*. Boston, 1700.
———. *Pray for the Rising Generation*. Cambridge, Mass., 1678.
———. Preface, to *God Brings to the Desired Haven*, by Thomas Prince. Boston, 1717.
———. *Renewal of the Covenant*. Boston, 1677.
———. *Returning Unto God the Great Concernment of a Covenant People*. Boston, 1680.
———. "To the Reader." In *An Exhortation unto Reformation*, by Samuel Torrey. Cambridge, Mass., 1674.
Munro, James. "Encouragements from the History of the Church under the Old and under the New Testament Dispensation, to expect, pray and labor for the Revival of Religion." In *Revival: Scriptural and Historical,* by John G. Lorimer and James Munro, pp. 83–142. Strathpine North, Australia: Covenanter Press, 1977.
[Murray, James.] *The Example of St. Paul, Represented to Ministers and Private Christians*. London, 1726.
A Narrative of the Revival of Religion in the County of Oneida . . . in the Year 1826. Utica, 1826.
Neal, Daniel. *A Sermon Preach'd to the Societies for Reformation of Manners*. London, 1722.
Norton, John. *An Essay Tending to Promote Reformation*. Boston, 1708.
———. "Sion the Outcast." In *Three Choice and Profitable Sermons*. Cambridge, Mass., 1664.
Owen, John. *The Works of John Owen*. Edited by William H. Goold. 16 vols. New York, 1851–53.
Pemberton, Ebenezer. *A Sermon Preached at the Ordination of Reverend Mr. Walter Wilmot*. Boston, 1738.
———. *A Sermon Preach'd before the Commission of the Synod of Philadelphia, April 20th, 1735*. New York, 1735.
Phillips, Samuel. *Three Plain Practical Discourses*. Boston, 1728.
Price, Samuel. *A Sermon Preach'd to the Societies for Reformation of Manners*. London, 1725.
Prince, Thomas. *Earthquakes the Work of God*. Boston, 1727.
———. *God Brings to the Desired Haven*. Boston, 1717.
———. *Six Sermons*. Edinburgh, 1785.
Proposals for a National Reformation of Manners. London, 1694.
Rawson, Grindal. *The Necessity of a Speedy and Thorough Reformation*. Boston, 1709.
Robe, James. *Mr. Robe's First-(Fourth) Letter to . . . James Fisher Concerning His Review of the Preface to a Narrative of the Extraordinary Work at Kilsyth*. 4 pts. Glasgow and Edinburgh, 1742, 1743.
———. *Sermons in Three Parts*. 1749. London, 1763.
[———.] *A Short Narrative of the Extraordinary Work at Cambuslang*. Glasgow, 1742.

[———.] *A Short Narrative of the Extraordinary Work at Cambuslang in Scotland.* Boston and Philadelphia, 1742.

[———.] *A Faithful Narrative of the Extraordinary Work of the Spirit of God, at Kilsyth.* Glasgow, 1789. Second title in *Narratives of the Extraordinary Work of the Spirit of God, at Cambuslang, Kilsyth, &c. Begun 1742*, by James Robe et al. Glasgow, 1790.

Rules Agreed upon to Be Observed, with Relation to the Encouragement of Young Men, Who Are Enclined to Give Themselves up to the Work of the Ministry. London, 1732.

S. H. [Samuel Hooker?] "To the Christian Reader." In *The Way of Israel's Welfare*, by John Whiting. Boston, 1686.

The Sense of the United Nonconforming Ministers, in and about London, concerning Some of the Erroneous Doctrines and Irregular Practices of Mr. Richard Davis. London, 1693.

Sewall, Joseph. *Christ Victorious.* Boston, 1733.

———. *The Duty of a People.* Boston, 1727.

———. *Repentance the Sure Way.* Boston, 1727

A Short Account of the Remarkable Conversions at Cambuslang. In a Letter from a Gentleman in the West-Country to his Friend at Edinburgh. Glasgow, 1742.

Smith, Josiah. *The Character, Preaching, etc. of the Rev. Mr. George Whitefield.* Boston, 1740.

Some, David. *The Methods to Be Taken by Ministers for the Revival of Religion.* London, 1730.

Sprague, William B. *Lectures on Revivals of Religion.* Albany, 1832; 2d ed. New York, 1833.

State of the Society in Scotland for Propagating Christian Knowledge. Edinburgh, 1741.

Stewart, Alexander. *An Account of the Late Revival of Religion.* 3d ed. Edinburgh, 1802.

Stoddard, Solomon. *An Appeal to the Learned.* Boston, 1709.

———. "The Benefit of the Gospel." In *The Efficacy of the Fear of Hell, to Restrain Men from Sin.* Boston, 1713.

[———.] *Cases of Conscience About Singing Psalms.* Boston, 1723.

———. *The Defects of Preachers Reproved.* New London, 1724.

———. *The Efficacy of the Fear of Hell, to Restrain Men from Sin.* Boston, 1713.

———. *Examination of the Power of the Fraternity.* Appended to *The Presence of Christ with the Ministers of the Gospel.* Boston, 1718.

———. *Falseness of the Hopes of Many Professors.* Boston, 1708.

———. *A Guide to Christ.* Boston, 1714.

———. *The Inexcusableness of Neglecting the Worship of God.* Boston, 1708.

———. *A Treatise Concerning Conversion.* Boston, 1719.

Strong, Nathan; Flint, Able; and Steward, Joseph, comps. *The Hartford Selection of Hymns.* Hartford: John Babcock, 1799.

Taylor, Abraham. *Of Spiritual Declensions.* London, 1732.

Tennent, Gilbert. *The Espousals, or a Passionate Perswasive to a Marriage with the Lamb of God*. New York, 1735.
———. *The Necessity of Religious Violence*. New York [1735].
———. *A Solemn Warning to a Secure World*. Boston, 1735.
———, Tennent, William, and Blair, Samuel. *Sermons on Sacramental Occasions by Diverse Ministers*. Boston, 1739.
The Testimony and Advice of an Assembly of Pastors of the Churches in New-England, At a Meeting in Boston July 7. 1743. Occasion'd By the late happy Revival of Religion in many Parts of the Land. . . . Boston, 1743.
The Testimony and Advice . . . With a Recommendation of It by the Revd. Dr. Watts. London, 1744.
The Testimony of a Number of Ministers. Boston, 1745.
The Testimony of the Pastors of the Churches in the Province of the Massachusetts-Bay in New-England, at their Annual Convention in Boston, May 15. 1743. Against several Errors in Doctrine, and Disorders in Practice. . . . Boston, 1743.
Torrey, Samuel. *An Exhortation unto Reformation*. Cambridge, Mass., 1674.
———. *Man's Extremity, God's Opportunity*. Boston, 1695.
———. *A Plea for the Life of Dying Religion*. Boston, 1683.
———, and Flint, Josiah. "To the Reader." In *The Necessity of the Pouring Out of the Spirit*, by William Adams. Boston, 1679.
Turell, Ebenezer. *The Life and Character of the Reverend Benjamin Colman*. Boston, 1749.
Tyler, Bennett, comp. *New England Revivals, as they existed at the Close of the Eighteenth, and the Beginning of the Nineteenth Centuries. Compiled Principally from Narratives First Published in the Conn. Evangelical Magazine*. Boston: Massachusetts Sabbath School Society, 1846.
Watts, Isaac. "The Atonement of Christ." In *Sermons on Various Subjects, Divine and Moral*, sermons 34–36. 3 vols. London, 1721, 1723, 1729.
———. *Discourses of the Love of God, and Its Influence on All the Passions: With a Discovery of the Right Use and Abuse of Them in Matters of Religion*. London, 1729.
———. *Divine Songs Attempted in Easy Language for the Use of Children*. London, 1715.
———, Guyse, John, Hubbard, John, Jennings, David, Neal, Daniel, and Price, Samuel. *Faith and Practice Represented in Fifty-Four Sermons*. 3d ed. Edinburgh, 1792.
———. *An Humble Attempt towards the Revival of Practical Religion*. London, 1731.
———. *Hymns and Spiritual Songs*. London, 1707.
———. "The Inward Witness of Christianity." In *Sermons on Various Subjects, Divine and Moral*. 3 vols. London, 1721, 1723, 1729.
———. *Psalms of David Imitated in the Language of the New Testament, and Applied to the Christian State and Worship*. London, 1719.
———. *The Strength and Weakness of Human Reason*. London, 1731.

———. *The Works of the Reverend and Learned Isaac Watts, D.D.* 6 vols. London, 1810–11.

Webster, Alexander. *Divine Influence the True Spring of the Extraordinary Work at Cambuslang*. Edinburgh, 1742.

Wesley, John. "Ernest Appeal to Men of Reason and Religion." In *The Works of John Wesley*. Edited by Gerald R. Cragg. Oxford: Clarendon Press, 1975.

———. *The Journal of the Rev. John Wesley*. Edited by Nehemiah Curnock. Standard ed. New York: Eaton & Mains [1909].

———. *Modern Christianity: Exemplified at Wednesbury and Other Adjacent Places in Staffordshire*. Newcastle, 1745. Also known as *Sufferings of the Primitive Methodists at Wednesbury*.

Whitby, Daniel. "Treatise of the True Millennium." In *A Paraphrase and Commentary on the New Testament*. Vol. 2. 1703. 2d ed., London, 1706.

Whitefield, George. *George Whitefield's Journals*. London: Banner of Truth Trust, 1960.

———. *Sermons on the Following Subjects*. Philadelphia, 1746.

———. *Three Letters . . . Concerning Archbishop Tillotson*. Philadelphia, 1740.

Whiting, John. *The Way of Israel's Welfare*. Boston, 1686.

Whitman, Samuel. *A Discourse of God's Omniscience*. New London, 1733.

———. *The Happiness of the Godly*. New London, 1727.

———. *Practical Godliness*. New London, 1714.

Willard, Samuel. *The Only Sure Way to Prevent Threatened Calamity*. Third title in *The Child's Portion*. Boston, 1684.

———. *Reformation the Great Duty*. Boston, 1694.

———. "To the Reader." In *Man's Extremity, God's Opportunity*, by Samuel Torrey. Boston, 1695.

Williams, Eleazar. *Sensible Sinners Invited to Come to Christ*. New London, 1735.

Williams, William. *The Duty and Interest of a People*. Boston, 1736.

———. *The Great Concern of Christians*. Boston, 1723.

———. *The Great Duty of Ministers*. Boston, 1726.

———. *The Great Salvation Revealed and Offered in the Gospel*. Boston, 1717.

———. *The Honor of Christ Advanced by the Fidelity of Ministers*. Boston, 1728.

———. *The Office and Work of Gospel Ministers*. Boston, 1729.

———. *A Painful Ministry*. Boston, 1717.

Williams, William, Jr. *Divine Warnings*. Boston, 1728.

Willison, John. *The Balm of Gilead for Healing a Diseased Land*. London, 1742.

———. *The Church's Danger and the Minister's Duty*. Edinburgh, 1733.

[———.] *A Fair and Impartial Testimony, Essayed in Name of a Number of Ministers . . . of the Church of Scotland*. Edinburgh, 1744.

———. *A Letter from Mr. John Willison, Minister at Dundee, to Mr. James Fisher, Minister at Glasgow. Containing Serious Expostulations with Him Concerning His Unfair-Declaring in His Review of Mr. Robe's Preface, &c*. 2d ed. Edinburgh, 1743.

———. *The Practical Works of the Rev. John Willison*. Glasgow, Edinburgh, and London, 1844.

———. Preface. To *Distinguishing Marks of a Work of the Spirit of God*, by Jonathan Edwards. Edinburgh, 1742.

———. *A Scripture Prophecy of the Increase of Christ's Kingdom and the Destruction of Antichrist*. Appended to *The Balm of Gilead*. London, 1742.

———. *A Sermon Preached Before His Majesty's High Commissioners*. Edinburgh, 1734.

Woodward, Josiah. *An Account of the Rise and Progress of the Religious Societies*. 3d ed. London, 1701.

[———.] *An Account of the Progress of the Reformation of Manners*. 12th ed. London, 1704.

SECONDARY SOURCES

Allport, Floyd H. *Theories of Perception and the Concept of Structure*. New York: John Wiley & Sons, 1955.

Armstrong, Anthony. *The Church of England, the Methodists and Society, 1700–1850*. Totowa, N.J.: Rowman and Littlefield [1973].

Bahlman, Dudley W. R. *The Moral Reformation of 1688*. New Haven, Conn.: Yale University Press, 1957.

Bailyn, Bernard. "Religion and Revolution: Three Biographical Studies." *Perspectives in American History* 4 (1970): 85–169.

Baker, Eric W. *A Herald of the Evangelical Revival: A Critical Inquiry into the Relation of William Law to John Wesley and the Beginning of Methodism*. London: Epworth Press, 1948.

Bebb, Evelyn Douglas. *Nonconformity and Social and Economic Life, 1660–1800: Some Problems of the Present as They Appeared in the Past*. London: Epworth Press, 1935.

Becker, Laura L. "Ministers vs. Laymen: The Singing Controversy in Puritan New England, 1720–1740." *New England Quarterly* 55 (1982): 79–94.

Bennett, G. V. *White Kennett, 1660–1728, Bishop of Petersborough*. London: S.P.C.K., 1957.

Benz, Ernst. "Ecumenical Relations between Boston Puritanism and German Pietism: Cotton Mather and August Hermann Francke." *Harvard Theological Review* 54 (1961): 159–93.

———. "The Pietist and Puritan Sources of Early Protestant World Missions (Cotton Mather and A. H. Francke)." *Church History* 20 (1951): 28–55.

Bercovitch, Sacvan. *The American Jeremiad*. Madison, Wis.: The University of Wisconsin Press, 1978.

Bloch, Ruth H. *Visionary Republic: Millennial Themes in American Thought, 1756–1800*. Cambridge: Cambridge University Press, 1985.

Bolam, C. Gordon, Goring, Jeremy, Short, H. L., and Thomas, Roger. *The English Presbyterians: From Elizabethan Puritanism to Modern Unitarianism*. Boston: Beacon Press, 1968.

Boles, John B. *The Great Revival, 1787–1805: The Origins of the Southern Evangelical Mind*. Lexington, Ky.: University Press of Kentucky, 1972.

Bonomi, Patricia U. *Under the Cope of Heaven: Religion, Society, and Politics in Colonial America*. New York: Oxford University Press, 1986.

Bonomi, Patricia U., and Eisenstadt, Peter R. "Church Adherence in the Eighteenth-Century British American Colonies." *William and Mary Quarterly*, 3d ser. 39 (April 1982): 245–86.

Breen, Timothy H. *The Character of the Good Ruler: A Study of Puritan Political Ideas in New England, 1630–1730*. New Haven, Conn.: Yale University Press, 1977.

Bridenbaugh, Carl. *Mitre and Sceptre: Transatlantic Faiths, Ideas, Personalities, and Politics, 1689–1775*. New York: Oxford University Press, 1962.

Brown, Callum G. *The Social History of Religion in Scotland since 1730*. London and New York: Methuen, 1987.

Bruce, Steve. "Social Change and Collective Behaviour: The Revival in Eighteenth-Century Ross-shire." *British Journal of Sociology* 34 (1983): 554–72.

Buchan, David. *The Ballad and the Folk*. London and Boston: Routledge & Kegan Paul, 1972.

Buchanon, John B. "Puritan Philosophy of History from Restoration to Revolution." *Essex Institute Historical Collections* 104 (1968): 329–48.

Bumsted, J. M., and Van de Wetering, John E. *What Must I Do to Be Saved? The Great Awakening in Colonial America*. Hinsdale, Ill.: Dryden Press, 1976.

Bumsted, John. "Religion, Finance, and Democracy in Massachusetts: The Town of Norton as a Case Study." *Journal of American History* 57 (1971): 817–31.

Burleigh, J. H. S. *A Church History of Scotland*. London: Oxford University Press, 1960.

Bushman, Richard L. *From Puritan to Yankee: Character and the Social Order in Connecticut, 1690–1765*. Cambridge, Mass.: Harvard University Press, 1967.

———, ed. *The Great Awakening: Documents on the Revival of Religion, 1740–1745*. New York: Atheneum, 1970.

———. "Jonathan Edwards as Great Man: Identity, Conversion, and Leadership in the Great Awakening." *Soundings* 52 (1969): 15–46.

Butler, Dugald. *John Wesley and George Whitefield in Scotland*. Edinburgh and London, 1898.

Byres, T. J. "Scottish Peasants and Their Song." *Journal of Peasant Studies* 3 (1976): 236–51.

Caldwell, Patricia. *The Puritan Conversion Narrative: The Beginnings of American Expression*. Cambridge: Cambridge University Press, 1983.

Carpenter, Edward. *The Protestant Bishop, Being the Life of Henry Compton, 1632–1713, Bishop of London*. London: Longmans, Green and Co., 1956.

Carwardine, Richard. *Transatlantic Revivalism: Popular Evangelicalism in Britain and America, 1790–1865*. Westport, Conn.: Greenwood Press, 1968.

Caulkins, Francis Manwaring. *History of Norwich, Connecticut*. Hartford, 1874.

Cavanagh, F. A. "Griffith Jones." *Journal of Adult Education*. Vol. 1, nos. 1 and 2 (London 1926 and 1927).

Clark, George, Sir. *English History: A Survey*. Oxford: Clarendon Press, 1971.

Clark, Ian D. L. "From Protest to Reaction: The Moderate Regime in the Church of

Scotland, 1752–1805." In *Scotland in the Age of Improvement: Essays in Scottish History in the Eighteenth Century.* Edited by N. T. Phillipson and Rosalind Mitchison. Edinburgh: Edinburgh University Press, 1970.

Clark, Stephen Morrieson. "Jonathan Edwards: The History of Redemption." Ph.D. dissertation, Drew University, 1986.

Clarke, W. K. Lowther. *Eighteenth-Century Piety*. London: S.P.C.K., 1944.

———. *A History of the S.P.C.K.* London: S.P.C.K., 1959.

Coalter, Milton J., Jr. *Gilbert Tennent, Son of Thunder: A Case Study of Continental Pietism's Impact on the First Great Awakening in the Middle Colonies.* Westport, Conn: Greenwood Press, 1986.

Coleman, D. C. *The Economy of England, 1450–1750*. London and New York: Oxford University Press, 1977.

Cook, Edward M., Jr. "Social Behavior and Changing Values in Dedham, Massachusetts, 1700 to 1775." *William and Mary Quarterly*, 3d ser. 27 (1970): 546–80.

Cooper, W. J. *Scottish Revivals*. Dundee, 1918.

Cowing, Cedric B. *The Great Awakening and the American Revolution: Colonial Thought in the 18th Century*. Chicago: Rand McNally, 1971.

———. "Sex and Preaching in the Great Awakening." *American Quarterly* 20 (1968): 624–44.

Cragg, Gerald R. *Puritanism in the Period of the Great Persecution, 1660–1688*. Cambridge: Cambridge University Press, 1957.

Crawford, Thomas. *Society and the Lyric: A Study of the Song Culture of Eighteenth-Century Scotland.* Edinburgh: Scottish Academic Press, 1979.

Currie, Robert. "A Micro-theory of Methodist Growth." *Proceedings of the Wesley Historical Society* 36 (1967): 65–73.

Currie, Robert; Gilbert, Alan; and Horsley, Lee. *Churches and Churchgoers: Patterns of Church Growth in the British Isles since 1700*. Oxford: Clarendon Press, 1977.

Dale, R. W., and Dale, A. W. W. *History of English Congregationalism*. 2d ed. London: Hodder and Stoughton, 1907.

Dallimore, Arnold A. *George Whitefield: The Life and Times of the Great Evangelist of the Eighteenth-Century Revival*. Vol. 1. London: Banner of Truth Trust, 1970. Reprint. 1971.

Davidson, James West. *The Logic of Millennial Thought: Eighteenth-Century New England*. New Haven, Conn.: Yale University Press, 1977.

Davies, Horton. *Worship and Theology in England: From Watts and Wesley to Maurice, 1690–1850*. Princeton, N.J.: Princeton University Press, 1961.

Davis, Arthur Paul. *Isaac Watts: His Life and Works*. New York: The Dryden Press, 1943.

de Jong, J. A. *As the Waters Cover the Sea: Millennial Expectations in the Rise of Anglo-American Missions, 1640–1810*. Kampen, Netherlands: J. H. Kok, 1970.

Demos, John. "Families in Colonial Bristol, Rhode Island: An Exercise in Historical Demography." *William and Mary Quarterly*, 3d ser. 25 (1968): 40–57.

Dictionary of Welsh Biography, Down to 1940. Honorable Society of Cymmrodorion. London: B. H. Blackwell, 1959.

Drummond, Andrew L., and Bulloch, James. *The Scottish Church, 1688–1843: The Age of the Moderates*. Edinburgh: Saint Andrew Press, 1973.

Durden, Diane Susan. "Transatlantic Communications and Literature in the Religious Revivals, 1735–1745." Ph.D. dissertation, University of Hull, 1978.

Durden, Susan. "A Study of the First Evangelical Magazines, 1740–1748." *Journal of Ecclesiastical History* 27 (1976): 255–75.

Durie, Alastair J. *The Scottish Linen Industry in the Eighteenth Century*. Edinburgh: John Donald, 1979.

Dwight, Sereno E. *The Life of President Edwards*. New York, 1820.

Elliot-Binns, L. E. *The Early Evangelicals: A Religious and Social Study*. Greenwich, Conn.: Seabury Press, 1953.

Elliott, Emory. *Power and the Pulpit in Puritan New England*. Princeton, N.J.: Princeton University Press, 1975.

Escott, Harry. *Isaac Watts, Hymnographer: A Study of the Beginnings, Development, and Philosophy of the English Hymn*. London: Independent Press, 1962.

Evans, Eifion. *Howell Harris, Evangelist, 1714–1773*. Cardiff: University of Wales Press, 1974.

Everitt, Alan. "Nonconformity in Country Parishes." *British Agricultural History Review* 18, suppl. (1970): 178–99.

Fawcett, Arthur. *The Cambuslang Revival: The Scottish Evangelical Revival of the Eighteenth Century*. London: Banner of Truth Trust, 1971.

Fiering, Norman S. "Will and Intellect in the New England Mind." *William and Mary Quarterly*, 3d ser. 29 (1972): 515–58.

Flaherty, David H. "Law and the Enforcement of Morals in Early America." In *Law in American History*. Edited by Donald Fleming and Bernard Bailyn, pp. 203–53. Vol. 5 of *Perspectives in American History*. Boston: Little, Brown, 1971.

Foote, Henry Wilder. *Three Centuries of American Hymnody*. Cambridge, Mass.: Harvard University Press, 1940.

Forster, Peter G. "Secularization in the English Context: Some Conceptual and Empirical Problems." *Sociological Review*, n.s., 20 (1972): 153–68.

Garrett, Clarke. *Spirit Possession and Popular Religion: From the Camisards to the Shakers*. Baltimore: Johns Hopkins University Press, 1987.

Gay, Peter. *A Loss of Mastery: Puritan Historians in Colonial America*. New York: Random House, 1968.

Gewehr, Wesley M. *The Great Awakening in Virginia, 1740–1790*. 1930. Reprint. Gloucester, Mass.: Peter Smith, 1965.

Gibb, Andrew. *Glasgow: The Making of a City*. London: Croom Helm, 1983.

Gilbert, Alan D. *Religion and Society in Industrial England: Church, Chapel, and Social Change, 1740–1914*. London: Longman, 1976.

Goodwin, Gerald J. "The Myth of 'Arminian-Calvinism' in Eighteenth-Century New England." *New England Quarterly* 41 (1968): 213–37.

Goring, Jeremy. "The Break-Up of the Old Dissent." In *The English Presbyterians: From Elizabethan Puritanism to Modern Unitarianism*. By C. Gordon Bolam et al., pp. 175–218. Boston: Beacon Press, 1968.

Goulding, James G. "The Controversy between Solomon Stoddard and the Mathers:

Western Versus Eastern Massachusetts Congregationalism." Ph.D. dissertation. Claremont University, 1971.

Graham, Henry Grey. *The Social Life of Scotland in the Eighteenth Century*. 1899. Reprint. London: A & C Black, 1928.

Green, V. H. H. *The Young Mr. Wesley: A Study of John Wesley and Oxford*. New York: St. Martin's Press, 1961.

Greene, Jack P. "Search for Identity: An Interpretation of the Meaning of Selected Patterns of Social Response in Eighteenth-Century America." *Journal of Social History* 3 (1970): 189–220.

Greven, Philip J., Jr. *Four Generations: Population, Land, and Family in Colonial Andover, Massachusetts*. Ithaca, N.Y.: Cornell University Press, 1970.

———. *The Protestant Temperament: Patterns of Child-rearing, Religious Experience, and the Self in Early America*. New York: Alfred A. Knopf, 1977.

———. "Youth, Maturity, and Religious Conversion: A Note on the Ages of Converts in Andover, Massachusetts, 1711–1749." *Essex Institute Historical Collections* 108 (1972): 119–34.

Griffiths, Olive M. *Religion and Learning: A Study in English Presbyterian Thought*. Cambridge: Cambridge University Press, 1935.

Gura, Philip F. "Sowing the Harvest: William Williams and the Great Awakening." *Journal of Presbyterian History* 56 (1978): 326–41.

Halévy, Élie. *The Birth of Methodism in England*. Edited and translated by Bernard Semmel. Chicago: University of Chicago Press, 1971. Originally published in *Revue de Paris*, 1 and 15 August 1906, pp. 519–39, 841–67.

Hall, A. Tindal. *Church and Society, 1600–1800*. London: Society for Promoting Christian Knowledge, 1968.

Hall, David D. *The Faithful Shepherd. A History of the New England Ministry in the Seventeenth Century*. 1972. Reprint. New York: W. W. Norton & Co., 1974.

———. *Worlds of Wonder, Days of Judgment: Popular Religious Belief in Early New England*. New York: Alfred A. Knopf, 1989.

Hambrick-Stowe, Charles E. *The Practice of Piety: Puritan Devotional Practice in Seventeenth-Century New England*. Chapel Hill, N.C.: University of North Carolina Press, 1982.

Hamilton, Henry. *An Economic History of Scotland in the Eighteenth Century*. Oxford: Clarendon Press, 1963.

Harmelink, Herman, III. "Another Look at Frelinghuysen and His 'Awakening.'" *Church History* 37 (1968): 423–38.

Harper, George W. "Clericalism and Revival: The Great Awakening in Boston as a Pastoral Phenomenon." *New England Quarterly* 57 (1984): 554–66.

Hatch, Nathan O. *The Sacred Cause of Liberty: Republican Thought and the Millennium in Revolutionary New England*. New Haven, Conn.: Yale University Press, 1977.

Heimert, Alan. *Religion and the American Mind: From the Great Awakening to the Revolution*. Cambridge, Mass.: Harvard University Press, 1966.

Henderson, G. H. *The Burning Bush: Studies in Scottish Church History*. Edinburgh: Saint Andrew Press, 1957.

Henretta, James A. *The Evolution of American Society, 1700–1800: An Interdisciplinary Analysis*. Lexington, Mass.: D.C. Heath & Co. 1973.

———. "The Morphology of New England Society in the Colonial Period." *Journal of Interdisciplinary History* 2 (1971): 379–98.

Hiner, N. Ray. "Adolescence in Eighteenth-Century America." *History of Childhood Quarterly* 3 (1975): 253–80.

Historical Catalogue of the First Church in Hartford 1633–1885. Hartford, Conn. 1885.

Holifield, E. Brooks. *The Covenant Sealed: The Development of Puritan Sacramental Theology in Old and New England, 1570–1720*. New Haven, Conn.: Yale University Press, 1974.

Hoskins, William G. *The Midlands Peasant: The Economic and Social History of a Leicester Village*. London: Macmillan & Co., 1957.

Houston, R. A. "Women in the Economy and Society of Scotland, 1500–1800." In *Scottish Society 1500–1800*. Edited by R. A. Houston and I. D. Whyte, pp. 118–47. Cambridge: Cambridge University Press, 1989.

Hughes, Hugh J. *Life of Howell Harris, the Welsh Reformer*. London, 1892.

Hutton, Holden. *The English Church from the Accession of Charles I. to the Death of Anne (1625–1714)*. London: Macmillan, 1913.

Irwin, Joyce. "The Theology of 'Regular Singing.'" *New England Quarterly* 51 (1978): 176–92.

Isaac, Rhys. *The Transformation of Virginia, 1740–1790*. Chapel Hill, N.C.: University of North Carolina Press, 1982.

Isaacs, Tina. "The Anglican Hierarchy and the Reformation of Manners." *Journal of Ecclesiastical History* 33 (1982): 391–411.

Johnson, David. *Music and Society in Lowland Scotland in the Eighteenth Century.* London: Oxford University Press, 1972.

Johnson, Paul E. *A Shopkeeper's Millennium: Society and Revivals in Rochester, New York, 1815–1837*. New York: Hill and Wang, 1978.

Johnson, Thomas H. *The Printed Writings of Jonathan Edwards, 1703–1758: A Bibliography*. Princeton, N.J.: Princeton University Press, 1940.

Jones, David. *Life and Times of Griffith Jones*. London: S.P.C.K., 1902.

Jones, M. G. *The Charity School Movement: A Study of Eighteenth Century Puritanism in Action*. 1938. Reprint. London and Edinburgh: Frank Cass and Co., 1964.

Jones, R. Tudor. *Congregationalism in England, 1662–1962*. London: Independent Press, 1962.

Keller, Charles Roy. *The Second Great Awakening in Connecticut*. New Haven, Conn.: Yale University Press, 1942. Reprint. Archon Books, 1968.

King, John Owen. *The Iron of Melancholy: Structures of Spiritual Conversion in America from the Puritan Conscience to Victorian Neurosis*. Middletown, Conn.: Wesleyan University Press, 1983.

Konig, David. *Law and Society in Puritan Massachusetts: Essex County, 1629–1692*. Chapel Hill, N.C.: University of North Carolina Press, 1979.

Landsman, Ned C. "Evangelists and Their Hearers: Popular Interpretation of Revivalist Preaching in Eighteenth-Century Scotland." *Journal of British Studies* 28 (1989): 120–49.

———. *Scotland and Its First American Colony, 1683–1765*. Princeton, N.J.: Princeton University Press, 1985.

Lang, Andrew. *A History of Scotland from the Roman Occupation*. Vol. 4. New York: Dodd, Mead, and Co., 1907.

Lee, Umphrey. *John Wesley and Modern Religion*. Nashville: Cokesbury Press, 1936.

Lockridge, Kenneth A. *A New England Town, the First Hundred Years: Dedham, Massachusetts, 1636–1736*. New York: W. W. Norton, 1970.

Lodge, Martin E. "The Crisis of the Churches in the Middle Colonies, 1720–1750." *Pennsylvania Magazine of History and Biography* 95 (1971): 195–220.

Lowry, Charles Wesley, Jr. "Spiritual Antecedents of Anglican Evangelicalism." In *Anglican Evangelicalism*. Edited by Alexander C. Zabriskie. Philadelphia: Church Historical Society, 1943.

Lucas, Paul R. "'An Appeal to the Learned': The Mind of Solomon Stoddard." *William and Mary Quarterly*, 3d ser. 30 (1973): 257–92.

MacEwen, Alexander R. *The Erskines*. New York: Charles Scribner's Sons [1900].

MacFarlan, D[uncan.] *The Revivals of the Eighteenth Century, Particularly at Cambuslang*. London and Edinburgh [1845]. Reprint. Wheaton, Ill.: Richard Owen Roberts, 1980.

MacInnes, John. *The Evangelical Movement in the Highlands of Scotland, 1688–1800*. Aberdeen: The University Press, 1951.

McKerrow, John. *History of the Secession Church*. 1839. 3d ed. London, n.d.

Maclear, James F. "'The Heart of New England Rent': The Mystical Element in Early Puritan History." *Mississippi Valley Historical Review* 42 (1956): 621–52.

———. Review of *Triumph of the Laity, Scots-Irish Piety and the Great Awakening, 1625–1760*, by Marilyn Westerkamp. *American Historical Review* 94 (1989): 1165–66.

McLeish, John. *Evangelical Religion and Popular Education: A Modern Interpretation*. London: Methuen, 1969.

Macleod, John. *Scottish Theology in Relation to Church History since the Reformation*. Edinburgh: Publications Committee of the Free Church of Scotland, 1943.

McLoughlin, William G. *Modern Revivalism: Charles Grandison Finney to Billy Graham*. New York: Ronald Press, 1959.

Mantoux, Paul. *The Industrial Revolution in the Eighteenth Century: An Outline of the Beginnings of the Modern Factory System in England*. Rev. ed. New York: Harper & Row, 1965.

Manual of the First Congregational Church of Norwich, Conn. Norwich, Conn., 1868.

Mathews, Donald G. *Religion in the Old South*. Chicago: University of Chicago Press, 1977.

Maxson, Charles Hartshorn. *The Great Awakening in the Middle Colonies*. 1920. Reprint. Gloucester, Mass.: Peter Smith, 1958.

Mead, Sidney E. "The Nation with the Soul of a Church." *Church History* 36 (1967): 262–83.

Mechie, Stewart. "The Theological Climate in Early Eighteenth Century Scotland." In *Reformation and Revolution: Essays presented to The Very Reverend Principal Emeritus Hugh Watt, D.D., D.Litt. on the Sixtieth Anniversary of his Ordination*, pp. 258–72. Edinburgh: The Saint Andrew Press, 1967.

Meek, James. "Parish of Cambuslang." In *The Statistical Account of Scotland*. Compiled by Sir John Sinclair, Vol. 5. Edinburgh: William Creech, 1793, pp. 241–74.

Middlekauff, Robert. *The Mathers: Three Generations of Puritan Intellectuals, 1596–1728*. New York: Oxford University Press, 1971.

———. "Piety and Intellect in Puritanism." *William and Mary Quarterly*, 3d ser. 22 (1965): 457–70.

Miller, Glenn T. "God's Light and Man's Enlightenment: Evangelical Theology of Colonial Presbyterianism." *Journal of Presbyterian History* 51 (1973): 97–115.

Miller, John C. "Religion, Finance, and Democracy in Massachusetts." *New England Quarterly* 6 (1933): 29–58.

Miller, Perry. "From the Covenant to the Revival." In *The Shaping of American Religion*. Edited James Ward Smith and A. Leland Johnson, pp. 322–68. Princeton, N.J.: Princeton University Press, 1961.

———. *Jonathan Edwards*. New York: W. Sloane Associates, 1949.

———. "Jonathan Edwards and the Great Awakening." In *Errand into the Wilderness*. By Perry Miller. Cambridge, Mass: Harvard University Press, 1956. Reprint. New York: Harper & Row, 1964, pp. 153–66.

———. "Jonathan Edwards' Sociology of the Great Awakening." *New England Quarterly* 21 (1948): 50–77.

———. *The New England Mind: From Colony to Province*. 1953. Reprint. Boston: Beacon Press, 1961.

———. *The New England Mind: The Seventeenth Century*. 1939. Reprint. Boston: Beacon Press, 1954.

Milner, Thomas. *The Life, Times, and Correspondence of the Rev. Isaac Watts, D.D.* London, 1845.

Mitchell, Mary Hewit. *The Great Awakening and Other Revivals in the Religious Life of Connecticut*. Vol. 26. New Haven, Conn.: Tercentenary Commission of the State of Connecticut, 1934.

Mitchison, Rosalind. *Lordship to Patronage: Scotland 1603–1745*. London: Edward Arnold, 1983.

Monk, Robert C. *John Wesley: His Puritan Heritage*. Nashville: Abbingdon Press, 1966.

Moran, Gerald F. "Conditions of Religious Conversion in the First Society of Norwich, Connecticut, 1718–1744." *Journal of Social History* 5 (1972): 331–43.

———. "The Puritan Saint." Ph.D. dissertation, Rutgers University, 1974.

Moran, Gerald F., and Vinovskis, Maris A. "The Puritan Family and Religion: A Critical Reappraisal." *William and Mary Quarterly*, 3d ser. 39 (1982): 29–63.

Morgan, Edmund S. *The Puritan Family: Religion & Domestic Relations in Seventeenth-Century New England*. 1944; New ed., rev. and enl. New York: Harper & Row, 1966.

———. *Visible Saints: The History of a Puritan Idea*. Ithaca, N.Y.: Cornell University Press, 1963.

Muirhead, Ian A. "The Revival as a Dimension of Scottish Church History." *Records of the Scottish Church History Society* 20 (1978): 179–96.

Murrin, John. "No Awakening, No Revolution? More Counterfactual Speculations." *Reviews in American History* 11 (1983): 161–71.

Nash, Gary B. *The Urban Crucible: Social Change, Political Consciousness, and the Origins of the American Revolution*. Cambridge, Mass.: Harvard University Press, 1979.

The New Schaff-Herzog Encyclopedia of Religious Knowledge. 13 vols. New York and London: Funk and Wagnalls, 1908–14. Reprint. Grand Rapids, Mich.: Baker, 1949–50.

North, Catherine M. *History of Berlin, Connecticut*. New Haven, Conn.: Tuttle, Morehouse & Taylor, 1916.

Nuttall, Geoffrey Fillingham. "Continental Pietism and the Evangelical Movement in Britain." In *Pietism und Reveil*. Edited by J. Van den Berg and J. P. van Dooren, pp. 207–36. Leiden: E. J. Brill, 1978.

———. *Richard Baxter and Philip Doddridge: A Study in a Tradition*. London: Oxford University Press, 1951.

Onuf, Peter S. "New Lights in New London: A Group Portrait of the Separatists." *William and Mary Quarterly*, 3d ser. 37 (1980): 627–43.

Orcutt, Samuel. *History of the Towns of New Milford and Bridgewater, Connecticut, 1703–1882*. Hartford, Conn., 1882.

Outler, Albert C., ed. *John Wesley*. New York: Oxford University Press, 1964.

Overton, John Henry. *Life in the English Church (1661–1714)*. London, 1885.

Oxford English Dictionary.

Patrick, Millar. *Four Centuries of Scottish Psalmody*. London: Oxford University Press, 1949.

Payne, Earnest A. "Doddridge and the Missionary Enterprise." In *Philip Doddridge 1702–1751: His Contribution to English Religion*. Edited by Geoffrey F. Nuttall. London: Independent Press, 1951.

Pettit, Norman. *The Heart Prepared: Grace and Conversion in Puritan Spiritual Life*. New Haven, Conn.: Yale University Press, 1966.

Pocock, J. G. A. *Politics, Language and Time: Essays on Political Thought and History*. New York: Atheneum, 1973.

———. *Virtue, Commerce, and History: Essays on Political Thought and History, Chiefly in the Eighteenth Century*. Cambridge: Cambridge University Press, 1985.

Pope, Robert G. *The Half-Way Covenant: Church Membership In Puritan New England*. Princeton, N.J.: Princeton University Press, 1969.

———. "New England versus the New England Mind: The Myth of Declension." *Journal of Social History* 3 (1969–70): 95–108.

Porter, Noah. *Half-Century Discourse*. Hartford, Conn., 1857.

Portus, Garnet V. *Caritas Anglicana*. London: A. R. Mowbray & Co., 1912.

Pratt, Anne Stokely. *Isaac Watts and His Gifts of Books to Yale College*. New Haven, Conn.: Yale University Library, 1938.

Probert, John C. C. *The Sociology of Cornish Methodism to the Present Day*. Cornish Methodist Historical Association, no. 17. Redruth, 1971.

Ragan, Allen E. *A History of Tuscullum College, 1794–1944*. Bristol, Tenn.: Tuscullum Sesquicentennial Committee, 1945.

Real-Encyklopedia fur Protestantische Theologie und Kirche. Edited by Johann Jakob Herzog. 22 vols. Hamburg: R. Besser, 1854–68. 3d ed. Edited by Albert Hauck. 24 vols. Leipzig: J. C. Hinrichs, 1896–1913.

Rees, Thomas. *History of Protestant Nonconformity in Wales: From its Rise in 1633 to the Present Time*. 2d ed., rev. London, 1883.

Roetger, R. W. "The Transformation of Sexual Morality in 'Puritan' New England: Evidence from New Haven Court Records, 1639–1698." *Canadian Review of American Studies* 15 (1984): 243–57.

Royster, Charles. *A Revolutionary People at War: The Continental Army and American Character, 1775–1783*. Chapel Hill, N.C.: University of North Carolina Press, 1979.

Rupp, Gordon. *Religion in England, 1688–1791*. Oxford: Clarendon Press, 1986.

Rutter, Robert Sherman. "The New Birth: Evangelicalism in the Transatlantic Community During the Great Awakening, 1739–1745." Ph.D. dissertation, Rutgers University, 1982.

Schafer, Thomas A. "Solomon Stoddard and the Theology of the Revival." In *A Miscellany of American Christianity: Essays in Honor of H. Shelton Smith*. Edited by Stuart C. Henry, pp. 328–61. Durham, N.C.: Duke University Press, 1963.

Scheick, William J. *The Writings of Jonathan Edwards: Theme, Motif, and Style*. College Station: Texas A & M University Press, 1975.

Schmidt, Leigh Eric. *Holy Fairs: Scottish Communions and American Revivals in the Early Modern Period*. Princeton, N.J.: Princeton University Press, 1989.

———. "Scottish Communions and American Revivals: Evangelical Ritual, Sacramental Piety, and Popular Festivity from the Reformation through the Mid-Nineteenth Century." Ph.D. dissertation, Princeton University, 1987.

———. "'A Second and Glorious Reformation': The New Light Extremism of Andrew Croswell." *William and Mary Quarterly*, 3d ser. 43 (1986): 214–44.

Schmotter, James W. "Ministerial Careers in Eighteenth-Century New England: The Social Context, 1700–1760." *Journal of Social History* 9 (1975): 249–67.

Schute, Michael N. "A Little Great Awakening: An Episode of the Enlightenment." *Journal of the History of Ideas* 37 (1976): 589–602.

Scott, Hew, ed. *Fasti Ecclesiae Scoticanae*. Edinburgh, 1915.

Sefton, Henry. "'Neu-lights and Preachers Legall': some observations on the beginnings of Moderatism in the Church of Scotland." In *Church, Politics and Society: Scotland 1408–1929*. Edited by Norman Macdougall, pp. 186–96. Edinburgh: John Donald, 1983.

Semmel, Bernard. *The Methodist Revolution*. New York: Basic Books, 1973.

Shepherd, T. B. *Methodism and the Literature of the Eighteenth Century*. New York: Haskell House, 1966.

Sher, Richard B. *Church and University in the Scottish Enlightenment: The Moderate Literati of Edinburgh*. Edinburgh: Edinburgh University Press, 1985.

Sher, Richard, and Murdoch, Alexander. "Patronage and Party in the Church of Scotland, 1750–1800." In *Church, Politics and Society: Scotland 1408–1929*. Edited by Norman Macdougall, pp. 197–220. Edinburgh: John Donald, 1983.

Shiels, Richard D. "The Connecticut Clergy in the Second Great Awakening." Ph.D. dissertation, Boston University, 1976.

Shipton, Clifford K. *Sibley's Harvard Graduates*. Vol. 5. Boston: Massachusetts Historical Society, 1937.

Shorter, Edward. *The Making of the Modern Family*. New York: Basic Books, 1975.

Simonson, Harold P. "Jonathan Edwards and His Scottish Connections." *Journal of American Studies* 21 (1987): 353–76.

Smith, Daniel Scott. "Parental Power and Marriage Patterns: An Analysis of Historical Trends in Hingham, Massachusetts." *Journal of Marriage and the Family* 35 (1973): 419–28.

Smith, Daniel Scott, and Hindus, Michael S. "Premarital Pregnancy in America, 1640–1971: An Overview and Interpretation." *Journal of Interdisciplinary History* 5 (1975): 537–70.

Smith, Timothy L. *Revivalism and Social Reform: American Protestantism on the Eve of the Civil War*. Abington Press, 1957. Reprint. New York: Harper & Row, 1965.

Smout, T. C., "Born Again at Cambuslang: New Evidence on Popular Religion and Literacy in Eighteenth Century Scotland." *Past and Present* 97 (1982): 114–27.

Smylie, John Edwin. "National Ethos and the Church." *Theology Today* 20 (1963): 313–21.

Stoeffler, F. Ernst. *The Rise of Evangelical Pietism*. Leiden: E. J. Brill, 1971.

Stout, Harry S. *The New England Soul: Preaching and Religious Culture in Colonial New England*. New York: Oxford University Press, 1986.

———. "Religion, Communications, and the Ideological Origins of the American Revolution." *William and Mary Quarterly*, 3d ser. (1977): 519–41.

Stout, Harry S., and Onuf, Peter S. "James Davenport and the Great Awakening in New London." *Journal of American History* 70 (1983): 556–78.

Strout, Cushing. *The New Heavens and New Earth: Political Religion in America*. New York: Harper & Row, 1974.

Sweet, William Warren. *Revivalism in America: Its Origin, Growth and Decline*. New York: Charles Scribner's Sons, 1944.

Sykes, Norman. *Church and State in England in the Eighteenth Century*. Cambridge: Cambridge University Press, 1934.

———. *Edmund Gibson, Bishop of London, 1669–1748*. London: Oxford University Press, 1926.

Tanis, James. *Dutch Calvinistic Pietism in the Middle Colonies: A Study in the Life and Theology of Theodorus Jacobus Frelinghuysen*. The Hague: Martinus Nijhof, 1967.

Thomas, Roger. "Presbyterians in Transition." In *The English Presbyterians: From*

Elizabethan Puritanism to Modern Unitarianism. By C. Gordon Bolam et al., pp. 113–74. Boston: Beacon Press, 1968.
Thompson, Roger. "Adolescent Culture in Colonial Massachusetts." *Journal of Family History* 9 (1984): 127–44.
Tracy, Patricia J. *Jonathan Edwards, Pastor: Religion and Society in Eighteenth-Century Northampton*. New York: Hill & Wang, 1980.
Trinterud, Leonard J. *The Forming of an American Tradition: A Re-examination of Colonial Presbyterianism*. Philadelphia: Westminster Press [1949].
Tyerman, Luke. *The Life of the Rev. George Whitefield*. 2d ed. London, 1890.
Van de Wetering, John E. "*The Christian History* of the Great Awakening." *Journal of Presbyterian History* 44 (1966): 122–9.
Verduin, Kathleen. "'Our Cursed Natures': Sexuality and the Puritan Conscience." *New England Quarterly* 56 (1983): 220–37.
Waddington, John. *Congregational History: In Relation to Contemporaneous Events, Education, the Eclipse of Faith, Revivals, and Christian Missions*. London, 1876.
Walker, George Leon. *History of the First Church in Hartford, 1633–1883*. Hartford, Conn., 1884.
Walker, James. *Theology and Theologians of Scotland 1560–1750*. 1872. 2d ed., rev. 1888. Reprint. Edinburgh: Knox Press, 1982.
Walsh, James. "The Great Awakening in the First Congregational Church of Woodbury, Connecticut." *William and Mary Quarterly*, 3d ser. 28 (1971): 543–62.
Walsh, J[ohn] D. "Élie Halévy and the Birth of Methodism." *Transactions of the Royal Historical Society*, 5th ser. 25 (1975): 1–20.
———. "Joseph Milner's Evangelical Church History." *Journal of Ecclesiastical History* 10 (1959): 174–87.
———. "Origins of the Evangelical Revival." In *Essays in Modern Church History, in Memory of Norman Sykes*. Edited by G. V. Bennett and J. D. Walsh, pp. 148–53. New York: Oxford University Press, 1966.
Ward, W. R. "Orthodoxy, Enlightenment and Religious Revival." *Studies in Church History* 17 (1982): 275–96.
———. "Power and Piety: The Origins of Religious Revival in the Early Eighteenth Century." *Bulletin of the John Rylands University Library* 63 (1980): 231–52.
———. "The Relations of Enlightenment and Religious Revival in Central Europe and in the English-speaking World." In *Reform and Reformation: England and the Continent, c1500-c1700*. Edited by Derek Baker, pp. 281–305. Oxford: Basil Blackwell, 1979.
Watts, Michael R. *The Dissenters, I: From the Reformation to the French Revolution*. Oxford: Clarendon Press, 1978.
Webster, Richard. *A History of the Presbyterian Church in America, from Its Origin until the Year 1760*. Philadelphia, 1857.
Wellwood, Henry Moncreiff. *Account of the Life and Writings of John Erskine*. Edinburgh, 1818.
Westerkamp, Marilyn J. *Triumph of the Laity: Scots-Irish Piety and the Great Awakening*. New York: Oxford University Press, 1988.

White, Eugene E. *Puritan Rhetoric: The Issue of Emotion in Religion*. Carbondale and Edwardsville, Ill.: Southern Illinois University Press, 1972.

Williams, Owen C. *The Puritan Experience: Studies in Spiritual Autobiography*. New York: Schocken Books, 1972.

Williams, W. Moses. *The Friends of Griffith Jones: A Study in Educational Philanthropy*. London: Honorable Society of Cymmrodorion, 1939.

Willingham, William F. "Religious Conversion in the Second Society of Windham, Connecticut." *Societas* 6 (1976): 109–19.

Wilson, James A. *A History of Cambuslang, A Clydesdale Parish*. Glasgow: Jackson, Wylie and Co., 1929.

Wilson, John F. "Jonathan Edwards as Historian." *Church History* 46 (1977): 5–18.

Winslow, Ola Elizabeth. *Meetinghouse Hill, 1630–1783*. New York: Macmillan Co., 1952.

Youngs, J. William T. *God's Messengers: Religious Leadership in Colonial New England, 1700–1750*. Baltimore: Johns Hopkins University Press, 1976.

Zehrer, Karl. "The Relationship between Pietism in Halle and Early Methodism." *Methodist History* 17 (1979): 211–24.

Zuckerman, Michael. *Peaceable Kingdoms: New England Towns in the Eighteenth Century*. New York: Alfred A. Knopf, 1970.

INDEX